Library of Congress Cataloging in Publication Data

Lenk, John D
  Logic designer's manual.

  Includes index.
  1. Logic circuits. 2. Computers—Circuits.
3. Electronic circuit design. I. Title.
TK7868.L6L46   621.3819'535   76-52957
ISBN 0-87909-450-8

To Irene, Karen, Mark, Brandon, and Lambie

© 1977 by Reston Publishing Company
A Prentice-Hall Company
Reston, Virginia   22090

10 9 8 7 6 5 4 3 2 1

Printed in the United States of America

Logic Designer's Manual

John D. Lenk

Consulting Technical Writer

Reston Publishing Company
A Prentice-Hall Company
Reston, Virginia

Logic Designer's Manual

John D. Lenk

Consulting Technical Writer

Reston Publishing Company
A Prentice-Hall Company
Reston, Virginia

Library of Congress Cataloging in Publication Data

Lenk, John D
  Logic designer's manual.

  Includes index.
  1. Logic circuits. 2. Computers—Circuits.
3. Electronic circuit design. I. Title.
TK7868.L6L46   621.3819'535   76-52957
ISBN 0-87909-450-8

To Irene, Karen, Mark, Brandon, and Lambie

© 1977 by Reston Publishing Company
A Prentice-Hall Company
Reston, Virginia   22090

10 9 8 7 6 5 4 3 2 1

Printed in the United States of America

Contents

Preface

The *Logic Designer's Manual* is an outgrowth of the author's popular *Handbook of Logic Circuits*. Most of today's logic design is accomplished with complete logic circuits, generally found in IC (integrated circuit) and/or plug-in module form. In prior years, it was necessary for the designer to implement such circuits as decoders, flip-flops, counters, registers, distributors, arithmetic units, and so on. Today, all these circuits are available in IC form, thus eliminating the need to make up circuits using basic logic gates. However, to use packaged logic circuits effectively, today's designer must be able to interconnect off-the-shelf logic circuits to form logic systems. The purpose of the *Logic Designer's Manual* is to fill that need.

This manual is written for logic IC users, rather than for designers of logic ICs. That is, the manual is written on the basis of using existing, commercial logic ICs to solve design and application problems. Typical users include design specialists who want to develop logic systems with available ICs, or technicians who must service logic equipment containing logic ICs. Experimenters and hobbyists can also make good use of this approach to logic IC applications.

There are two very common, although not necessarily accurate, concepts concerning logic ICs. First, it is assumed that the basic approach to logic design involves informing an IC manufacturer of design parameters and requirements for a particular logic system, and instructing the manufacturer to fabricate an IC (or group of ICs, or possibly a microprocessor) which meets these exact requirements. While this approach is satisfactory for some highly specialized logic systems, and particularly where cost is no factor, the approach may generally be wasteful and often unnecessary. On the other hand, it is often assumed that existing commercial logic ICs are limited in application, that such ICs are designed with only one or two uses in view.

There is a middle ground. Except for certain very special circuits, there are a number of commercial logic ICs that can be adapted to meet most logic circuit and system requirements. Also, new logic IC modules are being developed by various manufacturers. Likewise, although most off-the-shelf logic ICs are manufactured with certain specific uses in mind, these ICs are certainly not limited to only those uses. Thus, the approach found in this manual serves a two-fold purpose: (1) to acquaint the reader with logic ICs, in general, so that the user can select commercial units to meet his particular requirements, and (2) to show the reader the many other uses for existing logic ICs not found on the manufacturer's datasheet.

Chapter 1 is an introduction to logic design, which includes the basics of logic circuits, the logic symbols in common use throughout the industry, the basic principles of logic equations, and corresponding functions. The first chapter also provides detailed procedures for the design of both combinational and sequential networks. This chapter summarizes the entire subject of logic design, on the assumption that many readers will be students who are not familiar with all phases of the logic field.

Chapter 2 describes logic circuit elements available in IC form (decoders, distributors, counters, registers, and so on). This chapter describes what is available as well as how the circuits operate, and how they are used in basic systems.

The remainder of the text, Chapters 3 through 10, covers design applications of the basic circuits and elements discussed in Chapter 2. These chapters cover such subjects as data selectors, decoders, counters, registers, analog/digital and digital/analog converters, arithmetic units, memories, interface circuits, noise problems, and miscellaneous logic devices. The circuits described represent a cross-section of the entire logic field.

The author considers logic design to be an art rather than an exact science. He recognizes that there are many alternate solutions or designs to the problems described here. However, the designs presented in this manual are proven with time and will get the job done. To use the manual effectively, the reader is invited to study both the alphabetical index (at the end of the book) and the table of contents. In either or both lists, the reader will often find one or more designs listed that meet his exact needs. If not, the circuits described in this manual will at least provide a starting point for special design requirements.

The author has received much help from many organizations and individuals prominent in the field of logic design. He wishes to thank them all, and wants to express special thanks to the following: Cambridge Thermionic Corporation; Digital Equipment Corporation; Hewlett-Packard; Honeywell, Inc.; International Telephone & Telegraph; Litton Industries; Motorola Semiconductor Products, Inc.; Radio Corporation of America; and Texas Instruments Incorporated.

The author also wishes to express his appreciation to Mr. Joseph A. Labok of Los Angeles Valley College, and to Mr. Richard L. Castellucis of Southern Technical Institute for their help and encouragement.

John D. Lenk

1

Introduction to Logic Design

Today's logic designer must work with logic equations, circuits implemented by interconnecting basic logic gates, and complete logic circuits in IC module form. No matter what logic elements are used, many problems in logic design can be solved by working out the solution in equation form first, and then making circuits (or interconnecting IC elements) to match the equations. As an example, if a circuit is to be *minimized* (reduced to the least number of logic elements), the equipment is reduced to the simplest form "on paper," and the simplified equation is converted to a circuit.

Even when simplification is not involved, it is easier to write an equation than to wire a circuit. A good example of this is a logic module that has six inputs and one output, in which the output must be present when three (and only three) of the inputs are present. With such a problem, the equation is written to express the relationship of the six inputs to the output. Then the equation is converted to a corresponding circuit.

Today's designer must be familiar with logic equations (also known as logical algebra, Boolean algebra—after the English mathematician George Boole—computer algebra, or computer logic). The designer must be able to manipulate logical equations (usually to simplify them), and then to convert the equations into practical circuits. This task is greatly simplified if the designer can write an equation for a given circuit as a first step to improving the circuit. For example, it may be necessary to convert a circuit from positive logic to negative logic in order to accommodate an external circuit function.

In addition to a practical working knowledge of logical algebra, the designer must be familiar with the symbols used in logic circuits, the binary number system (typical to most logic systems), the basic logic circuit forms, mapping of logic circuits, and certain other problems common to all logic design. These subjects are discussed *in summary form* throughout this chapter.

1–1   DEFINING LOGICAL ALGEBRA

Conventional algebra is the symbolic expression for relationships of number variables. Logical algebra is a method of symbolically expressing the relationship between logic variables. Logical algebra differs from conventional algebra in two respects:

1. Arithmetic operations are not performed in logical algebra.

2. The symbols used in logical algebra (usually letters) do not represent numerical values.

Logical algebra is ideally suited to any system of intelligence based on *two opposite states,* such as "on" or "off." Thus, logical algebra is specifically suited to express the opening and closing of electrical switches, the presence or absence of electrical pulses, or the polarity or amplitude relationship of pulses. Logical algebra is also compatible with the *binary number system* (a two-state arithmetic system), and the various special logic codes discussed in Secs. 1–2 and 1–3.

1–2   BINARY NUMBER SYSTEM

In the binary system, all numbers can be made up by using only ones and zeros, rather than zero through nine, as in the decimal system. Consequently, instead of requiring ten different values to represent one digit, logic circuits using the binary method need only two values for each digit. In logic circuits, these values are easily indicated by the presence or absence of a signal (or pulse), or by positive and negative signals, or even by two different voltage levels.

In binary, the value of each digit is based on 2, and the

powers of 2. In a binary number, the extreme right-hand digit is multiplied by 1, the second-from-the-right digit is multiplied by 2, the third-from-the-right digit is multiplied by 4, and so on. This can be displayed as follows:

$$2^8 \quad 2^7 \quad 2^6 \quad 2^5 \quad 2^4 \quad 2^3 \quad 2^2 \quad 2^1 \quad 2^0$$

$$256 \quad 128 \quad 64 \quad 32 \quad 16 \quad 8 \quad 4 \quad 2 \quad 1$$

In binary, if the digit is zero, its value is zero. If the digit is one (1), its value is determined by its position from the right. For example, to represent the number 77 in binary form, the following combination of zeros and ones is used:

256	128	64	32	16	8	4	2	1
0	0	1	0	0	1	1	0	1

$$64 + 0 + 0 + 8 + 4 + 0 + 1 = 77$$

which means 1001101 in pure binary form $= 77$.

1–2.1   Converting Decimal Numbers Into Binary Numbers

A decimal number can be converted into a binary number in two ways. The obvious way is to make up a chart showing the power of two, as has just been done, and to then count the necessary number of ones and zeros to make up the desired decimal number.

For example, assume that the number 33 is to be converted. The number 33 is more than 32 (sixth position from the right) but less than 64 (seventh position from the right). This means that you need a combination of six digits (ones or zeros—probably both).

Start with the sixth position, or 32. Since you want number 32, write a one in the sixth position. Then move to the fifth position, or 16. Thirty-two plus 16 is greater than the desired 33, so use a zero for the fifth position. The fourth position is 8, and 8 plus 32 is more than 33, so you use a 0 for the fourth position. The same is true of the third position, or 4, and the second position, or 2, so both of these positions use a 0. The first (right-hand) position is 1, and 1 plus the sixth position (or 32) makes the desired 33, so both of these positions require a 1. Thus, the pure binary equivalent of 33 is 100001. This is shown in Figure 1-1, along with some other examples for tabular conversion of decimal and binary numbers.

Binary

Decimal	2^5	2^4	2^3	2^2	2^1	2^0
	32	16	8	4	2	1
0	0	0	0	0	0	0
1	0	0	0	0	0	1
2	0	0	0	0	1	0
3	0	0	0	0	1	1
4	0	0	0	1	0	0
5	0	0	0	1	0	1
6	0	0	0	1	1	0
7	0	0	0	1	1	1
8	0	0	1	0	0	0
9	0	0	1	0	0	1
10	0	0	1	0	1	0
20	0	1	0	1	0	0
30	0	1	1	1	1	0
→ 33	1	0	0	0	0	1
40	1	0	1	0	0	0
47	1	0	1	1	1	1
50	1	1	0	0	1	0

| 33 = 1 | 0 | 0 | 0 | 0 | 1 |

Figure 1-1. Tabular conversion of decimal and binary numbers.

The alternate method for converting from decimal to binary is to divide the decimal number by two as many times as is necessary to lower the quotient to a number less than two (1 or 0), using the *remainders* for each step of division as the binary numbers. This is shown in Figure 1-2.

Successive dividers	Original number and dividends	Remainder (binary number)
2	33	1
2	16	0
2	8	0
2	4	0
2	2	0
2	1	1 — 1 0 0 0 0 1

| 33 = 1 0 0 0 0 1 |

Figure 1-2. Converting decimal numbers to binary numbers by division.

For example, again assume that 33 is to be converted. Thirty-three divided by 2 is 16, with a remainder of 1. This 1 is the right-hand or first-position digit (also known as the least significant digit, or LSD).

Sixteen divided by 2 is 8 with a remainder of 0. This 0 is the second-position digit.

Eight divided by 2 is 4 with a remainder of 0. This 0 is the third-position digit. Two divided by 2 is 1 with a remainder of 0. This 0 is the fourth-position digit.

One divided by 2 is considered as 0 (since the whole number 1 can not be divided by 2), and there is a remainder of 1. This 1 is the sixth position. The 1 is also the left-hand or last-position digit (known as the most significant digit, or MSD).

Thus, the binary count for the decimal number 33 is 100001.

1-2.2   Adding Binary Numbers

The rules for adding binary numbers are:

$$0 + 0 = 0$$

$$0 + 1 = 1$$

$$1 + 0 = 1$$

$$1 + 1 = 0, \text{ with a carry of } 1.$$

Assume that the previous decimal 33 (binary 100001) is added to decimal 77 (binary 1001101):

```
  1001101
+  100001
 ─────────
  1101110  = decimal 110
```

There is a special rule for adding columns of binary numbers. If the number of the ones in any single column is greater than 2, divide the number of ones by 2. The number of times that 2 will divide into the number of ones is the amount of ones carried to the next column. If the number of ones divides evenly by 2, then the column total is written as 0. If the column does not divide out evenly, but has a remainder of 1, then the column total is written as 1.

For example:

```
   101
   101
   101
  ─────
  1111  = decimal 15.
```

In the first (right-hand) column there are three ones. This is more than 2, so it is divided by 2. Two goes into 3 once, with a remainder of 1. Therefore, the remainder of 1 is written for the first-position (right-hand) digit, and the 1 is carried into the second position. This results in:

$$
\begin{array}{r}
1 \\
101 \\
101 \\
101 \\
\hline
1
\end{array}
$$

In the second column, there is 1 plus 0, plus 0, plus 0, or simply 1. This is less than 2, so the special rule does not apply. Instead, $1 + 0 = 1$, and the second-position (middle) total is 1. This results in:

$$
\begin{array}{r}
101 \\
101 \\
101 \\
\hline
11
\end{array}
$$

In the third (left-hand) column, there are three 1s. This is more than 2, so it is divided by 2. Two goes into 3 once, with a remainder of 1. Therefore, the remainder of 1 is written for the third position, and the 1 is carried into the fourth position. This results in:

$$
\begin{array}{r}
1 \\
101 \\
101 \\
101 \\
\hline
111
\end{array}
$$

In the fourth position, there is a 1 (carried over) plus blanks or zeros. This is less than 2, so the special rule does not apply. Instead, $1 + 0 = 1$, and the fourth-position total is 1. This results in:

$$
\begin{array}{r}
101 \\
101 \\
101 \\
\hline
1111
\end{array} = \text{decimal } 15.
$$

1–2.3   Subtracting Binary Numbers

The rules for subtracting binary numbers are:

$$0 - 0 = 0$$
$$1 - 1 = 0$$
$$1 - 0 = 1$$
$$0 - 1 = 1, \text{ with a borrow of 1.}$$

Assume that the previous decimal 33 (binary 100001) is subtracted from decimal 77 (binary 1001101):

$$
\begin{array}{r}
1001101 \\
-100001 \\
\hline
101100
\end{array} = \text{decimal 44.}
$$

1–2.4   Multiplying Binary Numbers

The rules for multiplying binary numbers are:

$$1 \times 1 = 1$$
$$0 \times 1 = 0$$
$$1 \times 0 = 0$$
$$0 \times 0 = 0$$

Assume that the decimal 9 (binary 1001) is multiplied by decimal 3 (binary 11).

$$
\begin{array}{r}
1001 \\
\times\ 11 \\
\hline
1001 \\
1001 \\
\hline
11011
\end{array} = \text{decimal 27.}
$$

Note that multiplication is a form of adding and shifting. In logic circuits, there are many systems for multiplication, but they can be classified into two groups. In one group, multiplication is done by adding (with an adder, Chapters 2 and 5) and shifting (with a register, Chapter 4). The other method involves repeated addition. For example, 3×9 is, in effect, 9 added together three times (or 3 added together nine times). Multiplication of binary numbers is discussed further in Chapter 6.

1–2.5 Dividing Binary Numbers

In binary, division is a form of subtraction. For example, if decimal 12 is divided by decimal 3, then you can subtract 3 from 12, counting how many times it is subtracted until there is nothing left, or until the remainder is less than 3. Dividing 3 (binary 11) into 12 (binary 1100) results in:

$$
\begin{array}{r}
100 \text{ or decimal 4} \\
11\overline{)1100} \\
11 \\
\hline
00
\end{array}
$$

Dividing 5 (binary 101) into 12 (binary 1100) results in:

$$
\begin{array}{r}
10 \text{ or decimal 2} \\
101\overline{)1100} \\
101 \\
\hline
10 \text{ or decimal 2 remainder.}
\end{array}
$$

In logic circuits, division is usually done with subtractors (Chapter 2) and shift registers (Chapter 4). However, binary division can be done with logic subtraction circuits (operating at high speeds) and simple storage registers.

1–2.6 Binary Fractions

To express fractions by binary numbers, proceed initially as in decimals. For example,

$$
\frac{3}{5} = \frac{11}{101}
$$

This means that the radix uses *negative* powers or exponents. For example, binary 0.0011 ($= 0.375$ decimal) represents:

$$
(0 \times 2^{-1}) + (1 \times 2^{-2}) + (1 \times 2^{-3})
$$

$$
0 + \frac{1}{4} + \frac{1}{8} = \frac{3}{8} = 0.375
$$

Figure 1-3 converts a number of common decimal fractions to binary equivalents.

Fraction	Decimal	Power of two	Binary equivalent
$\frac{1}{2}$	0.5	2^{-1}	0.1
$\frac{1}{4}$	0.25	2^{-2}	0.01
$\frac{1}{8}$	0.125	2^{-3}	0.001
$\frac{1}{16}$	0.0625	2^{-4}	0.0001
$\frac{1}{32}$	0.03125	2^{-5}	0.00001
$\frac{1}{64}$	0.015625	2^{-6}	0.000001
$\frac{1}{128}$	0.0078125	2^{-7}	0.0000001
$\frac{1}{256}$	0.00390625	2^{-8}	0.00000001
$\frac{1}{512}$	0.001953125	2^{-9}	0.000000001

Figure 1-3.   Fractions, decimals, and binary equivalents.

1-3   BINARY CODED DECIMAL AND OTHER SPECIAL LOGIC CODES

The binary coded decimal (BCD) system combines the advantages of the binary system (the need in logic circuits for only two states, one or zero) and the convenience of the familiar decimal representation. In the BCD system, a number is expressed in normal decimal coding, but each digit in the number is expressed in binary form.

For example, the number 37 in BCD form appears as:

	Tens Digit	Units Digit
Decimal	3	7
BCD	0011	0111

Note that *four bits* of information are needed for each digit. In general, four bits yield 2^4 or 16 possible combinations. (However, in certain logic codes, not all combinations are used. When combinations are unused, they are often known as *forbidden codes*.)

Also note that there are many codes used in logic design other than the straight BCD. Figure 1-4 shows some typical logic codes. Note that most of the codes require four bits for each digit. The 2-out-of-5 (five bits) and biquinary (seven bits) codes are exceptions.

Decimal	Binary	Octal	Hexa decimal	BCD	2421	5421	XS3	Reflected grey	2 out of 5	Biquinary 5043210
0	0000	0	0	0000	0000	0000	0011-0011	0000	00011	0100001
1	0001	1	1	0001	0001	0001	0011-0100	0001	00101	0100010
2	0010	2	2	0010	0010	0010	0011-0101	0011	00110	0100100
3	0011	3	3	0011	0011	0011	0011-0110	0010	01001	0101000
4	0100	4	4	0100	0100	0100	0011-0111	0110	01010	0110000
5	0101	5	5	0101	1011	1000	0011-1000	0111	01100	1000001
6	0110	6	6	0110	1100	1001	0011-1001	0101	10001	1000010
7	0111	7	7	0111	1101	1010	0011-1010	0100	10010	1000100
8	1000	10	8	1000	1110	1011	0011-1011	1100	10100	1001000
9	1001	11	9	1001	1111	1100	0011-1100	1101	11000	1010000
10	1010	12	A	0001-0000	0001-0000	0001-0000	0100-0011	1111		
11	1011	13	B	0001-0001	0001-0001	0001-0001	0100-0100	1110		
12	1100	14	C	0001-0010	0001-0010	0001-0010	0100-0101	1010		
13	1101	15	D	0001-0011	0001-0011	0001-0011	0100-0110	1011		
14	1110	16	E	0001-0100	0001-0100	0001-0100	0100-0111	1001		
15	1111	17	F	0001-0101	0001-1011	0001-1000	0100-1000	1000		

Figure 1-4. Typical codes used in logic circuits.

1–4   THE STATES AND QUANTITIES OF LOGIC ALGEBRA

Only two discrete states are considered to exist in logical algebra.   Any pair of conditions (different from each other) could be chosen. However, in logic design, the states are usually described as "true" or "false," or possibly as "up" or "down" (or "high" or "low"), referring to the presence or polarity of pulses. A designer might choose any dissimilar value or state to represent "true," and any other value to represent "false." The resulting design would be normally consistent in the use of these chosen values.

Every logical quantity is single-valued.   That is, no quantity may be simultaneously both "true" and "false." Thus, every logical quantity must exist in one or the other of the chosen states ("true" and "false"). No other value is possible.

Every logical quantity has an opposite.   If the quantity is "true," then the inverse, or complement, is "false." If the quantity is "false," then the opposite is "true."

Any quantity that is "true" is equal to any other quantity that is "true."   Any "false" quantity is equal to any other "false" quantity. Thus, there can be only one "true" quantity and one "false" quantity.

A logical quantity may be either variable or constant.   If variable, the quantity may switch between the "true" and "false" states from time to time, but *only between* these two extremes. If constant, the logical quantity must remain either "true" or "false."

Logical quantities may be represented in many ways.   To help in the understanding of the operation of logic circuits, quantities are often represented by the position of a toggle switch (open or closed), or by the presence or absence of light (lamp "on" or lamp "off"). In practical logic circuits, quantities are usually represented by the presence or absence of pulses, the polarity of pulses, or the amplitude of pulses. For example, in one system, the presence of a pulse represents the "true" state or quantity, whereas the absence of a pulse represents "false." Another system might use the exact opposite representation ("false" by the present of pulse: "true" by absence of pulse).

1–5   LOGICAL ALGEBRA NOTATION

Letters of the alphabet are used to represent the condition or state of a logic variable in logical algebra. For example, the letter A might represent the condition of a logic variable. In this case, the values A could assume would be only "true" or "false," since these are the only values any logic variable may represent. (In conventional algebra, the letter A could represent any value, from minus to plus infinity.)

The inverse, or complement, of the letter A is represented by the symbol $\overline{A}$ (A-bar), A' (A-prime), or A^* (A-asterisk or A-star). (The "prime" and "asterisk" signs are the most popular notations in text for the inverse condition, since they are available on most typewriters. The opposite is true in logic diagrams or illustrations.)

No matter what system of notation is used to show the complement, or inverse, condition, A and A^* cannot have the same value at the same time. Thus, if A is "false" at any given time, then A^* must be "true," and vice versa.

If two or more logic variables are present at the same time, one might be represented by A, another by B, and so on. If at this time B happens to be "true" and A happens to be "true," then $A = B$, since a "true" always equals a "true." Simultaneously, the opposite of B (or B^*) and the opposite of A (A^*) are equal to each other, and are "false." A moment later, the variable represented by A may change state, and A then becomes "false" and A^* (the inverse or complement) becomes "true."

A^* (now equal to "true") = B (still true) and

A (now "false") = B^* (still false)

It is important to remember that A, A^*, B, and so on, are simply *symbols* used to represent logic variables and that, at any time, any symbol may be "true" and a moment later the same symbol may be "false."

Generally, the symbol or letter T is reserved to mean the "true" state, and F is reserved for the "false" state. An exception is where F is used to indicate *function* or *output* of a logic circuit.

Often, "1" represents "true," and "0" represents "false." The 1 and 0 are in common use since they are similar to binary numbers, and are easy to write. Note that 1 in logical notation is not necessarily equal to the number 1, and 0 is not necessarily equal to the number 0, unless proved to be so. The symbols 1 and 0 only denote "true" and "false" *states* in logical variables.

For example, a 1 means *yes, assertion, up, high, enable*, or *true*. Consequently, 0 means *no, negation, down, low, disable* (or *inhibit*), or *false*. Use of the words *true* and *false* (or *up* and *down*) does not imply that one state is more important than the other, or that one state is necessarily the "normal" state. The states are conditions, and both states are equally significant and are used equally in a two-state system.

Using a practical example, a designer might define as original conditions:

True state—represented by T or 1, and equal to −5 V.

False state—represented by F or 0 and equal to 0 V.

Under these conditions, the binary number 10011 (decimal 19) present in the output of a logic design might be represented electronically on five different lines, as shown in Figure 1-5. Line A could be given the logic notation of A, line B the logic notation B, and so on. If two lines are required for each output, each pair of lines would be given the logic notations A and A^*, B and B^*, and so on.

When line A has a true voltage (−5 V) on it, $A = T = 1$, and $A^* = F = 0$. When line A changes to zero volts (false), $A = F = 0$, and $A^* = T = 1$.

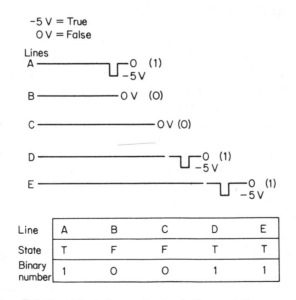

Line	A	B	C	D	E
State	T	F	F	T	T
Binary number	1	O	O	1	1

Figure 1-5.   Relationship between logic algebra notation, practical circuit voltage levels, and binary number system.

1–6   OPERATIONS IN LOGICAL ALGEBRA

The basic operations in logical algebra are:

> The AND operation.
>
> The OR operation.
>
> The NOT (or invert) operation.

In addition, there are five common operations (or functions) in logical algebra that are produced by combining the functions of the three basic operations. These common operations are:

> The NOR operation.
>
> The NAND operation.
>
> The AND/OR (or AO) operation.
>
> The EXCLUSIVE OR operation.
>
> The EXCLUSIVE NOR operation.

1–7   SYMBOLS IN LOGICAL ALGEBRA

As in conventional algebra, certain symbols (or connectives) are used to indicate the type of operation that is to be performed in logical algebra. Unfortunately, the use of connective symbols is not standard throughout the industry. However, the following symbols are generally accepted and understood:

The AND operation is represented by a dot between the variables $(A \cdot B)$. In many cases, the dot is omitted but is implied. Thus *ABC* reads: *A and B and C.*

The OR operation is represented by a plus sign between the variables $(A + B)$. This reads: *A or B.*

The NOT operation is represented by a log bar over the logical quantity to be complemented or inverted. This is similar to the inverse or complement notation discussed in Sec. 1–5. That is, $\overline{A} = A^*$. However, when the complement is subjected to the NOT operation, then $\overline{A}^* = A$.

If the term subjected to the NOT operation contains an OR or an AND operation in addition to the variables, the OR or AND operation is also complemented (or inverted). That is, if the variables are connected by an AND, the AND is changed to an OR, in addition to changing the variables. For example:

$$\overline{A^* + B} = AB^* \text{ and } \overline{A^*B} = A + B^*$$

Thus, the complement of the AND operation is an OR, and the complement of OR is an AND.

The equal sign (=) found in conventional algebra is carried over into logical algebra, and with the same meaning (equals, or, is the result of). However, the equal sign (=) with two bars is a *conditional equal* in logical algebra. That is, the conditions must be stated for the equations to be correct. For example, $A = B$, where A is true and B is true. The equal sign (≡) with three bars is an unconditional equality. For example, $A \equiv B$ means that A is the same as B, under all conditions.

Figure 1-6 shows typical symbols, equivalent symbols, and

Symbol or notation	Description
•, ∩, ∧, X	AND
+, ∪, V	OR
$\overline{X}$, ~X, X–, X′	Not X
=	Equivalence
⊕	Exclusive OR
$\overline{\oplus}$	Exclusive NOR
|, ⋏	NAND
↓, ⋎	NOR
A + B	A or B
$\overline{A}$	Not A
A ⊕ B	A exclusive – or B
A = B	A equals B, conditionally
A ≡ B	A is identical to B
I	True element
O	False element
A ⊃ B	A implies B
A → B	A implies B
A ⊂ B	A belongs to B
A ∪ B	Union of A and B
A ∩ B	Intersection of A and B

Figure 1-6.   Boolean symbols and notation.

forms of notation used in Boolean algebra. Figure 1-7 lists the equations commonly used in Boolean algebra. Note that the equations are grouped according to function, where practical.

$$A \cdot T = A \qquad A + (A \cdot B) = A$$
$$A \cdot F = F \qquad A \cdot (A + B) = A$$
$$A \cdot A = A \qquad A + (\overline{A} \cdot B) = A + B$$
$$A \cdot \overline{A} = F \qquad A \cdot B + A \cdot \overline{B} = A$$

Simplifying theorems

$$\overline{\overline{A}} = A$$

$$A + T = T \qquad \overline{A + B + C} = \overline{A} \cdot \overline{B} \cdot \overline{C}$$
$$A + F = A \qquad \overline{A \cdot B \cdot C} = \overline{A} + \overline{B} + \overline{C}$$

DeMorgan's theorems

$$A + A = A$$
$$A + \overline{A} = T$$

$$A + B = B + A$$
$$A \cdot B = B \cdot A$$

Communicative laws

$$A \cdot (B + C) = (A \cdot B) + (A \cdot C)$$
$$A + (B \cdot C) = (A + B) \cdot (A + C)$$

Distributive laws

$$(A + B) + C = A + (B + C)$$
$$(A \cdot B) \cdot C = A \cdot (B \cdot C)$$

Associative laws

Figure 1-7.   Equations commonly used in Boolean algebra.

1–8   TRUTH TABLES

The truth table is probably the most useful (and the simplest) tool for analyzing problems in logical algebra. As shown in Figure 1-8, the truth table consists of one vertical column for each of the logic variables involved in a given problem. The horizontal lines or rows of the truth table are filled with all possible true-false combinations the variables can assume with respect to each other.

For example, two variables can assume four different combinations at any of four different times: that is, both true at the same time; both false at the same time; and one true and one false, or vice versa. There are no other possible combinations. The additional column (or columns) will contain the output or results produced by the combinations of variables. For example, the output column (also known as the "function" column by some logic designers) can contain the result, or "output," when the variables are ANDed together, ORed together, or any particular output that must occur for each combination stated in the design problem.

Truth tables can be made up using any combination of letters, numbers, or symbols. Generally, the letters T and F are used in truth tables that accompany explanations of basic logic symbols.

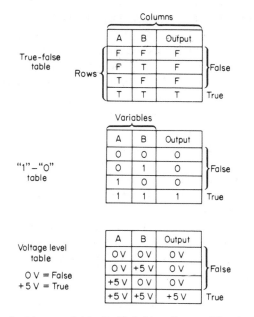

Figure 1-8.   Typical two-variable truth tables (for positive-logic AND gate).

Truth tables found on practical working logic diagrams (or logic data-sheets) generally use the numbers 1 and 0 (for true and false, respectively).

1–8.1   Summary of Truth Tables

Figure 1-9 summarizes the truth tables for the six basic logic circuit elements AND, OR, NAND, NOR, EXCLUSIVE OR, and EXCLUSIVE NOR.

A	B	AND	NAND	OR	NOR	Exclusive OR	Exclusive NOR
0	0	0	1	0	1	0	1
0	1	0	1	1	0	1	0
1	0	0	1	1	0	1	0
1	1	1	0	1	0	0	1

Figure 1-9.   Summary of truth tables. *(Courtesy of Cambridge Thermionic Corporation)*

1–9   POSITIVE AND NEGATIVE LOGIC

Although any system of letters and numbers can be used to represent true and false in logic equations, voltage (or current) *levels* are used in practical logic circuits. For example, true could be $+12$ V, false -12 V; true could be the presence of current, false the absence of current. Generally, voltage levels, rather than current levels, are used to define the true and false states.

Because logic circuits can work equally well with positive or negative voltages, it is necessary to define if the logic is positive or negative. In some (but not all) logic diagrams, a plus or minus sign may be used within any logic symbol to define the *true state* for that element. The sign should be used for all logic elements in which true and false levels are meaningful. As an alternative, the logic diagram can state in a note that all logic is positive true or that it is all negative true. Unfortunately, not all manufacturers or logic designers use this notation.

When used, the plus sign within a logic symbol means that the *relatively positive level* of the two logic voltages at which the circuit operates is said to be true. This is defined as *positive logic*. Note that the true voltage level does not have to be absolutely positive (that is, above ground or above the 0-V reference). The two voltage levels at which a logic circuit operates could be -1.7 V and -0.9 V (which is the case for ECL logic described in later sections of this chapter). A plus sign within the logic symbol (or a note specifying positive true) would indicate that the -0.9 V is true, and the -1.7 V is false, since the -0.9 V is closer to positive than -1.7 V.

A minus sign within the symbol (or a note specifying negative true) would indicate that the -1.7 V level is true (since it is more negative) and that the -0.9 V is therefore false. This is also defined as *negative logic*. Note that the voltage levels shown in Figure 1-5 use negative logic. That is, the true state is at -5 V, whereas the false state is at zero V

1–9.1   Logic Pulses

Note that the voltages shown in Figure 1-5 are actually electrical pulses. It is important that the reader understand the relationship of positive and negative logic to pulses. The logic designer, like everyone else who works with switching functions, finds it advantageous to use pulses to describe what goes on inside a logic circuit.

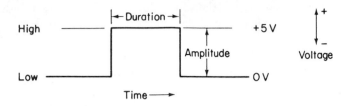

Figure 1-10.   Simplified pulse definitions (positive pulse).

This is especially convenient since logic circuits usually (but not always) work with pulses.

Some terms associated with electrical pulses should be defined here. Note that these definitions are simplified for the purpose of explanation.

The pulse of Figure 1-10 is a positive pulse because it goes from a lower level to a higher level, and back to a lower level. The pulse of Figure 1-11 is a negative pulse because it goes from a higher level to a lower level, and back to a higher level. Lower and higher levels represent *voltage levels.* Lower represents the lower voltage level or the *down* voltage. Higher represents the higher voltage or the *up* voltage level. The difference between the lower and higher level represents the *pulse amplitude.* For example, if the upper level of the pulse in Figure 1-10 is +5 V, with the lower level 0 V, the amplitude is

$$+5 - 0 = +5 \text{ V}.$$

If the upper level of the pulse in Figure 1-11 is 0 V, with the lower level −7 V, the amplitude is:

$$0 - 7 \text{ V} = -7 \text{ V}.$$

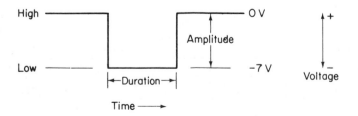

Figure 1-11.   Simplified pulse definitions (negative pulse).

With positive logic, the upper level of the pulses in Figures 1-10 and 1-11 is the true condition, with the lower level the false condition.

With negative logic, the lower level of the pulses is the true condition, with the upper level the false condition.

With either system, the length of time that a positive pulse stays up or a negative pulse stays down is called the *pulse duration.*

1–9.2   Logic Symbols for Positive and Negative Logic

Figure 1-12 shows how positive and negative logic is applied to logic symbols. Note that both MIL–STD–806B and MIL–STD–806C are shown. These are the military standard logic symbols in

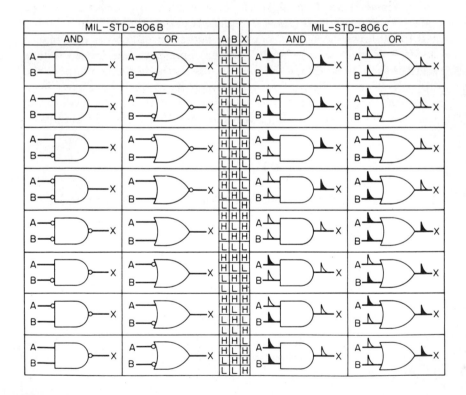

Figure 1-12.   Relationship between positive and negative logic symbols.

most common use. Also note that a solid arrow on a MIL–STD–806C symbol is the equivalent of *no inversion dot* on MIL–STD–806B. Likewise, an open arrow on MIL–STD–806C is the same as an inversion dot on MIL–STD–806B.

A solid arrow (no inversion dot) indicates that a 1 has been assigned to the higher level (voltage) in positive logic.

An open arrow (inversion dot) indicates that a 1 has been assigned to the lower level (voltage) in negative logic.

1–9.3   Logic Equivalences

Positive logic defines the 1 or true state as the most positive voltage level, whereas negative logic defines the most negative voltage level as the 1 or true state. Because of the difference in definition of states, it is possible for some logic elements to have *two equivalent outputs or functions,* depending upon definition. For example, the same basic logic gate may have a NOR function in positive logic, and a NAND function if negative logic is used. Figure 1-13 shows a comparison of several common logic functions. Note that a positive AND is equivalent to a negative OR, a positive OR is equivalent to a negative AND, and so on. Also note that a positive EXCLUSIVE OR is equivalent to a COINCIDENCE gate.. (The COINCIDENCE gate is the same as an EXCLUSIVE NOR, and the latter term is used by many logic designers.)

1–10   THE *AND* FUNCTION

The AND function (or operation) can be defined as follows:

The function (or output in the case of a gate)
is true when all the ANDed logical equations are true.

The function is false when one or more of the
ANDed quantities is false.

Figure 1-14 shows the truth table, logic symbol, and circuit diagrams of a basic AND gate. The diagrams show both positive and negative logic. The circuits shown are often referred to as "diode logic," since only switching diodes (and no transistors) are used.

With positive logic, when both inputs A and B are up (as indicated by the corresponding positive pulses), then and only then

Inputs		AND	OR	NAND	NOR	EXOR	COIN*
A	**B**						
Lo	Lo	Lo	Lo	Hi	Hi	Lo	Hi
Lo	Hi	Lo	Hi	Hi	Lo	Hi	Lo
Hi	Lo	Lo	Hi	Hi	Lo	Hi	Lo
Hi	Hi	Hi	Hi	Lo	Lo	Lo	Hi
A	**B**	OR	AND	NOR	NAND	COIN*	EXOR

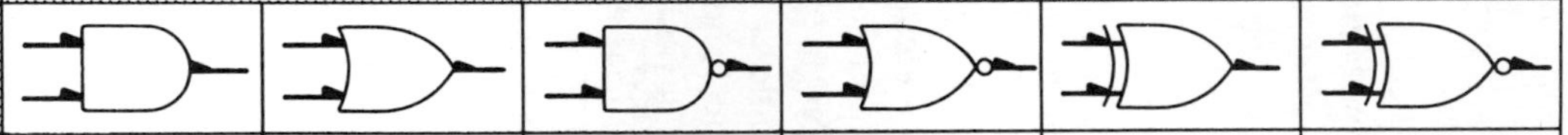

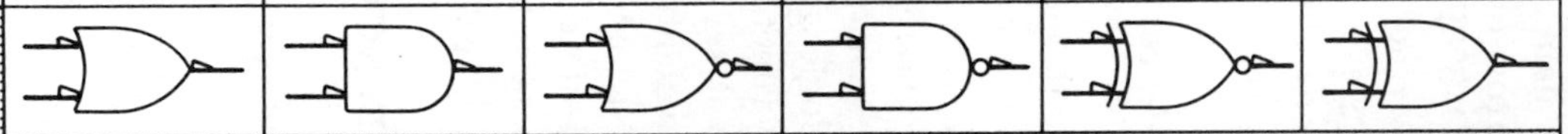

Positive logic

Negative logic

*Coincidence

Figure 1-13. Comparative positive and negative logic functions. (*Courtesy of Motorola*)

22

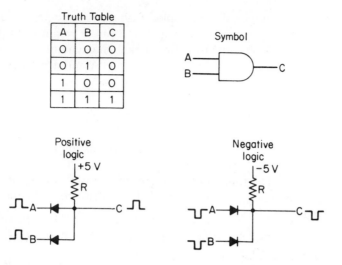

Figure 1-14.   AND gate truth table, logic symbol, and circuit diagrams.

the output goes up. (Both diodes are reverse-biased, current flow through resistor R is halted, and the output line rises to some voltage near that of the source of +5 V.) If any one or both inputs are down, then the logic AND function is not satisfied, and the output goes down. (One or more of the diodes are forward-biased, current flows through resistor *R*, and the output line drops.)

Note that resistor *R* is often referred to as the *pull-up resistor* on some gate datasheets.

In practical circuits, AND gates may appear in multiple-input form (usually about six inputs is maximum). No matter how many inputs are involved, *all inputs must be true* for the output to be true.

Although all inputs must be true for an AND gate to produce an output, the inputs need not necessarily be of the same polarity. That is, the gate circuit can be arranged to produce a true output when mixed high and low inputs are applied. This is shown in Figure 1-15. An inverted input is represented on the symbol by a dot (or small circle) on that input which is inverted from the *normal logic* on

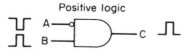

Figure 1-15.   AND gate with mixed inputs.

the rest of the diagram. For example, with positive logic, an *inversion dot* on an input indicates that a negative input is required to produce a true condition. Thus, if it is assumed that the symbol of Figure 1-15 uses positive logic, a low is required at input *A* and a high at input *B* to produce a true output.

1–11   THE *OR* FUNCTION

The OR function (or operation) can be defined as follows:

> The function (or output in the case of a gate) is true when one or more of the logical quantities is true.

> The function is false only when all the variables are false.

Figure 1-16 shows the truth table, logic symbol, and circuit diagrams of a basic OR gate. The diagrams show both positive and negative logic. Note that the positive OR gate is identical to the negative logic AND gate (Figure 1-14), and vice versa, as discussed in Sec. 1–9.3.

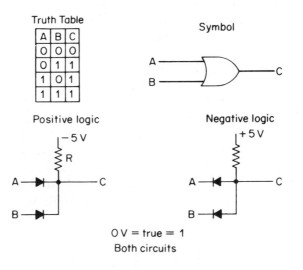

Figure 1-16.   OR gate truth table, logic symbol, and circuit diagrams.

With the positive logic OR gate, the output goes up with either (or both) *A* and *B* inputs up. Assume that true is 0 V and false is −5 V. With inputs *A* and *B* at −5 V (false) both diodes are reverse-biased. No current flows through resistor *R*, and the output line assumes a voltage near that of the source, or −5 V (false). If input *A* goes to 0 V (true), the corresponding diode is forward-biased (appearing as a short), and the output line drops to 0 V (true). The same condition occurs if input *B* goes to 0 V (true), or if both inputs *A* and *B* go to 0 V.

In practical circuits, OR gates may appear in multiple-input form (usually about six inputs is maximum). No matter how many inputs are involved *any one true input* will produce a true output.

The inputs need not be of the same polarity. The OR gate can be arranged to produce a true output when mixed high and low inputs are applied. This is shown in Figure 1-17. On the symbol, a dot indicates an input that is inverted from normal logic. For example, with positive logic, an inversion dot on an input indicates that a negative input is required to produce a true condition. Thus, if it is assumed that the symbol of Figure 1-17 uses positive logic, a low is required at input *A* or a high at input *B* to produce a true output.

Positive logic

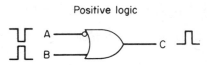

Figure 1-17.   OR gate with mixed inputs.

1–12   THE *AND–OR* FUNCTION

The AND–OR function is not basic in the same sense as the AND and OR functions. However, the AND–OR function is used so frequently in logic design (particularly in a version where the OR output is inverted) that the function can be treated as a basic operation.

Figure 1-18 shows the truth table, logic symbol, and circuit diagrams of a basic AND–OR gate. The diagram shows positive logic only. However, negative logic can be accomplished by reversing the diodes and voltage polarities.

Note that the AND–OR function is made up of two AND gates at the input, and one OR gate at the output. Operation of the individual gates is identical to that described in Secs. 1–10 and 1–11.

Truth Table

A	B	C	D	E
0	0	0	0	0
0	1	0	0	0
1	0	0	0	0
1	1	0	0	1
0	0	1	0	0
0	0	0	1	0
0	0	1	1	1

Symbol

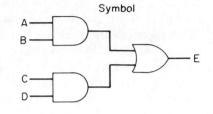

Positive logic

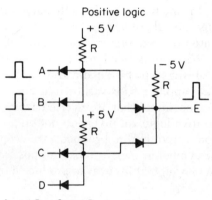

A <u>and</u> B <u>or</u> C <u>and</u> D = output

Figure 1-18.   AND-OR function truth table, logic symbol, and circuit diagram.

1–13   THE *NOT* FUNCTION

The NOT function is often referred to as negation, or a complementing function. The negative of a quantity may also be called the inverse, converse, or opposite. The NOT circuit is not a gate. However, the NOT function is required to produce the NAND and NOR gate functions (described in Secs. 1–14 and 1–15).

The output of a NOT circuit is the inverse of the input. For example, if the input is a positive-going pulse, the output will be a negative-going pulse. The NOT function is sometimes combined with amplification.

The basic NOT circuits are shown in Figure 1-19. Note that these circuits are essentially single-stage, common-emitter transistor amplifiers. In a negative-logic system, a PNP transistor is used, since a negative voltage at the base causes the transistor to conduct, creating a ground (0 V) output that is the opposite of the negative input

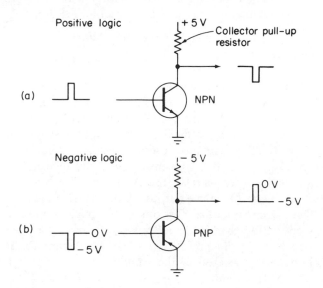

Figure 1-19. Basic NOT circuits.

voltage. For example, assume that the logic is negative (Figure 1-19b), that true is represented by −5 V, and false is 0 V. If the input is true, the −5 V at the base will cause the transistor to conduct. The output (collector) will then rise to 0 V (or near 0 V), producing a false output. If the input is false, the 0 V at the base will not turn the transistor on, and the output will remain at some voltage near the source of −5 V (or true).

The transistor circuit may also provide some amplification. The outputs of diode switching circuits have an amplitude that is not as high as their input pulses. The output pulses are always attenuated in a diode circuit. Also, since diodes are passive devices, the output of a diode switching circuit does not have the capability to drive too many other diode circuits. The transistor NOT circuit can provide pulse amplitude restoration, and driving capability, in addition to inversion of the pulses.

1–14   THE *NAND* FUNCTION

The NAND operation is a combination of the NOT and AND operations. The term NAND is a contraction of NOT AND. The NAND operation can be defined as follows:

The function (or output in the case of a gate) is true when one or more of the variables is false.

The function is false only when all of the variables are true.

Figure 1-20 shows the truth table, logic symbol, and circuit diagram of a basic NAND circuit. Note that the circuit is that of a diode AND gate and a transistor NOT gate. The AND gate performs the usual AND function, producing a true output when both inputs are true. The NOT circuit (a common-emitter amplifier) inverts this true into a false. Thus, two true inputs will produce a false output. If either (or both) inputs is false, the AND gate produces a false output, which is inverted to a true output by the NOT circuit.

There are many other versions of the NAND gate, some using all transistors, rather than combinations of diodes and transistors. The NAND gate is generally preferred to the AND gate for most logic design applications because of the amplification factor provided by the transistor NOT circuit.

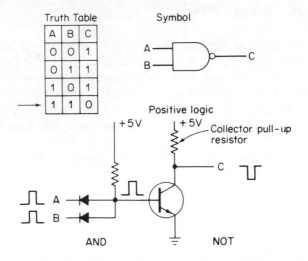

Figure 1-20. NAND function truth table, logic symbol, and circuit diagram.

1–15 THE *NOR* FUNCTION

The NOR operation is a combination of the NOT and OR operations. The term NOR is a contraction of NOT OR. The NOR operation can be defined as follows:

The function (or output in the case of a gate) is false when one or more of the logical variables is true.

The function is true only when none of the variables is true.

Figure 1-21 shows the truth table, logic symbol, and circuit diagram of a basic NOR gate. The circuit is that of a diode OR gate, and a transistor NOT circuit. The OR gate performs the usual OR function, producing a true output when either or both inputs are true. The NOT circuit inverts this true into a false. Thus, one or both true inputs will produce a false output. If both inputs are false, the OR gate produces a false output, which is inverted to a true output by the NOT circuit.

There are many other versions of the NOR gate, some using all transistors, rather than combinations of diodes and transistors. The NOR gate is generally preferred to the OR gate for most logic design applications because of the amplification factor provided by the transistor NOT circuit.

Truth Table

A	B	C
0	0	1
0	1	0
1	0	0
1	1	0

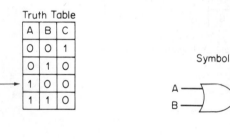

Symbol

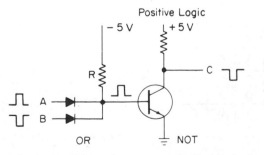

Figure 1-21. NOR function truth table, logic symbol, and circuit diagram.

1–16   THE *EXCLUSIVE OR* AND *EXCLUSIVE NOR* FUNCTIONS

An EXCLUSIVE OR gate is a special type of OR gate. This gate has only two inputs, and one output. The output is true if one, but not both, of the inputs is true. The converse statement is equally accurate: the output is false if the inputs are both true or both false. The EXCLUSIVE OR gate is independent of polarity, and is not generally spoken of as being either positive-true or negative-true. One method of producing an EXCLUSIVE OR gate, along with the symbol and truth table, are shown in Figure 1-22.

Figure 1-22 also illustrates the symbol and truth table for the EXCLUSIVE NOR gate, which operates in the same way as the EXCLUSIVE OR, except that the output is inverted. *Two false* inputs,

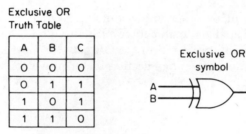

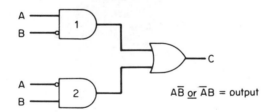

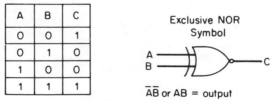

Figure 1-22.   EXCLUSIVE OR and EXCLUSIVE NOR truth tables, logic symbols, and circuit diagrams.

or *two true* inputs, produce a true output. (Thus, the EXCLUSIVE NOR is also called a COINCIDENCE function.) One false and one true input produce a false output.

1–17   BASIC LOGIC ELEMENTS AND SYMBOLS

Although there are many variations of logic elements and their symbols, there are only four *basic* classes or groups. These are gates, amplifiers (or inverters), switching elements, and delay elements.

1–17.1   Basic Gates

A gate is a circuit that produces an output on condition of certain rules governing input combinations. As shown in Figure 1-23, the basic gate symbol has input lines connecting to the flat side of the symbol, and output lines connecting to the curved or pointed sides. Since inputs and outputs are thus easily identifiable, the symbol

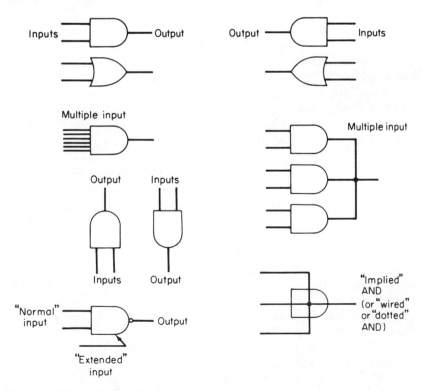

Figure 1-23.   Basic gate symbols.

can be shown facing left or facing right (or facing up or down), as necessary.

There may be two inputs to a gate. In some cases, multiple inputs (more than two) can be accomplished by increasing internal circuit components of the gate (such as adding an extra diode for each input). In other cases, it is necessary to connect the outputs of two (or more) gates together, in parallel. For example, the output of three two-input AND gates can be connected in parallel, resulting in a six-input AND gate. This is sometimes known as a *wired, dotted,* or *implied* AND function. Some designers use the symbol shown in Figure 1-23 when the term "implied AND" is used.

It is also possible for a gate to have an input other than the normal input. This is often referred to as an "extended input." For example, a NAND gate can be made up of a diode AND gate, followed by a transistor amplifier (which also inverts). An input can be connected directly to the transistor base, thus bypassing the diode AND function. Generally, a signal at an extended input produces an output, regardless of the condition (true or false) at all normal inputs.

1–17.2 Amplifiers

When amplifiers are used in logic circuits, the driving or input signals are normally pulsed. Consequently, the output of the amplifier is an amplified form of the input pulse. As shown in Figure 1-24, the amplifier symbol is an equilateral triangle, with the input applied to the center of one side and the output taken from the opposite point of the triangle. Like gates, the amplifier may be shown in any of the four positions.

When an amplifier is used as a separate element in logic circuits, it is assumed that the output is essentially the same as the input, but in amplified form. That is, a true input produces a true output, and vice versa. When inversion occurs, an inversion dot (or possibly an inverted pulse symbol) is placed at the output. Usually, the element is then termed an *inverter* rather than an amplifier, even though amplification may occur. Also, an amplifier (with or without inversion) can be called a *buffer* when it is used between two logic elements or circuits.

If a plus or minus sign is used in the symbol, this indicates the *input* polarity required to turn the amplifier on.

One amplifier or inverter symbol may represent any number of amplification stages or, optionally, separate symbols may be shown for each stage. Logic symbols, by themselves, do not necessarily imply a specific number of components, but rather relate to overall logic effect.

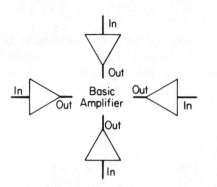

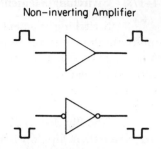

Non-inverting Amplifier

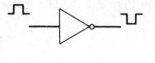

Phase-splitter

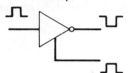

Inverting Amplifier or Inverter

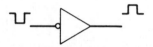

Differential Amplifier

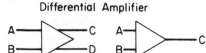

Figure 1-24.   Amplifier, inverter, and phase-splitter symbols.

Sometimes an amplifier is used as a *phase splitter;* that is, one input with dual outputs. One of the dual outputs is in phase with the input, whereas the other is out of phase with the input.

A similar case exists with *differential amplifiers,* which have dual inputs and dual outputs (although some differential amplifiers have dual inputs and single outputs).

Note that the present trend in logic design is to use NAND and NOR circuits (or AND–OR INVERT), rather than separate amplifiers or inverters. This is discussed further in Sec. 1–19.

1–17.3   Switching Elements

Switching elements used in logic design are some form of multivibrator: bistable (flip-flop, latch, Schmitt trigger), monostable (one-shot), and astable (free-running). According to the type of circuit, inputs cause the state of the circuit to switch, reversing the output; that is, an output formerly true will switch to false, and vice versa.

The flip-flop or latch is the most common logic circuit switching element. A flip-flop is bistable; that is, it takes an external signal to set the element and another signal to reset it. The flip-flop will remain in a given state until switched to the opposite state by the appropriate external signal.

The basic switching element and flip-flop symbols are shown in Figure 1-25. The various types of flip-flops and switching elements

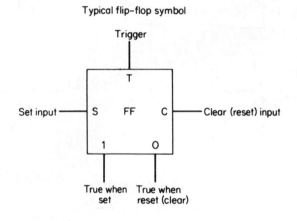

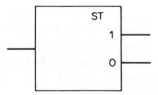

Figure 1-25.   Typical switching element symbols.

used in present-day logic circuits are discussed fully in other sections of this chapter, and in other chapters, as applicable.

1–17.4   Delay Elements

A delay element provides a finite time between input and output signals. The delay symbols, with examples of actual delay time, are shown in Figure 1-26. Many types of delay elements are used in logic design. Two frequently used delays are the *tapped delay* and delays effective only on the leading or trailing edges of pulses. Such delay elements, together with the theoretical waveforms, are shown in Figure 1-26.

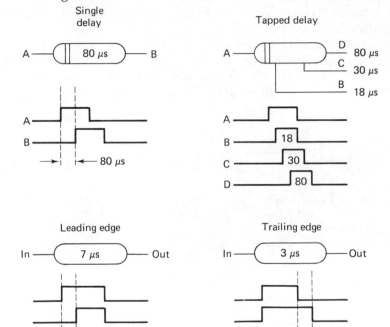

Figure 1-26.   Basic delay element symbols.

1–18   MODIFICATION AND IDENTIFICATION OF LOGIC SYMBOLS

The basic logic symbols are usually modified to express circuit conditions. Although designers may have their own set of modifiers, together with those of MIL-STD–806, the following modifiers are in general use.

1–18.1   Truth Polarity

As previously discussed, positive (+) or negative (−) indicators may be placed inside a symbol to designate whether the true state for that circuit is positive or negative, *relative* to the false state. This is frequently done with gates and switching elements, as shown in Figure 1-27. When all symbols on a particular diagram have the same polarity, a note to the effect that all logic is positive-true or all is negative-true may be used instead of having individual polarity signs in each symbol.

Note that positive-true logic is used throughout this manual, unless otherwise specified. Polarity signs used in amplifier symbols do not have any direct logic significance. Rather, the polarity signs are a troubleshooting aid, indicating the *polarity required to turn the amplifier on.*

As shown in Figure 1-27, the positive-true gate and positive flip-flop (FF) operate with true levels positive with respect to the false levels. Similarly, the negative-true gate and FF operate with true levels negative with respect to the false levels.

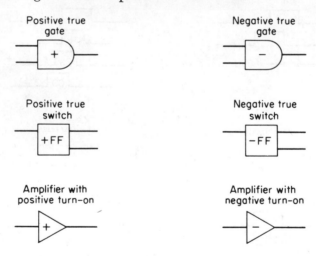

Figure 1-27.   Truth polarity in logic symbols.

1–18.2   Inversion

Generally, logic inversion is indicated by an inversion dot at inputs or outputs. (In some cases, inverted pulses are shown at the inputs and outputs.)

When the inversion dot appears on an input (generally only on gates and switching elements), the input will be effective when the input signal is of opposite polarity to that *normally required*. For example, if the switching element in Figure 1-28 is normally positive-true or is used on a diagram where all logic is positive-true, a negative-input (or the complemented input) at the inversion dot will set the circuit.

When the inversion dot appears at an output (generally only on gates and amplifiers), the output will be of opposite polarity to that *normally delivered*. For example, if the gate of Figure 1-28 is used in a positive-true logic circuit, the output will be negative. Likewise, the amplifier in Figure 1-28 will produce a positive output if the input is negative, and vice versa.

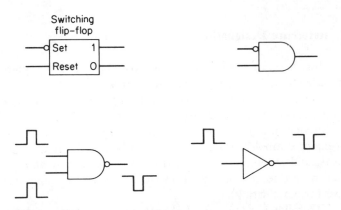

Figure 1-28.   Methods of indicating inversion in logic symbols.

1–18.3   Alternating-current Coupling

Capacitor inputs to logic elements are indicated by an arrow, as shown in Figure 1-29. In the case of gates and switching elements, the element responds only to a change of the ac coupled input in the true-going direction.

An inversion dot used in conjunction with the coupling arrow indicates that the element responds to a change in the false-going direction.

In the case of an amplifier, a pulse edge of the same polarity as given in the symbol turns the amplifier on briefly, then off as the capacitor discharges. The output is then a pulse of the same width as the amplifier on-time. With an inversion dot at the amplifier output, the output pulse is inverted.

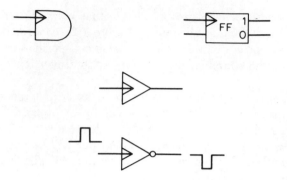

Figure 1-29.   Methods of indicating ac coupling in logic symbols.

1–18.4   Reference Designations

Most logic diagrams show logic elements as a complete component rather than the many components that make up the element. For example, an amplifier is shown as a triangle, rather than as several dozen resistors, capacitors, transistors, and so on. Thus, the amplifier has a reference designation of its own. On some logic diagrams, the logic element symbols are mixed with symbols of individual resistors, capacitors, and so on. Either way, the logic element symbol must be identified by a reference designation (to match descriptions in text or as a basis for parts listing).

The following logic symbol reference designations are in general use.

        Gates   G

        Amplifiers, inverters, and buffers   A

        Flip-flops and latches   FF

        One-shot   OS (sometimes SS for single-shot)

        Multivibrator   MV

        Schmitt trigger   ST

        Delay   D

Figure 1-30 shows some examples of how the reference designations are used. Note that the reference designations for switching are placed within the symbol. All other designations are (usually) placed beside the symbol. In the case of switching elements, the true-state sign can be used as a prefix to the designations.

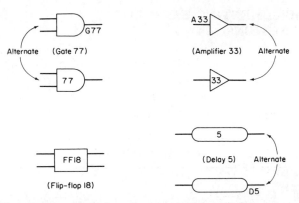

Alternate   (Gate 77)

(Amplifier 33)   Alternate

(Flip-flop 18)

(Delay 5)   Alternate

Figure 1-30. Reference designations on logic symbols.

When logic elements appear in integrated circuit (IC) or *microcircuit* form, the system of reference designations is usually changed. Typical examples of microcircuit logic reference designations are shown in Figure 1-31.

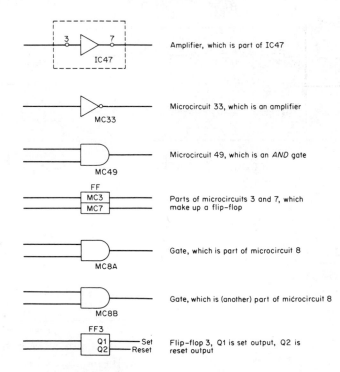

Amplifier, which is part of IC47

Microcircuit 33, which is an amplifier

Microcircuit 49, which is an *AND* gate

Parts of microcircuits 3 and 7, which make up a flip-flop

Gate, which is part of microcircuit 8

Gate, which is (another) part of microcircuit 8

Flip-flop 3, Q1 is set output, Q2 is reset output

Figure 1-31. Reference designations for microcircuit (integrated circuit or IC) logic symbols.

1–19   IMPLEMENTING BASIC LOGIC FUNCTIONS

The designer is often faced with the problem of producing (or implementing) a basic logic function using other logic functions. For example, a designer may be working with ICs or logic modules that contain a number of general-purpose NAND gates. (Many manu-

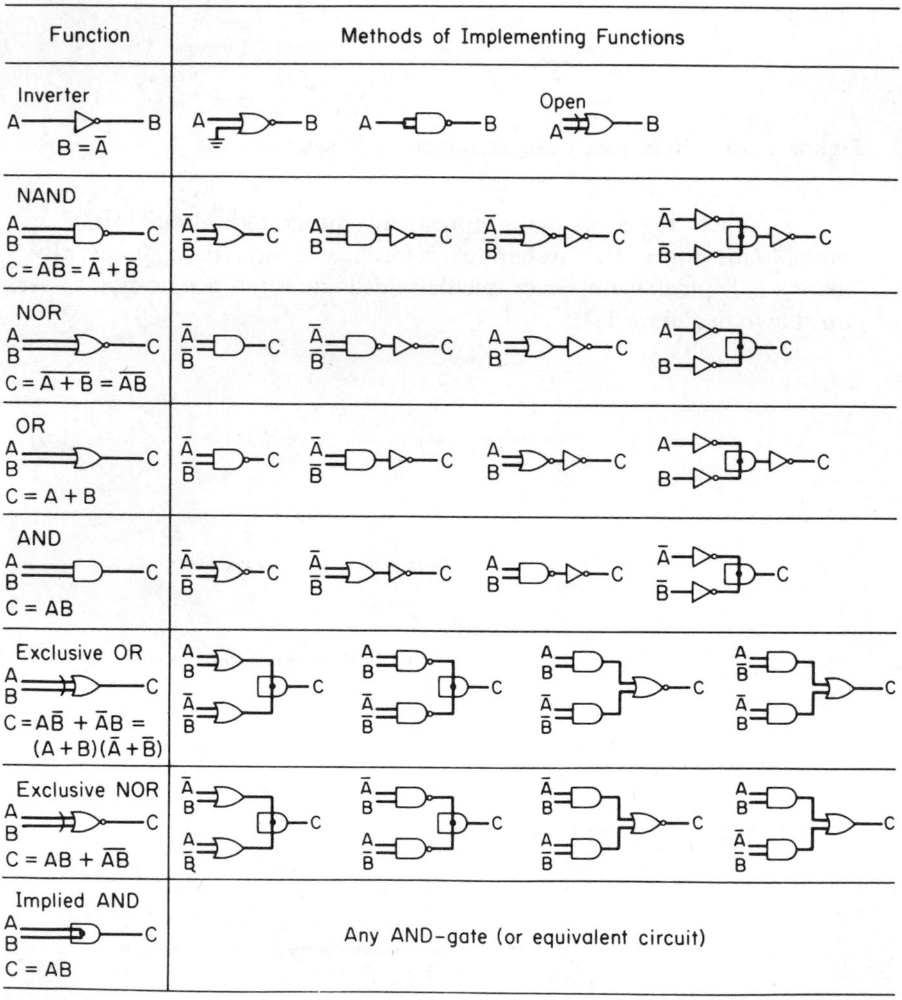

Function	Methods of Implementing Functions			
Inverter $B = \bar{A}$			Open	

Figure 1-32.   Implementing basic logic functions with other basic logic elements.

facturers produce such ICs and circuit boards as off-the-shelf items.) Although most of the design can be done using only NAND gates, assume that the design calls for the use of one inverter, or one NOR gate, and so on. It is possible to form all of the basic logic functions discussed thus far in this chapter by using only NAND gates.

Figure 1-32 shows the connections required to implement all of the basic logic functions, using other logic elements. For example, to form an inverter from a NOR gate, connect one input to ground or zero. To form the same inverter with a NAND gate, connect both inputs of the NAND gate together. With either connection, the output will be inverted from the input.

Figure 1-32 shows how to implement all of the basic logic functions using an assortment of other functions. Although the author does not necessarily recommend this method of implementing basic functions, he recognizes that the method is in common use, and does provide considerable convenience. Also, he recognizes that there are many alternate methods of implementing basic functions.

1–19.1   Parallel Outputs (Wired AND and Wired OR)

When gate outputs are connected in parallel an AND function or an OR function can result. These functions are often referred to as "wired AND" and "wired OR." If the true condition is represented by zero volts (or ground), then the parallel outputs or gates produce an OR function. If the true condition is represented by a voltage of any value or polarity, then the function is AND. The reason is that if a gate output is zero volts (ground) all the other outputs are, in effect, shorted to ground.

Any gates with active pull-up elements should not be connected for wired OR. When one gate has a 0 output and the other a 1 output, the resulting logic output is unpredictable.

1–20   WORKING WITH COMBINATIONAL AND SEQUENTIAL LOGIC NETWORKS

Today's logic designer has an infinite variety of combinational and sequential networks available in IC form. That is, the logic designer can now buy such networks as decoders, encoders, flip-flops, counters, registers, arithmetic units, and memories in complete, functioning packages that require only a power source. These packages are described in Chapter 2. However, on the assumption that it may be necessary for the reader to design a basic circuit, the remaining sections of this chapter are devoted to logic circuit design "from scratch." A review of these sections will also help the student understand the basics of logic design. We will start with some definitions.

Combinational logic networks are those which have no feedback or memory (or do not depend upon feedback or memory for their operation). The combinational network produces an output (or outputs) in response to the presence of two or more variables (or inputs). The nature of the output depends upon the combination of gates used. Adders, even and odd circuits, decoders, and subtractors are examples of combinational networks.

Sequential logic networks have a memory, and possibly a feedback, and must operate in a given sequence. The output of sequential network depends on the time relation or timing of the inputs and the feedback, as well as the combination of gates or other elements. Sequential networks usually involve flip-flops, and include such circuits as multivibrators, counters, registers, shift cells, multipliers, and dividers.

1–21 DESIGN PROCEDURE FOR BASIC COMBINATIONAL NETWORKS

Problems in logic design may be stated in many ways. Likewise, the logic designer may receive the problem in many forms. For example, in some cases, the designer need only convert an existing equation (already simplified) into a working circuit. This process is often referred to as implementing or mechanizing an equation. In other cases, the equation must first be simplified, thus making the circuit equally simple.

Note that all equations should not be reduced to the simplest terms, on an arbitrary basis. For example, an equation might contain three variables, one of which could be eliminated by proper manipulation of the equation. However, what if the designer must produce a module that will solve the equation, but the module has three inputs? The third input cannot be ignored.

In still other cases, the designer must write the equation from a truth table, or word statement, for the problem. No matter what the starting point, the logic designer should have a routine set of steps for the entire procedure. The designer should also use the same procedure repeatedly, until he no longer has to remember each step.

There are many procedures in common use by logic designers for writing, simplifying, and implementing equations. Some of these procedures involve the use of Veitch or Venn diagrams, or Karnaugh maps. The use of such diagrams and maps to simplify and implement logic circuits is discussed in Sec. 1–22. In this section, we shall use an even more basic procedure, based on simple truth tables and the rules of logic algebra. Once simplified to the desired terms, the equation can be implemented as a combinational network using basic logic gates.

1–21.1   Simplifying Logic Equations

Commonly used Boolean equations are given in Figure 1-7. The following rules and notes explain the equations.

The first (and probably most important) rule is: If two terms are identical except for one variable, and that variable is true in one term and inverted (or false) in the other term, the variable may be eliminated, and the two terms may be replaced by a common term. For example,

$$A^*B + AB \text{ can be replaced by } B.$$

For the purposes of this discussion, a term can be considered as any group of variables, separated from other groups of variables by the OR connective (plus sign).

Thus, in the expression $XYZ + AB + XYZ^* + BC$, the Z and Z^* can be eliminated, and the two XY terms combined into one. The resulting expression is $XY + AB + BC$.

A double negation converts an expression back to its original form. For example, $\overline{B^*} = B$, likewise $\overline{X^*Y^*Z^*} = XYZ$.

The expression $A^* + A$ is a true statement. This is because the expression indicates that A^* or A is to be considered.

The expression $A^* \cdot A$ is false. That is, $A^* \cdot A$ can never equal a true or 1. This is because the expression indicates the combination of a positive and a negative value, thus resulting in a cancellation.

The expressions $A + A$, AA, and $A + (AB)$ all equal A. Thus, any of these expressions (or their equivalents) in an equation can be reduced to A.

The equation $A + (A^*B) = A + B$, and can be so reduced in an equation.

The basic AND and OR functions are *commutative*. That is, the end results of the AND and OR connectives are not altered by the sequence that makes up the logical equation. For example, $A + B = B + A$, $XY = YX$.

Included in the commutative laws are De Morgan's laws:

$$\overline{AB} = \overline{A} + \overline{B} \text{ and } \overline{A} + \overline{B} = \overline{AB}$$

De Morgan's laws can be used to find the complement of any Boolean expression. For example, the negation of the AND ($\overline{AB}$) function (NOT–AND) is equal to the alternate denial ($\overline{A} + \overline{B}$), which expresses, in effect, that NOT–A or NOT–B is true.

In practical circuits, De Morgan's laws show the relationship of AND, OR, NAND, and NOR gates. If a NOT circuit follows an AND gate (the NAND function), the results are the same as inverting the $A + B$ inputs to an OR gate. Likewise, if a NOT circuit follows an OR gate (the NOR function), it is comparable to the circuit obtained if the $A \cdot B$ inputs to an AND gate are inverted.

The AND and OR functions are *associative*. The associative law states that the elements of a Boolean expression may be grouped as desired, so long as they are connected by the same sign. For example,

$$(X + Y) + Z = X + (Y + Z) = X + Y + Z$$
$$ABC = (AB)C = A(BC)$$

The AND and OR functions are also *distributive*. The distributive law has to do with the functional characteristics of logic connectives. If terms contain a common variable, even though the remainder of the terms are different, the common variables can be reduced to a single variable, and the terms can be recombined in simpler form, provided that the connective signs remain the same. For example,

$$(XY) + (XZ) = X(Y + Z)$$
$$(X + Y)(X + Z) = X + (YZ)$$

1–21.2 Sum of Products Versus Product of Sums

There are two classic solutions for the conversion of a truth table or word statement into a practical circuit (implementing the statement or truth table). These are *sum of products* (S of P) and the *product of sums* (P of S). Either solution can be applied to the same problem. This is shown in Figure 1-33, where two circuits satisfy the conditions of the same truth table.

Although the circuits of Figure 1 33 are workable, they have one obvious drawback. There is no amplification of the signals, since AND and OR gates are used. When the logic signals (pulses) must pass through many gates or other circuits, it is standard practice to provide some amplification, preferably at each gate, or at least for the group of gates that form a basic circuit.

Truth Table

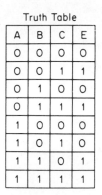

A	B	C	E
O	O	O	O
O	O	1	1
O	1	O	O
O	1	1	1
1	O	O	O
1	O	1	O
1	1	O	1
1	1	1	1

Sum of Products

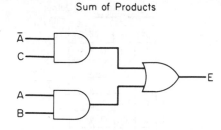

Product of Sums

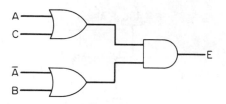

In Both Circuits

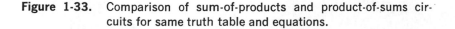

$$E = \bar{A}\bar{B}C + \bar{A}BC + AB\bar{C} + ABC$$
$$\bar{E} = \bar{A}\bar{B}\bar{C} + \bar{A}B\bar{C} + A\bar{B}\bar{C} + A\bar{B}C$$

Figure 1-33. Comparison of sum-of-products and product-of-sums circuits for same truth table and equations.

This amplification is usually provided by the use of NAND
and NOR gates, or by AOI (AND–OR INVERT) networks. The S-of-P
solution can best be implemented with NAND gates, whereas the
P-of-S solution uses OR gates. The AOI solution requires AND gates,
followed by a NOR gate.

1–21.3   The Basic S-of-P Solution

Figure 1-34 shows a truth table and a basic S-of-P circuit
(using AND and OR gates). The following steps describe the procedure for implementing the circuit, starting with the word statement.

A typical word statement of the problem could be: "The
circuit must accept three variables (or inputs); the circuit output
must be false when variables *ABC* are false, and when *AB* are true
and *C* is false, the circuit output must be true with all other combinations of variables."

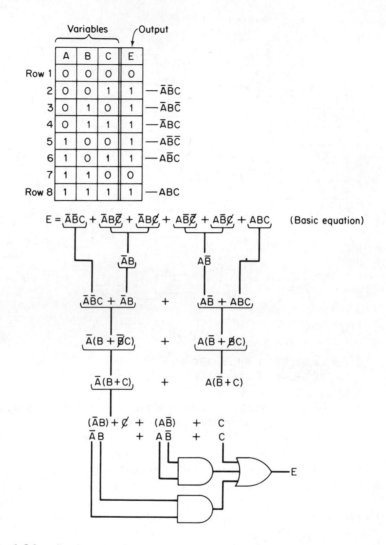

Figure 1-34. Basic sum-of-products solution (from truth table to circuit).

To convert a word statement into a truth table, make sure that each variable has a separate column in the table. Since there are three variables in this problem, the truth table must have three columns. There are always eight rows in a three-variable truth table, since each variable can be in only one of two states ($2^3 = 8$).

It is convenient to arrange the truth table the same way each time. Let the top row (or row number 1) have all zeros and the bottom row have all ones. Then let the second row from the top (row 2) show a binary count of 1 (or 001), the third row (row 3) show a binary count of 2 (010), and so on. The count of the next to bottom row (row 7 in this case) should be one less than the bottom row. In our case, the bottom row has a binary count of 7, whereas the next to last row has a count of 6, which is correct. In this way, the truth tables will be consistent, and the number of rows will be correct (all possible combinations of variables will be present).

In an S-of-P solution, only those rows that produce a true (1) output need be considered. In this case, rows 1 and 7 can be omitted, since they produce a false output.

An equation can be written using the remaining rows as shown in Figure 1-34. A circuit can be implemented from this basic equation. The circuit would require a separate AND gate for each of the six terms. The outputs of the AND gates would then be connected to a six-input OR gate.

However, it is possible to simplify the equation by using the steps shown in Figure 1-34. These steps are based on the laws and rules of Sec. 1–21.1, and should be self-explanatory. As shown, the final expression is reduced down to $A^*B + AB^* + C$.

With the equation reduced to the shortest desired terms, the equation is implemented as follows:

If a term has only one variable, that variable (or input) is connected directly to the OR gate.

If a term has more than one variable, all variables of the term are connected to the input of an AND gate. In turn, the AND gate outputs are connected to the OR gate input.

1–21.4   The Basic P-of-S Solution

Figure 1-35 shows a truth table and a basic P-of-S circuit (using OR and AND gates). The following steps describe the procedure for implementing the circuit.

The word statement and truth table for this problem are identical with those of Figure 1-34, and need not be repeated.

However, in a P-of-S solution, only those rows of the truth table that produce a false (0) output need be considered. In this case, rows 1 and 7 only need be used. All other rows can be ignored.

The basic equation written from rows 1 and 7 is shown in Figure 1-35. Note that the output of this equation is inverted (or false), since the false outputs are selected. To convert the output to true, invert *both* sides of the equation (*both* sides of the equal sign). A double inversion of the output makes the output true. An inversion of

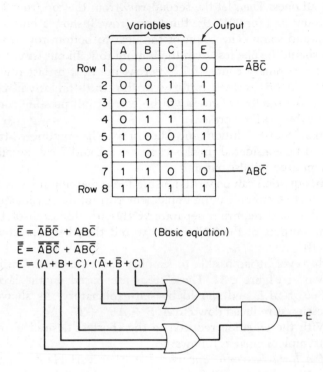

Figure 1-35. Basic product-of-sums solution (from truth table to circuit).

the terms changes them from two AND terms, separated by an OR connective, into two OR terms, separated by an AND connective.

With the equation reduced and changed, the equation is implemented as follows:

If a term has only one variable, that variable (or input) is connected directly to the AND gate input.

If a term has more than one variable, all variables of the term are connected to the input of an OR gate. In turn, the OR gate outputs are connected to the AND gate input.

1–21.5   Converting S of P to NAND

An S-of-P solution can be converted directly to a NAND network. That is, the OR gate and AND gates of an S-of-P network are replaced by NAND gates, as shown in Figure 1-36. However, one

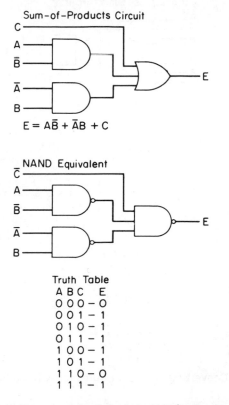

Figure 1-36.   Converting sum of products to NAND circuit.

point must be considered. If the S-of-P circuit includes a single-variable term connected to the OR gate, the sign of that variable must be changed. This applies only to single-variable terms. For example, in Figure 1-36, the single-variable C must be changed to $C^{\ast}$.

1–21.6   Converting P of S to NOR

A P-of-S solution can be converted directly to a NOR network. That is, the AND gate and OR gates of the P-of-S network are replaced by NOR gates, as shown in Figure 1-37. Again, one point must be considered. If the P-of-S circuit includes a single-variable term connected to the AND gate, the sign of that variable must be changed back to its original form. This applies only to single-variable terms (not shown in Figure 1-37).

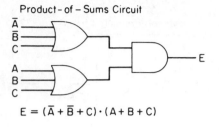

Product - of - Sums Circuit

$$E = (\bar{A} + \bar{B} + C) \cdot (A + B + C)$$

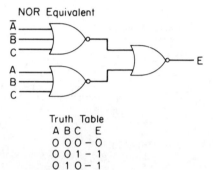

NOR Equivalent

Truth Table

A	B	C	E
0	0	0	0
0	0	1	1
0	1	0	1
0	1	1	1
1	0	0	1
1	0	1	1
1	1	0	0
1	1	1	1

Figure 1-37.   Converting product of sums to NOR circuit.

1–21.7   The AOI Solution

As discussed, the AOI solution is a network consisting of AND gates followed by a NOR gate. The AOI solution can be applied to the same problems covered by S-of-P and P-of-S solutions. There are two methods of arriving at an AOI network. These are shown in Figures 1-38 and 1-39. Note that the truth table in each of the illustrations is the same as the truth tables in the S-of-P and P-of-S solutions (Figures 1-36 and 1-37).

The AOI solution shown in Figure 1-38 is similar to the S-of-P solution, in that the basic equation is written using those combinations of variables that produce a true (1) output. The equation is then simplified, and the circuit is implemented as follows:

If a term has more than one variable, all variables of the term are connected to the input of an AND gate. In turn, the AND gate outputs are connected to a NOR gate input.

```
ABC    E
000 — 0
001 — 1 ✓
010 — 1 ✓
011 — 1 ✓
100 — 1 ✓
101 — 1 ✓
110 — 0
111 — 1 ✓
```

Basic equation: $\bar{A}\bar{B}C + \bar{A}B\bar{C} + \bar{A}BC + A\bar{B}\bar{C} + A\bar{B}C + ABC = E$

Simplified equation: $\bar{A}B + A\bar{B} + C = E$

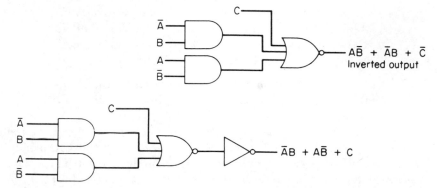

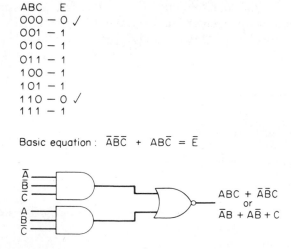

Figure 1-38. Basic AOI solution (with and without inverter output).

```
ABC    E
000 — 0 ✓
001 — 1
010 — 1
011 — 1
100 — 1
101 — 1
110 — 0 ✓
111 — 1
```

Basic equation: $\bar{A}\bar{B}\bar{C} + AB\bar{C} = \bar{E}$

Figure 1-39. Alternate AOI solution (based on P-of-S solution).

> *If a term has only one variable,* that variable
is connected directly to the NOR gate input.

The output of the NOR gate represents an inversion of the
final equation. This output must be inverted to satisfy the truth table.
Any form of inverter can be used.

The AOI solution shown in Figure 1-39 is similar to the S-of-P
solution, in that the basic equation uses those combinations of variables that produce a false (0) output. The equation is then simplified,
and the circuit is implemented as follows:

> *If a term has more than one variable,* all vari-
ables of the term are connected to the input of an
AND gate. In turn, the AND gate outputs are connected to a NOR gate input.

> *If the term has only one variable,* that vari-
able is connected directly to the NOR gate input.

The output of the NOR gate represents an inversion of the
false condition. As such, the output is equivalent to the true condition. That is, the circuit should perform to produce the true (1) condition of the truth table, even though it was based on the false (0)
condition.

1–21.8   Complementary Inputs

In the logic circuits described so far, it is assumed that both
true and complementary inputs (A and A^*, B and B^*, and so on) are
always available. This is rarely true. Many logic circuits must be
designed with only true (or only complementary) inputs.

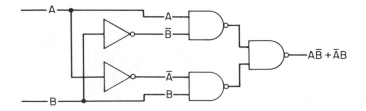

Figure 1-40.   Providing complementary inputs by means of inverters in
each input line.

The obvious solution is to use some form of inverter. As shown in Figure 1-32, inverters can be implemented using NAND, NOR, or EXCLUSIVE OR gates. If true inputs are available, connect an inverter between the true input and the circuit input that requires a complement. This is shown in Figure 1-40.

Another solution to the complementary input problem is to use a NAND gate as shown in Figure 1-41. The NAND gate will invert the true inputs, making them complementary inputs. Note that the complete circuit of Figure 1-41 is the equivalent of the EXCLUSIVE OR function.

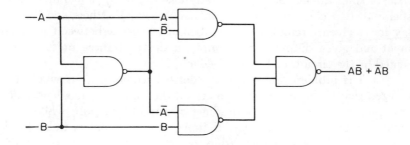

Figure 1-41.   Providing complementary inputs by means of a NAND gate.

1–21.9   Fan-in, Fan-out, Load, and Drive

The terms fan-in, fan-out, load, and drive are used interchangeably by logic designers and the manufactures of IC logic elements. To avoid confusion, the following definitions apply to circuits in this manual.

Fan-in.   The term fan-in is generally applied to IC logic circuits or elements. Fan-in is defined here as the number of independent inputs on an element. For example, a three-input NAND gate has a fan-in of 3.

Load.   Each input to a logic element represents a load to the circuit trying to drive the element. If there are several logic elements (say several gates) in a given circuit, the load is increased by one unit for each element. Load, therefore, means the number of elements being driven from a given input.

Fan-out.   Fan-out (primarily an IC term) is defined as the number of logic elements that can be driven directly from an output, without any further amplification or circuit modification. For example, the output of a NAND gate with a fan-out of three can be connected directly to three elements. The term *drive* can be used in place of fan-out.

1–21.10   Propagation Delay Time

Any solid-state element (diode, transistor, and so on) will offer some delay to signals or pulses. That is, the output pulse will occur some time after the input pulse. Thus, every logic element has some delay. This delay is known as *propagation delay time, delay time,* or simply *delay.* Some logic designers and IC logic manufacturers spell out a specific time (usually in nanoseconds). Others specify delay for a given circuit as the number of elements between a given input and given output. For example, a delay of three units is presented by the circuits of Figures 1-40 and 1-41.

The problems presented by delay become obvious when it is realized that all logic circuits operate on the basis of coincidence. For example, an AND gate will produce a true output only when both inputs are true, simultaneously. If input *A* is true, but switches back to false before input *B* arrives (due to some delay of the *B* input), the inputs will not occur simultaneously, and the AND gate output will be false. The problem is compounded when delay occurs in sequential networks, which depends upon timing as well as coincidence.

Many of the failures that occur in logic design (or the limitations of design) are the result of undesired delay. For this reason, the problems associated with delay are discussed frequently throughout this manual.

1–22   BASIC LOGIC DESIGN USING DIAGRAMS AND MAPS

Many designers use diagrams and maps to aid in the design of basic logic circuits. The Venn and Veitch diagrams, as well as Karnaugh maps, are the most used design tools. This section describes these tools and how to use them when designing circuits.

1–22.1   Venn Diagrams

The Venn diagram is a pictorial representation of logical expressions. Figure 1-42 shows some typical examples of Venn diagrams for basic equations and logic functions.

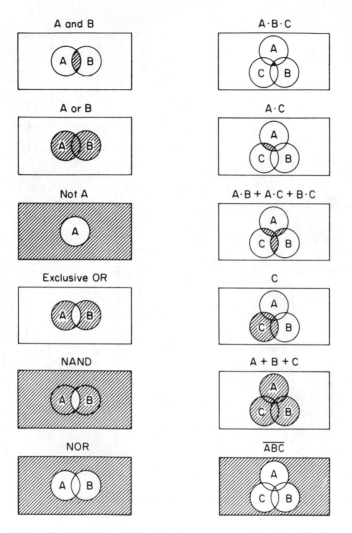

Figure 1-42.   Typical Venn diagrams.

1–22.2   Veitch Diagrams

Veitch diagrams provide a graphic means of representing logic equations. In effect, a Veitch diagram is a form of truth table. However, a Veitch diagram is also a tool that may be used to simplify logical equations. Veitch diagrams may be constructed for any number of variables. However, the diagrams become more difficult to use

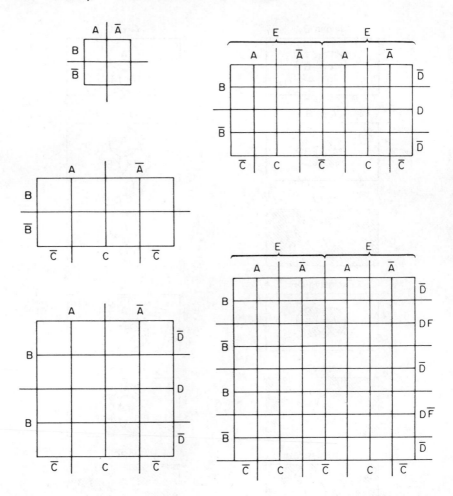

Figure 1-43.   Veitch diagrams for two through six variables.

as the number of variables increases. The Veitch diagram for two through six variables are shown in Figure 1-43. The author recognizes that there are many ways to draw a Veitch diagram. For the sake of uniformity, the author prefers the forms shown in Figure 1-43.

Since each variable has two possible states (true or false), the number of squares required in a Veitch diagram is 2^n, where n is the number of variables. Thus, for a three-variable Veitch diagram, there must be 2^3, or 8 squares.

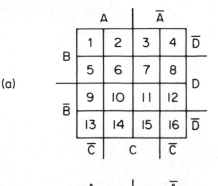

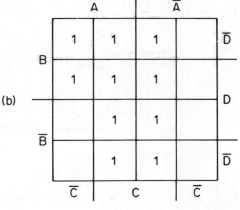

$$F = AB + \overline{A}BCD + A\overline{B}C + \overline{A}C + ACD$$

Figure 1-44.   Example of Veitch diagram for four variables.

1–22.3 Using Veitch Diagrams for Simplification

To illustrate the use of Veitch diagrams for simplification of logic equations, consider the following equation:

$$F = AB + \overline{A}BCD + A\overline{B}C + \overline{A}C + ACD.$$

Since there are four variables (A, B, C, and D), the diagram must contain 2^4 or 16 squares. The procedure is as follows:

1. Draw a 16-square diagram such as shown in Figure 1-44. (The numbers in the section of Figure 1-44a are for discussion only.)

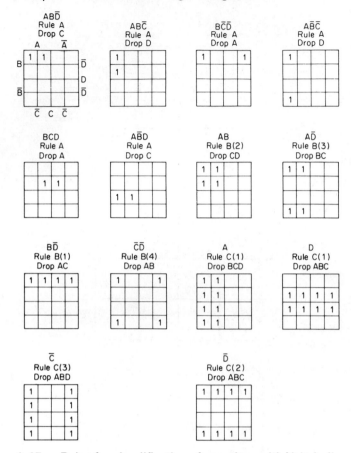

Figure 1-45. Rules for simplification of equations with Veitch diagrams.

2. Plot the function on the diagram. This is done by placing a one in each square that is represented by the terms of the equation. As examples:

$$F = AB \qquad \text{squares 1, 2, 5, and 6}$$

$$+ \ \overline{A}BC D \ \text{square 7}$$

$$+ \ A\overline{B}C \quad \text{squares 10 and 14}$$

$$+ \ \overline{A}C \qquad \text{squares 3, 7, 11, and 15}$$

$$+ \ ACD \quad \text{squares 6 and 10}$$

When ones are placed in the diagram, it should appear as shown in Figure 1–44b.

3. Obtain the simplified logical equation from the Veitch diagram, using the following four rules. The rules are illustrated in Figure 1–45.

Rule A: If ones are located in adjacent squares, or at opposite ends of any row or column, one of the variables may be eliminated. The variable that may be dropped is the one that appears in both states.

Rule B: Two variables may be dropped if there are all one's in:

1. Any row or column of squares.

2. Any block of four squares.

3. The four end squares of any adjacent rows or columns.

4. The four corner squares.

The two variables that may be dropped are those appearing in both states.

Rule C: Three of the variables may be dropped if there are all one's in:

1. Any two adjacent rows or columns.

2. The top and bottom rows.

3. Each square of both the right and left columns.

The three variables that may be dropped are those appearing in both states.

Rule D: To reduce the original equation to its simplest form, sufficient simplification must be made until all one's have been considered in the final equation. Any of the one's may be used more than once, and the largest combinations should be used.

Now the equation can be simplified: Squares 1, 2, 5, and 6 may be combined by using Rule B(2), and the variables $\overline{C}$, C, $\overline{D}$, and D may be eliminated, leaving only the term AB.

Squares 2, 3, 6, 7, 10, 11, 14, and 15 can be combined by using Rule C(1), and the variables $\overline{A}$, A, $\overline{B}$, B, $\overline{D}$, and D may be eliminated, leaving only the term C.

All of the one's have been used, so the logical equation can be written in its simplest form:

$$F = AB + C.$$

1–22.4 Karnaugh Maps

The Karnaugh map is a logic tool similar to the Veitch diagram in that Karnaugh maps can substitute for a truth table, or be used to simplify an equation. The Karnaugh map can also be used to convert logic equations or truth tables into corresponding logic circuit diagrams in a simple and orderly manner.

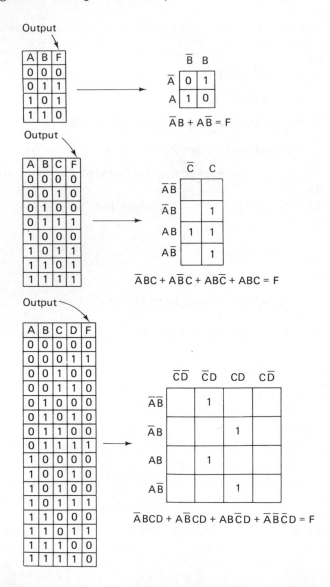

Figure 1-46. Karnaugh maps for two, three, and four variables.

Figure 1-46 shows two-, three-, and four-variable Karnaugh maps for corresponding truth tables and equations. Again, the author recognizes that many formats are used for Karnaugh maps, but he has found that the format used here is the simplest.

Note that each square of the Karnaugh map corresponds to a row of the truth table, and to one term of the equation. (If there are eight rows in the truth table, there must be eight squares in the Karnaugh map.) For example, row 1 of the three-variable truth table (the term $\overline{A}\overline{B}\overline{C}$) corresponds to the upper left-hand square of the Karnaugh map. Since the output of this row is 0 (or false), the corresponding map square contains a 0.

Each increase in the number of input variables doubles the number of squares required in the map. As in the case of Veitch diagrams, Karnaugh maps are not very practical for more than about four or five variables.

When you are interested in only outputs, the zeros can be omitted from the map. Also, the squares can be marked with letters T (for true) or F (for false) if desired. No matter what scheme is used, the rows and columns must be labeled so that only one variable changes at a time as you move from square to square, either horizontally or vertically. For example, in the lower left-hand square of the three-variable map ($A\overline{B}\overline{C}$), if you move one square up, the A and $\overline{C}$ remain the same, but the $\overline{B}$ changes to a B. Also, starting from the same lower left-hand square, if you move to the right one square, the A and $\overline{B}$ remain the same, but the $\overline{C}$ changes to a C.

1–22.5   Looping Adjacent Ones for Simplification

The arrangement of adjacent squares permits the Karnaugh map to be used to simplify equations, and to convert equations or truth tables directly to circuit diagrams.

The first step is to draw loops around adjacent ones. The following rules are illustrated in Figure 1-47.

1. On a two-variable map, any two adjacent squares can be looped.

2. On a three-variable map, any two or four adjacent squares can be looped.

3. On a four-variable map, any two, four, or eight squares can be looped.

Note that the top row is considered to be adjacent to the bottom row, and the extreme left-hand row is adjacent to the extreme right-hand row. Also note that when four ones are looped, they

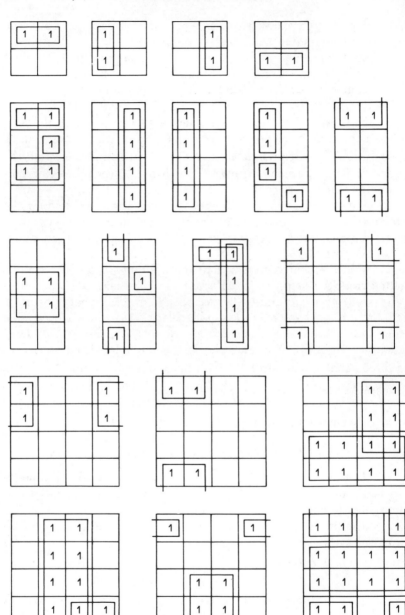

Figure 1-47.   Looping adjacent ones on Karnaugh maps for simplification of equations.

can be in one row, one column, or in a 2-by-2 square. When eight ones are looped, they must be in a 2-by-4 square.

Loops may overlap other loops. However, each loop must be treated as a separate term of the equation.

Any ones not covered by adjacent loops must be looped separately.

Note that Karnaugh maps often give the designer alternate methods of looping. This is shown in Figure 1-48. There are no fixed rules as to the best method of looping. However, as a general rule, always make the loops as large as possible. That is, first draw loops of 8, then 4, then 2. Small loops do not provide the simplest equation, nor do they produce the minimum solutions to circuit implementations.

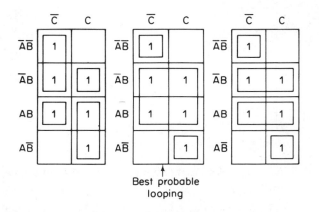

Best probable looping

Figure 1-48.   Alternate methods of looping on Karnaugh maps.

1–22.6 Simplifying Equations on Karnaugh Maps

Once the loops have been drawn on the map, each loop is converted to a term of the simplified equation. Some typical loops and corresponding equations are shown in Figure 1-49.

The following rule applies to loops of any size:

> When a variable appears in both true and complemented form within a loop, the variable can be eliminated. When the variable remains the same for all squares of the loop, that variable must be included in the simplified term of the equation.

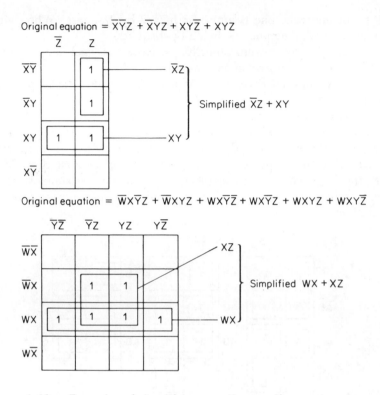

Original equation $= \overline{X}\overline{Y}Z + \overline{X}YZ + XY\overline{Z} + XYZ$

Simplified $\overline{X}Z + XY$

Original equation $= \overline{W}X\overline{Y}Z + \overline{W}XYZ + WX\overline{Y}\overline{Z} + WX\overline{Y}Z + WXYZ + WXY\overline{Z}$

Simplified $WX + XZ$

Figure 1-49.   Examples of simplifying equations on Karnaugh maps.

For example, in the three-variable map of Figure 1-49, there are two loops. Thus, the simplified equation will have two terms. In the upper loop, the $\overline{X}$ and Z variables remain the same for both squares. Thus, the simplified term must contain $\overline{X}Z$. The Y variable appears in both true and complemented form, and can thus be dropped. In the lower loop, the X and Y remain constant for both squares, resulting in a simplified term of XY (the $\overline{Z}$ and Z are dropped).

By combining the simplified terms (one for each loop) the final equation is $\overline{X}Z$ and XY.

As another example, in the four-variable map of Figure 1-49, there are two loops, requiring two terms in the final simplified equation.

In the four-in-a-row loop, the W and X variables remain constant (true only), while YZ variables appear in both true and complemented forms. This results in a final term of WX.

In the two-by-two loop, the X and Z variables appear in true-only form, while the WY variables appear in both true and complemented forms. This results in a final term of XZ.

The combined terms result in a final simplified equation $WX + XZ$.

1–22.7   Converting Karnaugh Maps to Circuit Diagrams

It is not necessary to simplify equations and then convert the equations to a circuit diagram when working with Karnaugh maps. It is possible to go directly from the mapping to a circuit diagram. The following are some examples of converting Karnaugh maps to basic logic functions (or circuits).

1–22.8   Converting Karnaugh Maps to NAND Circuits

The following paragraphs describe the procedure for converting a Karnaugh map directly to a NAND circuit.

The first step is to loop all ones using the rules previously discussed. This is shown in Figure 1-50. Note that the truth table is

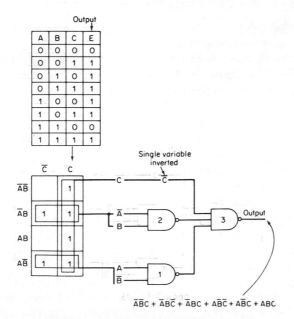

Figure 1-50.   Converting Karnaugh map to NAND circuit.

the same as that used in Figures 1-34 through 1-39. Three loops are drawn on the map. More loops could have been used, but this would have required more circuit elements and more inputs.

Each loop represents an input to the final NAND gate (shown as gate 3). If a loop contains only one variable in the true-only or complement-only form, that variable is connected directly to the final NAND gate, but in inverted form. For example, in the four-in-a-row loop, only the C variable remains constant (true-only), while AB appear in both true and complemented forms. This C is inverted to $\overline{C}$, and connected directly to the final NAND gate.

If a loop contains two or more variables that remain in true-only or complement-only form, those variables are connected to a separate NAND gate. In turn, the output of the separate gate (or gates) is connected to the final NAND gate. For example, in the bottom loop, only A and $\overline{B}$ variables remain constant. Thus A and $\overline{B}$ are connected to a separate NAND gate (gate 1), whereas the output of this gate is connected to the final NAND gate (gate 3). In the middle loop, only $\overline{A}$ and B variables remain constant. Thus $\overline{A}$ and B are connected to a separate NAND gate (gate 2), while the output of this gate is connected to the final NAND gate (gate 3).

1–22.9 Converting Karnaugh Maps to NOR Circuits

The following paragraphs describe the procedure for converting a Karnaugh map directly to a NOR circuit.

The first step is to loop all zeros (*not* the ones) using the previous rules. The zeros are looped since the NOR circuit is implemented from a product-of-sums solution, as discussed in Sec. 1–21.4. This is shown in Figure 1-51. Two loops are drawn on the map. Other looping arrangements could have been used. For example, the bottom four zeros could have been looped in a two-by-two square. However, the following is based on the looping arrangement of Figure 1-51.

Each loop represents an input to the final NOR gate (shown as gate 2). If a loop contains only one variable in a true-only or complement-only form, that variable is connected directly to the final NOR gate, with no inversion. For example, in the four-in-a-row loop, only the C variable remains constant (true only), while A and B appear in both true and complemented forms. Thus, C is connected directly to the final NOR gate.

If a loop contains two or more variables that remain in true-only or complement-only form, those variables are connected to a separate NOR gate, but in inverted form. In turn, the output of the separate gate (or gates) is connected to the final NOR gate. For example, in the two-square loop, only the A and $\overline{C}$ variables remain

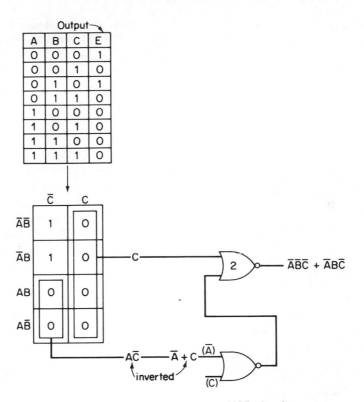

Figure 1-51.   Converting Karnaugh maps to NOR circuit.

constant. Thus, A and $\overline{C}$ are inverted to $\overline{A} + C$ and connected to a separate NOR gate (gate 1), while the output of this NOR gate is connected to the final NOR gate (gate 2).

When proving this circuit against the truth table, keep in mind that the final output must be true for the ones, not the zeros, even though mapping is based on the zeros. For example, the only combinations that must result in a true output are $\overline{A}\overline{B}\overline{C}$ and $\overline{A}B\overline{C}$. Also keep in mind that a NOR gate produces a true output only when both inputs are false.

1–22.10   Converting Karnaugh Maps to AOI Circuits

The following paragraphs describe the procedure for converting a Karnaugh map directly to an AOI gate circuit. In Sec. 1–22.7, it is pointed out that there are alternate solutions when using AOI gates. In brief, one solution is to base the circuit on false conditions (the zeros in the truth table). If the ones are used, the AOI gate output must be inverted. That is, the AOI NOR gate output must be followed by an inverter. No inversion is required if the AOI circuit is based on the zeros (since the NOR gate provides its own inversion). The choice of solutions is up to the designer.

Keep in mind that each row of the truth table used will require a separate AND gate (or a separate input to the NOR gate), and it is usually desirable to use the minimum number of gates. Thus, if there are fewer zeros (fewer combinations of variables that will produce a false output), base the circuit on zeros. If there are fewer ones, base the circuit on ones, even though an inverter is required at the output. (It is generally better to design the circuit with fewer gates and inputs, even if the inverter is required.)

The Karnaugh map conversions of true and false (one and zero) AOI circuits are shown in Figures 1-52 and 1-53, respectively.

As shown in Figure 1-52, three loops are drawn on the map.

Output⟶

A	B	C	E
0	0	0	0
0	0	1	1
0	1	0	1
0	1	1	1
1	0	0	1
1	0	1	1
1	1	0	0
1	1	1	1

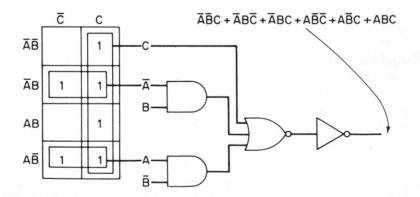

Figure 1-52.   Converting Karnaugh map to AOI gate (with output inverter).

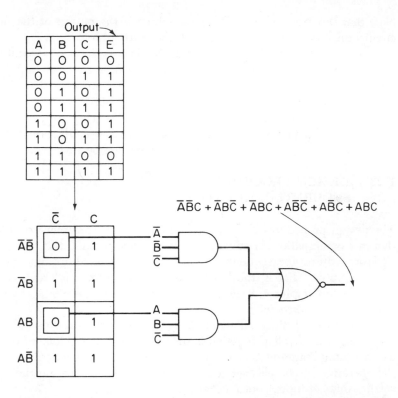

Figure 1-53. Converting Karnaugh map to AOI gate (without output inverter).

Each loop represents an input to the NOR gate. If a loop contains only one variable in true-only or complement-only form, that variable is connected directly to the NOR gate. For example, in the four-in-a-row loop, only C remains constant, and is connected directly to the NOR gate.

If a loop contains two or more variables that remain in the true-only or complement-only form, those variables are connected to a separate AND gate. In turn, the outputs of separate AND gates are connected to the NOR gate. For example, in the bottom loop, only A and $\bar{B}$ variables remain constant. Thus A and $\bar{B}$ are connected to the NOR gate. In the middle loop, only the $\bar{A}$ and B variables remain constant, and are thus connected to their corresponding AND gate. In turn, the output of this gate is connected to the NOR gate.

Note that the input of the NOR gate (which is the output of the AOI circuit) must be inverted to agree with the truth table.

As shown in Figure 1-53, each zero is looped separately, since the zeros are not adjacent. Each loop represents an input to the NOR gate. Again, single-variable loops are connected directly to the NOR gate (no such loops exist in Figure 1-53), while multiple-variable loops are connected through AND gates. However, with the mapping shown in Figure 1-53 (zero mapped), the NOR gate output (which is the output of the AOI circuit) need not be inverted.

1–23 DESIGN PROCEDURE FOR BASIC SEQUENTIAL NETWORK

The design procedures for sequential logic circuits are more complex than for combinational networks. Thus, in this manual, the basic design procedures for sequential circuits are discussed in the appropriate chapter. That is, the basic design for counters and registers is discussed in Chapter 4, basic design for arithmetic circuits is covered in Chapter 6, and so on.

However, since all sequential circuits involve the use of flip-flops, the basic flip-flop function is discussed in this section. Also, the use of timing diagrams is discussed here since it is impractical to show the operation of complex sequential circuits without showing the time relationships of inputs, outputs, feedback, and so on.

1–23.1 The Basic Flip-flop

Flip-flops used in logic design can be of the transistor type. In effect, a bistable transistor-type multivibrator is used. The output of one transistor is applied to the input of the other transistor, while the output of the second transistor is returned to the input of the first. An external input signal to either transistor will "set" the two transistors to opposite states. One transistor will have a high (or true, or one) output, whereas the other transistor will have the complemented output (low, false, or zero).

Logic circuit flip-flops can also be implemented by using logic gates. The simplest such flip-flop (the basic reset-set, or RS) is shown in Figure 1-54, along with the corresponding truth table. Note that NAND gates are used, and that the gates are cross-coupled (the output of one gate is connected to the input of the opposite gate). The presence of a pulse at either the SET or RESET inputs causes the cross-coupled gates to assume the corresponding state, as shown in the truth table. The gates will remain in the state until a pulse is applied at the correct input to change stages.

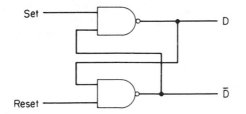

Truth Table

Previous state		Input condition		Result	
D	D̄	Set	Reset	D	D̄
L	H	L	H	H	L
H	L	H	L	L	H
L	H	H	H	No change	
H	L	H	H	No change	
H	L	L	H	No change	
L	H	H	L	No change	
L	H	L	L	H	*
H	L	L	L	H	*

Figure 1-54.   Basic RS flip-flop implemented with NAND gates.

The circuit shown in Figure 1-55 is a clocked-*RS* flip-flop. This circuit changes states only when both the input pulse and a *clock pulse* are present simultaneously. (The clock pulse is also known as a *gate pulse* or *trigger pulse*).

1–23.2 Timing Diagrams

Operation of the circuit in Figure 1-55 is dependent upon the time relation of pulses, as well as the presence of inputs. This relationship can be shown by means of a truth table (as is shown in Figure 1-54). However, a timing diagram such as that shown in Figure 1-55 is generally a simpler and more effective means of showing time relationships.

For example, in the circuit of Figure 1-55, the states will change only on the *positive-going edge of the clock pulse* and with the appropriate input present.

Timing diagrams have one major drawback. They do not show what will happen if the inputs are abnormal. The classic example of this occurs when both SET and RESET inputs appear simultaneously. The circuit may move to either state, or remain in

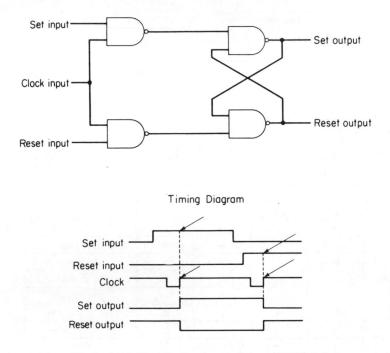

Figure 1-55.   Clocked RS flip-flop implemented on NAND gates.

its previous state, or may move back and forth between states (known as a *race* condition). Some designers use *flowcharts* (also known as flow tables) to show the theoretical operation of sequential networks. The author has no quarrel with flowcharts, except that the final test of a circuit is the performance under actual operating conditions. The author prefers to make an actual timing test with the circuit in the breadboard state. Both normal and abnormal inputs may be applied, and the results may be noted.

IC Logic Devices

This chapter is devoted to logic circuit elements available in IC form. The chapter not only describes what is available to the logic designer, but how (in brief) the circuits operate. It is essential that logic designers understand these operating principles to use and interconnect the IC elements in logic systems properly. The IC logic elements discussed here represent a cross-section of the entire logic field, and do not necessarily represent the products of any particular manufacturer.

2–1 LOGIC FORMS

The following paragraphs describe the various logic forms in use today. Some of these forms appeared as discrete component circuits. However, most of the forms are the result of packaging logic elements as integrated circuits.

The author makes no attempt to promote one form over another, but simply summarizes the facts (capabilities and limitations) about each form. Thus, designers can make informed and intelligent comparisons of the logic forms, and select those that are best suited to their needs.

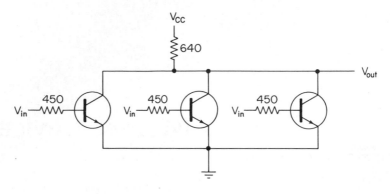

Figure 2-1.   Basic RTL gate.

2–1.1   Resistor-transistor Logic (RTL)

Resistor-transistor logic, or RTL, was derived from Direct Coupled Transistor logic, or DCTL, and was the first integrated circuit logic form introduced around 1960. The basic circuit is a direct translation from the discrete design into integrated form. This circuit was the most familiar to logic designers, easy to implement and, therefore, the first introduced by IC manufacturers.

The basic RTL gate circuit shown in Figure 2-1 is presented here to illustrate the basic building block type of logic. The most complex elements are constructed simply by the proper interconnection of this basic circuit. The small resistor added in the base circuit increases the input impedance to assure proper operation when driving more than one load. Without this resistor, when several base-emitter junctions are driven from the same output, the input with the lowest base-emitter junction forward bias could severely limit the drive current to the other transistor bases.

There are many electronic devices using RTL still in operation. Thus, RTL is of interest to the student and service technician. However, because of the problems discussed in Sec. 2–2, RTL is not generally used in the design of new logic circuits.

2–1.2   Diode-transistor Logic (DTL)

Diode-transistor logic, or DTL, is another logic form that was translated from discrete design into IC elements. DTL was very familiar to the discrete component logic designer in that the form used diodes and transistors as the main components (plus a minimum number of resistors). The diodes provided higher signal thresholds than could be obtained with RTL.

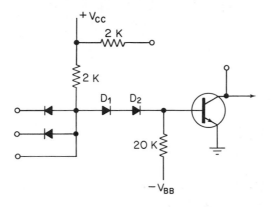

Figure 2-2.   Basic DTL gate.

Figure 2-2 shows the basic DTL gate. Note that it requires two power supplies (to improve turn-off time of the transistor inverter). As in the case of RTL, the DTL logic form is primarily of interest to students and technicians, but it is not generally used in the design of new devices.

2–1.3   High-threshold Logic (HTL)

HTL (high-threshold logic) is designed specifically for logic systems where electrical noise is a problem, but where operating speed is of little importance (since HTL is the slowest of all IC logic families). Because of this slow speed, and because other logic forms (such as the MOS logic described later in Sec. 2–1.6) offer similar noise immunity, HTL is generally not being used for design of current logic systems (although HTL is found in many existing systems, particularly where logic must be connected to industrial and other heavy duty equipment).

Figure 2-3 shows a typical HTL gate along with the transfer characteristics. Note that the gate is identified as an MHTL, or Motorola HTL. Also note that the MHTL gate is compared with an MDTL, or Motorola DTL, gate. This comparison is made since the HTL is essentially the same as the DTL, except that a Zener diode is used for D_1 in the HTL. The use of a Zener for D_1 (with a conduction point of about 6 or 7 V), and the higher supply voltage (a V_{CC} of about 15 V), produces the wide noise margins shown in the transfer characteristics of Figure 2-3. That is, the HTL gate will not operate unless the input voltage swing is large. Noise voltages below about 5 V will have no effect on the gate.

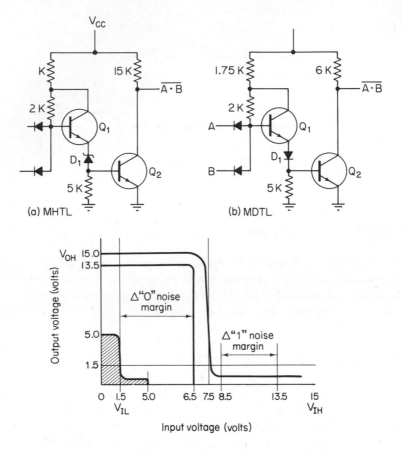

Figure 2-3. Motorola MHTL gate and transfer characteristics.

The problems of noise in logic systems are discussed further in Chapter 9.

2–1.4   Transistor-transistor Logic (TTL or T²L)

TTL (or T²L) has become one of the most popular logic families available in IC form. Most IC manufacturers produce at least one line of TTL, and often several lines. This fact gives TTL the widest range of logic functions.

Figures 2-4 and 2-5 show the schematics of two typical logic gates for comparison. Figure 2-4 shows a conventional high-speed TTL gate, whereas Figure 2-5 shows an MTTL, or Motorola TTL gate. One difference between the MTTL and TTL is the replacement of resistor R_4 (Figure 2-4) with resistors R_{4A} and R_{4B}, and transistor Q_6, as shown in Figure 2-5.

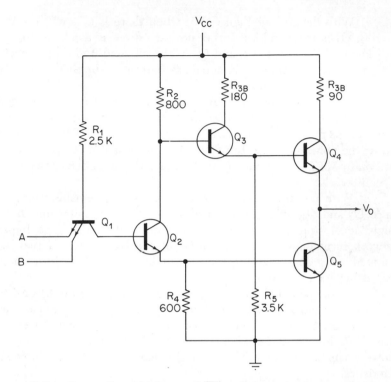

Figure 2-4. Conventional high-speed TTL gate.

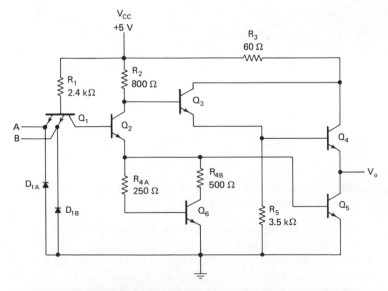

Figure 2-5. Improved Motorola MTTL gate. *(Courtesy of Motorola)*

With the gate of Figure 2-5, when there is a "low" on either A or B, Q_1 is forward-biased and no base drive is available for Q_2. This keeps Q_2, as well as Q_5 and Q_6 in the off condition. The collector of Q_2 is approximately at V_{CC}, and base current is supplied to Q_3 and Q_4, keeping Q_3 and Q_4 on.

Now assume that input A is "high" and input B gradually goes from a "low" to a "high." The base of Q_2 tracks the voltage at the input (the same as for conventional TTL) by the difference of $V_{BE} - V_{BC}$ of Q_1. At the point where Q_2 turns on in conventional TTL, Q_2 does not turn on in MTTL, since the equivalent of an open circuit exists at the Q_2 emitter.

With no current flow, the collector of Q_2 remains near V_{CC}. Since Q_3 and Q_4 act as emitter-followers, the output remains at the "high" level, approximately two V_{BE} drops below V_{CC}. The bypass network turns on when the potential at the base of Q_2 is two V_{BE} drops above ground, causing the output transistor Q_5 to turn on (as well as bypass transistor Q_6).

As the potential on input B increases further, output transistor Q_5 saturates, and point E is reached. The resistors in the bypass network are chosen so that the network conducts the same current as resistor R_4 in Figure 2-4 when transistor Q_5 is saturated. Figure 2-6 shows a comparison of MTTL and conventional TTL transfer characteristics.

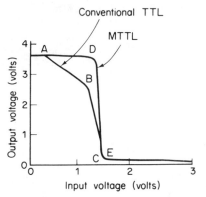

Figure 2-6.   Comparison of MTTL and conventional TTL characteristics. *(Courtesy of Motorola)*

Function of input diodes.   Due to high speeds of operation, TTL generates large values of current and voltage rates-of-change. A 1 V in approximately 1.3-ns (nanoseconds) rise time, and a 1 V in approximately 1-ns falltime produces dV/dt rates on the order of 10^9 V/s (volts per second). With these rates-of-change, undershoot exceeding 2 V can develop in the system. Such undershoot can cause two serious problems.

First, false triggering of the following stage is possible since a positive overshoot follows the large undershoot. This positive overshoot may act as a "high" signal, and turn on the following stage for a short period of time.

Second, if the unused inputs of a gate are returned to the supply voltage (as is usually recommended for TTL logic design), and a negative undershoot in excess of 2 V occurs, the reverse-biased emitters of the inputs may break down, and draw excessive current, generating noise in the system.

Typical TTL gates.   The basic gate in most TTL systems is the NAND gate. However, a full TTL logic line will include AND, OR, AOI, NOR, as well as EXCLUSIVE OR and NOR.

TTL logic elements.   In addition to basic gates, there is an almost unlimited variety of TTL logic elements, such as flip-flops, counters, registers, decoders, multiplexers, and so on. These elements are discussed in Chapters 3 through 10.

TTL power gates.   In some logic systems, there are fan-out requirements that exceed the capability of a standard gate. Power gates are designed to meet these requirements with a minimum of additional circuitry. A typical power gate (an MTTL AND gate) is shown in Figure 2-7. With this power gate, the output circuitry is designed to provide twice the fan-out of conventional gates (in this case, 20 standard gate loads, instead of 10).

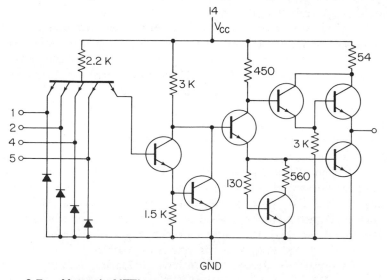

Figure 2-7.   Motorola MTTL power gate.

TTL line drivers.   IC line drivers are generally used as amplifiers to increase the fan-out capability of gates, without the use of power gates. Figure 2-8 shows the circuit of a NAND line driver;

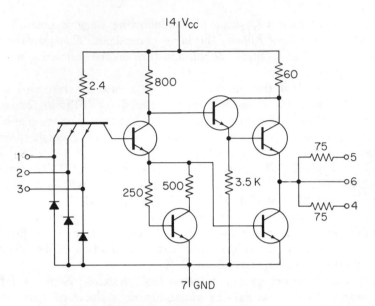

Figure 2-8.   Motorola MTTL terminated line driver.

Figure 2-9 shows a typical application of the circuit. Note that the gate output has 75-Ω resistors in series with the standard output (at pins 4 and 5), in addition to the direct output (at pin 6). These resistors provide for terminating the line.

Using an unterminated line driver, a line appears essentially as an open circuit at each end. Any pulse traveling down the line will see a reflection almost equal in magnitude to the original pulse. By terminating the loaded end, reflections and switching transients are minimized. For driving 93-Ω coaxial cable, or 120-Ω twisted pair, a good match can be made at the output of each resistor. For loads of 50 to 93Ω, the two resistive outputs are shorted together for better impedance matching. The nonresistive output can be used to drive gates in a normal manner.

Open collector gates. Most TTL gates have an active pull-up resistor at the output. As discussed in Sec. 1–19.1, this does not permit using a wired-OR operation. Special gates, such as the MTTL gate shown in Figure 2-10, is included in many TTL product lines to overcome this limitation. The output of the Figure 2-10 circuit can be used for wired-OR, or to drive discrete components.

Output impedance. Some form of Darlington output is used for most TTL lines. The Darlington output configuration provides extremely low output impedance in the "high" state. The low impedance results in excellent noise immunity, and allows high-speed

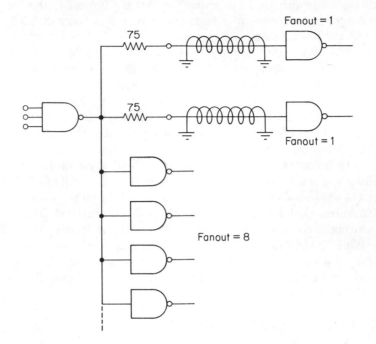

Figure 2-9.   Typical application of line driver. *(Courtesy of Motorola)*

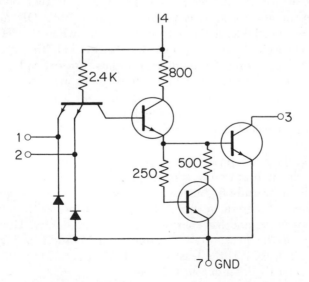

Figure 2-10.   Open collector Motorola MTTL gate for implementing the wired-OR function.

operation while driving large capacitive loads. Typically, the "high" state output impedance varies from about 10Ω at outputs of 3.5 V, to about 64Ω at lower output voltages.

The "low" state output impedance is typically 6Ω at an output of 0.2 V, and goes up to about 500Ω if the output increases to 0.5 V.

Totem-pole output.   One disadvantage of conventional TTL is the so-called "totem-pole" output. As shown in Figure 2-4, both output transistors are on during a portion of the switching time. Since the turn-off time of a transistor is normally greater than the turn-on time, the following occurs.

In going from a "high" state to a "low" state on the output, transistor Q_4 is initially on, and is in the process of turning off. Transistor Q_5, at the same instant in time, is off and is attempting to turn on. Transistor Q_5 turns on before transistor Q_4 can turn off. The result is a current spike through both transistors and the load resistor. The same effect takes place when the conditions are reversed, and transistor Q_4 turns on before transistor Q_5 can turn off. The active bypass network in the MTTL line (transistor Q_6), shown in Figure 2-5, helps to limit this problem.

2–1.5   Emitter-coupled Logic (ECL)

ECL, shown in Figure 2-11, operates at very high speeds (compared to TTL). Another advantage of ECL is that *both a true and complementary* output is produced. Thus, both OR and NOR functions are available at the output. Note that when the NOR functions of two ECL gates are connected in parallel, the outputs are ANDed, thus extending the number of inputs. For example, as shown in Figure 2-11, when two two-input NOR gates are ANDed, the results are the same as a four-input NOR gate (or a four-input NAND gate in negative logic). When the OR functions of two ECL gates are connected in parallel, the outputs are ANDed, resulting in an OR/ AND function.

The high operating speed is obtained since ECL uses transistors in the nonsaturating mode. That is, the transistors do not switch full-on or full-off, but swing above and below a given bias voltage. Delay times range from about 2 to 10 ns. ECL generates a minimum of noise, and has considerable noise immunity. However, as a tradeoff for the nonsaturating mode (which produces high speed and low noise), ECL is the least efficient. That is, ECL dissipates the most power for the least output voltage.

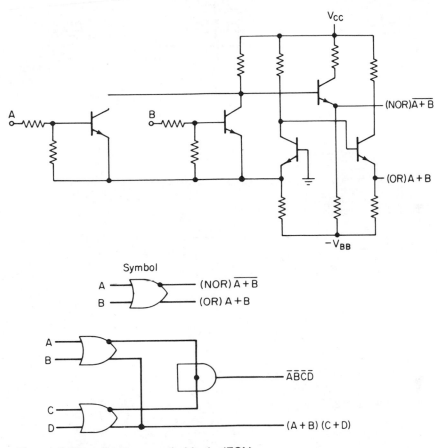

Figure 2-11.   Emitter-coupled logic (ECL).

A typical ECL gate is shown in Figure 2-12. The tables of Figure 2-12 illustrate the logic equivalences of the ECL family. As discussed in Sec. 1–9.3, it is possible for some logic elements to have two equivalent outputs or functions, depending upon logic definition (positive or negative logic). The ECL gate shown in Figure 2-12 can be considered as a NAND in negative logic, or a NOR in positive logic.

Saturated logic families such as TTL have traditionally been designed with the NAND function as the basic logic function. However, in positive logic, the basic ECL function is NOR. Thus, the designer may either design ECL systems with positive logic using the NOR, or design with negative logic using the NAND, whichever is

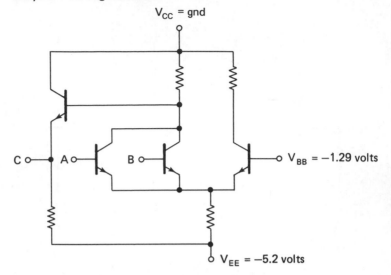

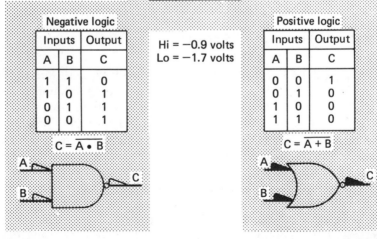

Figure 2-12. Basic MECL gate circuit and logic function in positive and negative nomenclature. *(Courtesy of Motorola)*

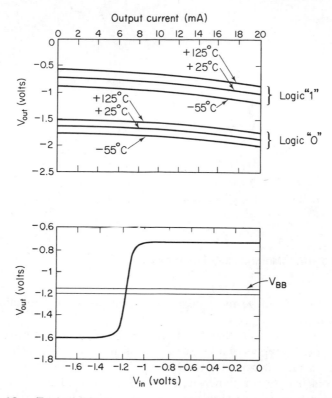

Figure 2-13.   Typical ECL transfer characteristics.

more convenient. On one hand, traditionally TTL designers are familiar with positive logic levels and definitions. On the other hand, they are familiar with implementing systems using NAND functions. The problems of logic equivalences are discussed and illustrated in Sec. 1–9.3.

For positive logic, a logical 1 for the circuit of Figure 2-12 is about -0.9 V, which corresponds to one base-emitter voltage (V_{BE}) drop below ground. Logical 0 is -1.7 V, which yields a nominal voltage switch of about 0.8 V. (However, ECL lines are available with logic swings up to about 2 V.) Some typical ECL transfer characteristics are shown in Figure 2-13.

Bias problems.   Unlike TTL (and other saturated logic), ECL requires a bias voltage V_{BB}. In the case of the Figure 2-12 gate, the V_{BB} bias is -1.29 V, when the supply voltage V_{EE} is -5.2 V (with V_{CC} at ground or 0 V). With such a gate, if the power supply voltage is increased (V_{EE} increased due to poor supply regulation, and so on),

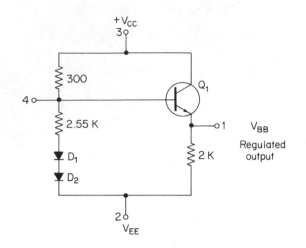

Figure 2-14.   Motorola MECL bias driver.

the 0-level will move more negative, while the 1-level remains essentially constant.

It is essential that the bias voltage V_{BB} track any variations in the power supply voltage. For this reason, some ECL manufacturers provide a *bias driver* with their ECL line. An example of this is the Motorola bias driver shown in Figure 2-14. The bias driver provides a temperature and voltage compensated reference for MECL logic. Any of the three MECL voltages may be grounded, but the common voltage of the bias driver must correspond to that of the logic system. If V_{BB} is obtained from the bias driver connected to the *same power supply* as the ECL logic element, the bias or reference voltage will track the supply voltage changes or temperature variations, thus keeping V_{BB} in the center of the logic levels.

Using ECL logic.   A fairly complete line of ECL devices is available in IC form. The use of these devices is described in Chapters 3 through 10. Here, we shall list the highlights or general rules for using ECL.

1. The maximum recommended ac fan-out for typical ECL is about fifteen input loads. Direct-current fan-out is about twenty-five loads. The ac fan-out is lower than dc fan-out because of the increase in rise time and fall time with high fan-out. Also, if high fan-outs and long leads are used, overshoot due to lead inductance becomes a problem.

2. A circuit such as the bias driver (Figure 2-14) will fan-out to about twenty-five loads. Note that a dual gate or half adder is equivalent to two gate input loads for a circuit such as the bias driver.

3. Each *J* or *K* input to a flip-flop is equivalent to one and one-half loads. (Flip-flops are discussed in Chapter 4.) For example, a *J* and *K* input tied together as a flip-flop clock input would be a load of three, allowing a gate (with an ac fan-out of 15) to drive five flip-flops. All other inputs are a load of unity (or one).

4. The output of two ECL gates may be tied together to perform the wired-OR function, in which case a maximum fan-out of 5 is allowed. If only one *pull-down* resistor (an emitter resistor in the output, rather than a collector pull-up resistor) is used, each additional common output is equivalent to one gate load. For example, if 6 gates are wired together with only one pull-down resistor connected, the fan-out would be (15 − 5), or a fan-out of 10 remaining.

5. All unused inputs should be tied to V_{EE} for reliable operation (assuming that the power connections are as shown in Figure 2-12). As seen from the gate input characteristics, the input impedance of a gate is very high when at a low level voltage. Any leakage to the input and/or wiring of the gate will gradually build up a voltage on the input. This may affect noise immunity of the gate or hinder switching characteristics at low repetition rates. Returning the unused inputs to V_{EE} insures no buildup of voltage on the input, and a noise immunity depended only upon the inputs used.

6. A recommended maximum of three input expanders should be used (assuming that each input expander provides 5 inputs). Thus, the recommended maximum input to any ECL gate is 15. If this is exceeded, the NOR output rise and fall times suffer noticeably because of the increased capacitance at the collector of the input transistors. For low frequencies, higher fan-ins may be used, if rise and fall times are of no significance.

7. Each gate in the IC package must have external bias supplies (except for certain ECL gates which have an internal bias scheme). ECL flip-flops do not normally require an external bias.

2–1.6   MOS Logic

MOS (metal oxide semiconductor) logic is somewhat different from that used in other logic lines such as TTL, ECL, and so on. MOS logic is based on using MOSFETs (MOS field effect transistors) instead of two-junction or bipolar transistors, as do other logic families. It is assumed that the reader has a basic understanding of MOSFETs (and other MOS devices). Such subjects are discussed in the author's *Manual for MOS Users* (Reston Publishing Company, Inc., Reston, Virginia, 1975), and will not be repeated here. However, before going into details of MOS logic, let us consider the "what" and "why" of MOS in digital applications.

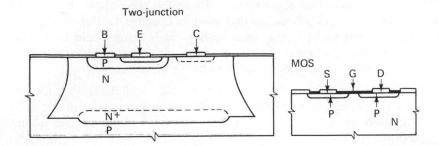

Figure 2-15. Comparison of two-junction and MOSFET size. *(Courtesy of Texas Instruments)*

MOS integrated circuits require only one-third of the process steps needed for two-junction IC. The most significant feature of MOS ICs is the large number of semiconductor circuit elements that can be put on a small chip. The size relationship of MOS and two-junction ICs is shown in Figure 2-15. This high circuit density means *large-scale integration* (LSI), instead of *medium-scale integration* (MSI) found in TTL, ECL, and so on. For example, it is possible to put 5,000 devices on a silicon chip only 150 × 150-mils square. Each transistor in a typical MOS/LSI array requires as little as 1 square mil of chip area, a great reduction over the two-junction IC transistor which requires about 50 square mils.

MOS/LSI has several advantages over two-junction MSI. These include: lower cost per circuit function, fewer subsystems to test, fewer parts to assemble and inspect, increased circuit complexity per package, lower power drain per function, and a choice of standard or custom products.

This last advantage is of particular importance to the designer. Most major MOS device manufacturers offer a complete line of standard logic ICs. Typically, the line will include gates, switches, registers, dividers, counters, generators, memories, coders, decoders, and general purpose logic. The great majority of design problems can be solved with these standard, off-the-shelf devices, as discussed in the remaining chapters of this manual.

In addition to the standard logic ICs, many MOS manufacturers offer a custom production service. That is, they will produce complete logic devices from the customer's logic drawings. In effect, the customer draws the desired logic, defines the inputs and outputs, and describes the test procedures. This information is then sent to various MOS manufacturers for bids on completed hardware. Some MOS suppliers also provide software for both standard and custom devices. Likewise, manufacturers will sometimes produce MOS logic elements from customer's software.

From this description, it would appear that the customer need only tell the manufacturer what is wanted, check the bids, and then wait (indefinitely) for the finished hardware. However, it is necessary that the customer or designer have a working knowledge of MOS logic devices in order to make an intelligent comparison of the manufacturer's services. For example, the designer should know the switching characteristics of MOS devices, and the basics of complementary logic.

The purpose of the following paragraphs and subsections is to acquaint the readers with MOS logic, in general, so that they can select commercial units to meet their particular circuit requirements, and so that they can understand the applications described in later chapters of this manual. If needs cannot be met with existing devices, or with the applications described, the following information will provide a sound basis for venturing into the uncertain world of custom MOS logic hardware.

2–1.7   Basic Complementary MOS Logic

Although MOS logic is not limited to the complementary inverter, or to the complementary fabrication technique (known as CMOS), the complementary principle forms the backbone of most current MOS logic ICs, both standard and custom. The basic complementary inverter circuit is shown in Figure 2-16. Note that this circuit is formed using an N-channel MOSFET and a P-channel MOSFET on a single semiconductor chip. As discussed in later chapters, some MOS logic devices are formed using only P-channel (PMOS) or N-channel (NMOS) fabrication techniques.

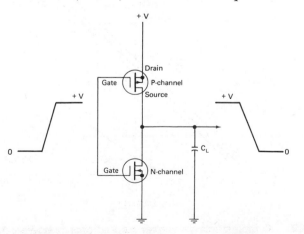

Figure 2-16.   Basic MOS device complementary (CMOS) inverter circuit.

The complementary inverter of Figure 2-16 has the unique advantage of dissipating almost no power in either stable state. *Power is dissipated only during the switching interval.* And, since MOSFETs are involved, the capacitive input lends itself to direct-coupled circuitry. (The input gate of a MOSFET acts as a capacitor.) No capacitors are required between circuits; this results in a savings in component count and in circuit wiring.

With the input at zero, the upper *P*-channel MOSFET is fully ON. The load capacitance C_L is charged to $+V$ through the *P*-channel MOSFET. Once C_L is charged, the only current flow is the I_{DSS} (drain-source current) of the *P*-channel. Typically, this is in the picoampere range, since the MOSFET is completely OFF with zero gate voltage.

The voltage drop across the *P*-channel MOSFET is simply the I_{DSS} of the *N*-channel MOSFET multiplied by the channel resistance of the *P*-channel. Thus, if the I_{DSS} is 1 pA, and the channel resistance is 300Ω, the voltage drop is about 3 nV.

With the input at $+V$, the *N*-channel is fully ON, and the *P*-channel is OFF. C_L discharges to ground and the *P*-channel MOSFET limits current flow to a few pA. The voltage drop across the *N*-channel MOSFET similarly is in the nV region.

Figure 2-17 shows the switching performance of the complementary inverter of Figure 2-16 as a function of fan-out. Here, fan-out

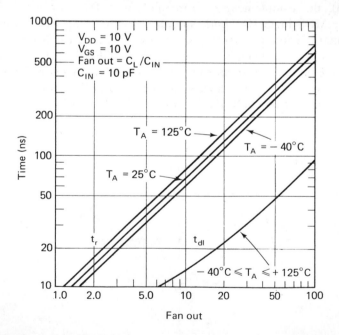

Figure 2-17.   Complementary inverter turn-on time temperature variation. *(Courtesy of Motorola)*

is defined as C_L/C_{IN}, where C_{IN} for the complementary pair is about 10 pF. Thus, the rise time (t_r) with a fan-out of 10 at $T_A = 25°C$ is 70 ns, and the delay time (t_d) is 13 ns. At a fan-out of 100, representing a load capacitance of 1000 pF, the rise time is about 650 nA, with a 100 ns delay time.

The temperature variation of the complementary inverter is also shown in Figure 2-17. At a fan-out of 10, the variation in rise time over the temperature range of $-40°C$ to $+125°C$ is only 20 nA. There is no appreciable change in the delay time over this temperature range. Turn-off is much the same as turn-on, since one device of the pair is always turning on.

Complementary MOS logic NAND gate. The MOS logic NAND gate is formed as shown in Figure 2-18. The P-channel devices are connected in parallel, and the N-channel complements are connected in series. The truth table for the three-input NAND gate is also given on Figure 2-18. Note that the output swings from 0 V to $+V$ (supply voltage). If a supply voltage is 10 V or 15 V, the difference

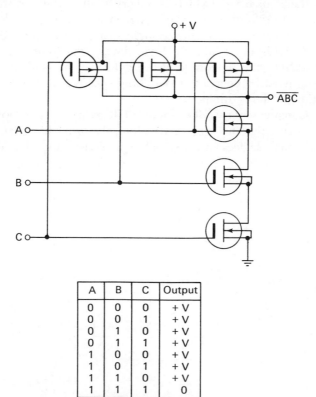

A	B	C	Output
0	0	0	+ V
0	0	1	+ V
0	1	0	+ V
0	1	1	+ V
1	0	0	+ V
1	0	1	+ V
1	1	0	+ V
1	1	1	0

Figure 2-18.   Three-input NAND gate using MOS devices. *(Courtesy of Motorola)*

between a logic 0 and a logic 1 is approximately 10 V or 15 V. Thus the MOS logic device has a much greater noise immunity than the HTL described in Sec. 2–1.3. For this reason, and because MOS uses much less power and operates at higher speeds, HTL is generally being replaced by MOS.

As discussed in Chapter 1, for the NAND function, the output is always high unless all three inputs are high. If any one or any pair of inputs is high, one or more of the *P*-channel devices will be held ON by the remaining low inputs and the common output, but will be at +V. When all three inputs are high, all three series *N*-channels will be ON, and the output is low.

Note that the zero output level is developed across three series elements. However, the leakage current from all three of the *P*-channel devices is in the pA range, resulting in nV output levels (even for very large gates). For example, assume a *P*-channel leakage of 20 pA, and an *N*-channel resistance of 130Ω. For a 50-input gate, the total leakage current is 1 nA (50 × 20 pA). Assuming a series output resistance of 6.5 K (which is a very high output resistance), the resultant output voltage is 6.5 μV (which is an extremely low output voltage, particularly if the normal logic swing is from 0 to 10 or 15 V).

As with any solid-state logic device, the limitation on width of the NAND gate (or how long the gate may be held in the 1 or 0 condition) is set by decreasing switching speeds and increasing power dissipation (as the width increases).

Complementary MOS logic NOR gate. The complementary NOR gate is shown in Figure 2-19. Here, the order has been reversed. The *P*-channel devices are connected in series, and the *N*-channels

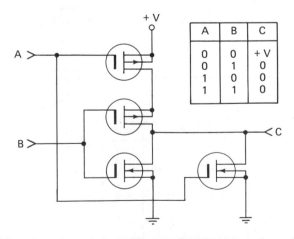

A	B	C
0	0	+ V
0	1	0
1	0	0
1	1	0

Figure 2-19. Complementary NOR gate using MOS devices. (*Courtesy of Motorola*)

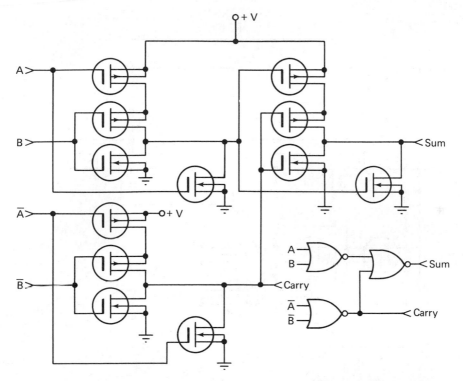

Figure 2-20. Half-adder using complementary MOS device NOR gates. *(Courtesy of Motorola)*

are in parallel. If any one of the inputs is high, one of the parallel N-channels will be ON, and the output will be low. The same is true if both inputs are high. Only when both inputs are low will both series P-channels be ON, allowing the output to become high. Thus, the conditions stated in the truth table are satisfied. The same comments regarding size of the NAND gate apply to the NOR gate.

    Complementary MOS logic half-adder. Figure 2-20 shows a half-adder using three MOS logic NOR gates. (Adders are discussed in Chapter 6.) The carry digit is taken from the NOR gate that handles the complemented inputs, while the sum digit is picked up from the NOR gate that sums outputs from the first two NOR gates.

    Referring to Figure 2-17, which shows switching times of the complementary pair, consider the carry digit to the next stage as a fan-out of 1, and the input to the sum digit NOR gate as a fan-out of 1. The carry NOR gate now faces a fan-out of 2. For a fan-out of 2 (according to Figure 2-17), the rise time at room temperature should be 15 ns. Thus, the propagation delay from the output to the carry digit is 15 ns.

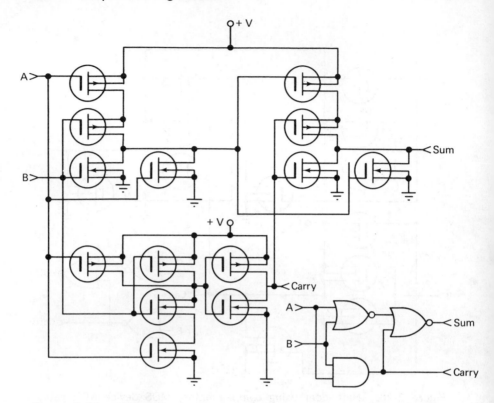

Figure 2-21. Half-adder (not requiring complementary inputs) using MOS devices. *(Courtesy of Motorola)*

Assume a fan-out of 2 for the sum output NOR gate, and a total propagation delay from the input to the sum digit of 30 ns. This allows 15 ns to the carry input, plus another 15 ns through the final NOR gate, for a total of 30 ns propagation delay.

Modified MOS logic half-adder. The modified half-adder can be modified to handle only one set of inputs by changing one NOR gate to an AND gate, as shown in Figure 2-21. (The AND gate is formed by a NAND gate, followed by an inverter, as described in Sec. 1–19.)

The additional stage increases the propagation delay somewhat. The fan-out from the NAND gate is 1 for a propagation delay of 10 ns, and the fan-out of the inverter is assumed to be 2 for an additional 15 ns. Total propagation time to the carry digit is now 25 ns, and to the sum digit, 40 ns. The additional stage adds 10 ns to the propagation delay times in the NOR gate circuit of Figure 2-20.

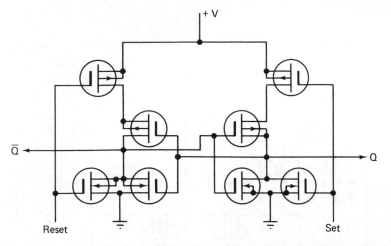

Figure 2-22.   Complementary MOS device *RS* flip-flop. *(Courtesy of Motorola)*

MOS logic RS flip-flop. An *RS* flip-flop implemented from complementary MOS devices is somewhat more complex than simply cross-coupling gates (as described in Sec. 1–23). One version of the *RS* flip-flop is given in Figure 2-22. (Flip-flops are discussed in Chapter 4.)

One of the major advantages of MOS logic is the ability to design highly reliable direct-coupled circuits. However, care must be taken to avoid sneak-paths that are an inadvertent hazard of direct-coupled circuits. The *RS* flip-flop is a case in point.

If the set and reset lines are applied directly to the basic flip-flop gates, the output is driven directly from the set and reset lines. An additional complementary pair is required for both the set and reset lines to isolate them from the load. Such an arrangement is shown in Figure 2-22 where the set and reset inputs are isolated from the basic flip-flop by complementary MOSFETs.

MOS logic JK flip-flop. With a suitable triggering scheme, the basic flip-flop can be made into a *JK* flip-flop as shown in Figure 2-23. In order to steer the trigger pulse for the $J = K = 1$ condition, a complementary pair is used to sense the output of the opposite side. Additional single device gates are used to gate the *J*, *K*, and clock inputs.

Assume that the *J* output is grounded, and that the $J = K = 1$ condition is to toggle the flip-flop. With the *J* output low, the *P*-channel of the right-hand sensing pair is biased ON. High inputs at the *J*

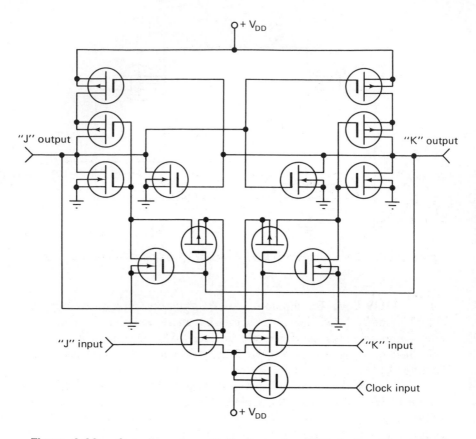

Figure 2-23.   Complementary *JK* flip-flop using MOS device. *(Courtesy of Motorola)*

and *K* inputs turn on their respective gates. When the clock is high (true), the $+V_{DD}$ is propagated through the clock gate, the *K* input, the right-hand sensing gate, and to the *K* output gate. This toggles the flip-flop.

The trigger signal does not propagate through the left-hand sensing gate since it is blocked. This scheme requires a *narrow clock pulse*. The clock pulse must not be present when the *K* output has switched since the left-hand sensing pair would then be biased ON.

The circuit of Figure 2-23 can be operated at frequencies up to about 5 MHz without undue precautions. A frequency of about 10 MHz is possible with careful layout and component selection.

2–1.8 Handling MOS Logic Devices

Damage due to static discharge can be a problem with MOS logic devices. Electrostatic discharges can occur when a MOS device is picked up by its case and the handler's body capacitance is discharged to ground through the series capacitances of the device. This requires proper handling, particularly when the MOS device is out of the circuit. In circuit, a MOS logic device is just as rugged as any other solid state component of similar construction.

MOS devices are generally shipped with the leads all shorted together to prevent damage in shipping and handling (there will be no static discharge between leads). Usually, a shorting spring, or similar device, is used for shipping. The spring *should not be removed until after* the device is soldered into the circuit. An alternate method for shipping or storing MOS devices is to apply a conductive foam between the leads. Polystyrene insulating "snow" is not recommended for shipment or storage of MOS logic devices. Such snow can acquire high static charges which could discharge through the device.

When removing or installing a MOS device, first turn the power off. If the MOS device is to be moved, your body should be at the same potential as the unit from which the device is removed and installed. This can be done by placing one hand on the chassis before moving the MOS device.

2–1.9 Protecting MOS Logic Devices

Because of the static discharge problem, manufacturers provide some form of protection for a number of their MOS logic devices. Generally, this protection takes the form of a diode incorporated as part of the substrate material. A diode (or diodes) can be fabricated as part of the monolith chip. The protection scheme used by Motorola in their complementary MOS ICs is shown in Figure 2-24.

The diode is designed to break down at a lower voltage than the MOS device junctions. Typically, the diode will break down at about 30 V, whereas the junctions break down at 100 V. The diode can break down without damage, provided the current is kept low (as is usually the case where there is a static discharge).

The *single diode* scheme provides protection by clamping positive levels to V_{DD}. Negative protection is provided by the 30 V reverse breakdown. The *resistor diode protection* method adds some delay, but provides protection by clamping positive and negative

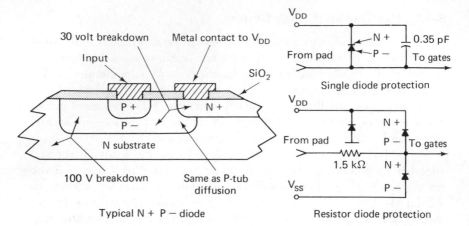

Figure 2-24. Gate protection for complementary MOS devices. *(Courtesy of Motorola)*

potentials to V_{DD} and V_{SS}, respectively. The resistor is included to provide additional circuit isolation.

2–1.10   State-of-the-art MOS Logic IC Devices

In the following paragraphs, we shall describe three current lines of MOS logic. These include: the McMOS, which is the trademark of Motorola Semiconductor Products, Inc.; complementary MOS devices; the COS/MOS, which is the Radio Corporation of America complementary MOS line; and the standard MOS/LSI line of Texas Instruments, Inc.

Keep in mind that these descriptions are for current equipment and are subject to change. That is, the details such as power requirements, propagation times, logic levels, and so on, may change in the future. However, the basic principles for MOS logic devices, both standard lines and custom products, will remain the same. A careful study of this section will enable you to interpret the future data of the manufacturers, as well as the data of other manufacturers.

2–1.11   Motorola Complementary MOS Logic ICs

Figure 2-25 shows how the *P*-channel and *N*-channel devices are connected to form the basic element of McMOS. Figure 2-26 gives the V_{IN} versus V_{OUT} transfer curve for the basic inverter at a power supply voltage of 10 V. Note that the logic pulse output is approximately equal to the supply voltage.

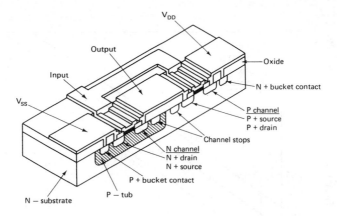

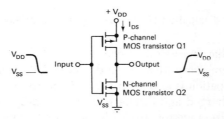

Figure 2-25.   Basic McMOS *P*-channel and *N*-channel devices connected to form an inverter. *(Courtesy of Motorola)*

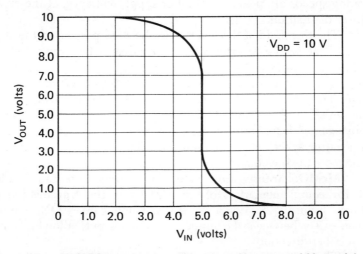

Figure 2-26.   McMOS inverter transfer curve. *(Courtesy of Motorola)*

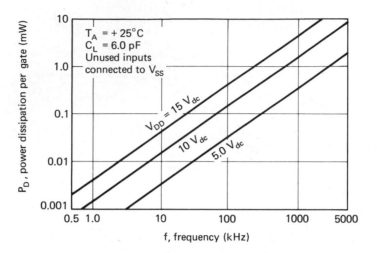

Figure 2-27.   Power dissipation of CMOS inverter. *(Courtesy of Motorola)*

Dynamic characteristics. Because the dynamic power dissipation of a complementary MOS logic device results from capacitive loading, power dissipation is also a function of the frequency at which the capacitance is charged and discharged. Figure 2-27 illustrates this relationship for a basic inverter. It can be seen that power dissipation is linear with frequency. Figure 2-28 shows the effects of capacitive loading and power supply voltage on propagation delays. Higher operating speeds are possible at higher supply voltages, at the expense of power dissipation.

The threshold of the inverter is about 45 percent of the supply voltage. That is, the inverter will switch over when the input logic signal (or pulse) is about 45 percent of the supply voltage. The output voltage then switches from zero to almost 100 percent of the supply voltage. Thus, the MOS logic can be operated over a wide range of supply voltages (5, 10, and 15 V are typical). Also since the output is equally isolated from both V_{DD} and V_{SS} terminals, McMOS can operate with negative as well as positive supplies. The only requirement is that V_{DD} be more positive than V_{SS}.

McMOS transmission gate. A second important building block for complementary McMOS circuits is the transmission gate shown in Figure 2-29. When the transmission gate is ON, a low resistance exists between the input and the output, allowing current flow in either direction.

The voltage on the input line must always be positive with respect to the substrate (V_{SS}) of the N-channel device, and negative with respect to the substrate (V_{DD}) of the P-channel device. The gate

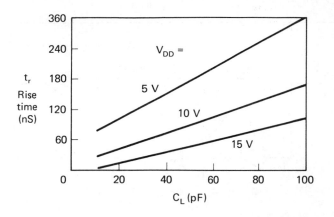

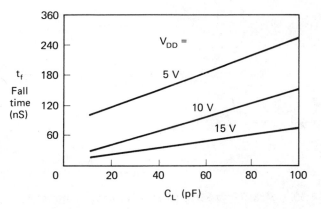

Figure 2-28.   Typical delay characteristics of two-input McMOS NOR gate. *(Courtesy of Motorola)*

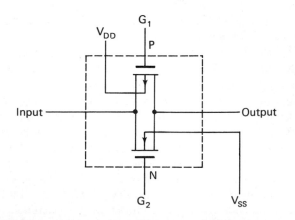

Figure 2-29.   Basic McMOS transmission gate. *(Courtesy of Motorola)*

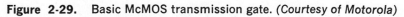

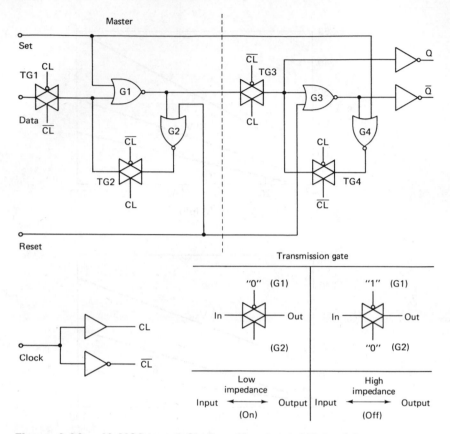

Figure 2-30.   McMOS type-*D* flip-flop. *(Courtesy of Motorola)*

is ON when the gate (G_1) of the *P*-channel is at V_{SS}, and the gate (G_2) of the *N*-channel is at V_{DD}. When G_2 is at V_{SS} and G_1 is at V_{DD}, the transmission gate is OFF, and a resistance of greater than 10^9 ohms exists between input and output.

The resistance between the input and output of a basic transmission gate in the ON condition is dependent upon the voltage applied at the input, the potential difference between the two substrates $(V_{DD} - V_{SS})$, and the load on the output. Typically, R_{ON} is defined as the input-to-output resistance with a 10 KΩ load resistor from the output to ground. With voltages between the two extremes, both devices are partially ON and the value of R_{ON} is caused by the parallel resistance of the *P*- and *N*-channel devices.

Transmission gate applications. An illustration of use for the basic transmission gate is given in the McMOS MC14013 flip-flop, shown in Figure 2-30. The flip-flop works on the master-slave principle and consists of four transmission gates, as well as four NOR gates, two inverters, and a clock buffer/inverter. (Flip-flops, including the master-slave principle, are discussed in Chapter 4.)

When the clock is at a logic 0, transmission gates (TG) 2 and 3 are OFF, and 1 and 4 are ON. In this case, the master is logically disconnected from the slave. With TG_4 ON, gates (G) 3 and 4 are cross-coupled and latched in a stable state.

Assuming that the SET and RESET inputs are low, the logic states of G_1 and G_2 are determined by the logic changes to a logic 1. Under these conditions, TG_2 and TG_3 turn ON, and TG_1 and TG_4 turn OFF. Gates G_1 and G_2 are cross-coupled through TG_2, and the gates latch into the state in which they existed at the time the clock changed from a 0 to a 1. With TG_3 ON, the logic state of the master section (output of gate G_1) is fed through an inverter to the Q output, and through G_3 and another inverter to the $\overline{Q}$ output.

When the clock returns to a logic 0, TG_3 turns OFF, and TG_4 turns ON. This disconnects the slave from the master and latches the slave into the state that existed in the master when the clock changed from a 1 to a 0. Thus, information is entered into the master on the positive edge of the clock. When the clock is high, the output of the master is transmitted directly through the slave to Q and $\overline{Q}$. When the clock changes back to a low state, the state of the master is stored by the slave which then provides the output.

2-1.12   RCA COS/MOS Complementary MOS Logic Devices

COS/MOS fundamentals can best be understood by reference to Figure 2-31. The typical characteristics are shown in Figure 2-32.

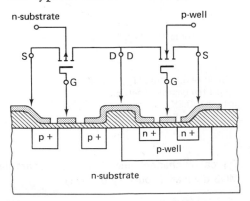

Figure 2-31.   Cross-section of COS/MOS device. *(Courtesy of RCA)*

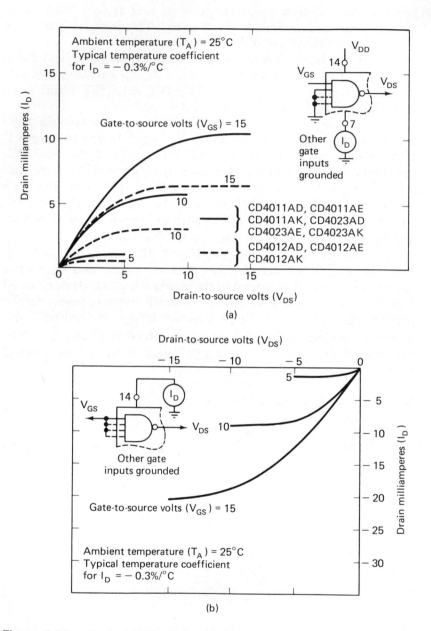

Figure 2-32. Typical *N*-channel and *P*-channel characteristics of COS/ MOS devices. *(Courtesy of RCA)*

104

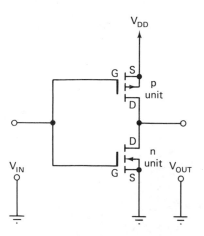

Figure 2-33.   Basic COS/MOS inverter. *(Courtesy of RCA)*

The basic logic inverter (or logic gate) formed by use of P- and N-type devices in series is illustrated in Figure 2-33.

Quiescent device dissipation. When the input lead is grounded, or otherwise connected to 0 V (logic 0), the N-device is cut off, and the P-device is biased on. As a result, there is a low-impedance path from the output to V_{DD}, and an open circuit to ground. The resultant output voltage is essentially V_{DD}, or a logic 1.

When this occurs, the N-channel device becomes a low impedance, while the P-channel device becomes an open circuit. The resultant output becomes essentially zero volts (logic 0).

Note that one of the devices is always cut off at either logic extreme, and that no current flows into the insulating gates, resulting in negligible inverter quiescent power dissipation (equal to the product of V_{DD} times the leakage current).

A cross-section of the COS/MOS inverter (as it is formed in an integrated circuit) on an N-type substrate is illustrated in Figure 2-31. Compare this with Figure 2-34. Note that the source-drain diffusions and the p-well diffusion form *parasitic diodes* (in addition to the desired transistors). (Keep in mind that the parasitic diodes are not to be confused with the protective diodes.) The parasitic diode elements are back-biased (across the power supply) and contribute, in part, to the device leakage current, and thus to the quiescent power dissipation.

Power dissipation of product line. The RCA COS/MOS product line consists of circuits of varying complexity (from the dual

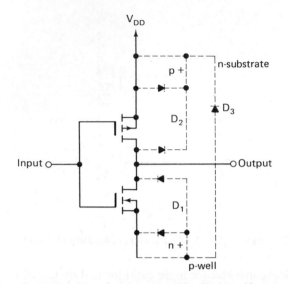

Figure 2-34.   Basic inverter showing parasitic diodes. *(Courtesy of RCA)*

four-input logic gates that contain sixteen MOS devices to the more complex sixty-four-bit static shift registers that contain over 1000 devices). The COS/MOS devices occupy different amounts of silicon area, and are composed of varying numbers of circuits formed from inverters. Consequently, each device in the family shows a particular magnitude of leakage current, depending on the total effect of device count and parasitic diode area.

For example, some logic gates are specified to operate with a typical power dissipation of 5 nW ($V_{DD} = 10$ V), but 7-stage counters or registers are specified to operate with a typical power dissipation of 5 μW ($V_{DD} = 10$ V). Published data includes both typical device quiescent-current levels and maximum levels ($V_{DD} = 5$ V and $V_{DD} = 10$ V).

Switching characteristics. The input/output characteristics for the COS/MOS inverter are shown in Figure 2-35. The signal extremes at the input and output are approximately zero volts (logic 0) and V_{DD} (logic 1). The switching point is shown to be typically 45 to 55 percent of the magnitude of the power supply voltage (regardless of the magnitude) over the entire range from 3 to 15 V (or 5 to 15 V). Note the negligible change in operating point from $-55°$C to $+125°$C.

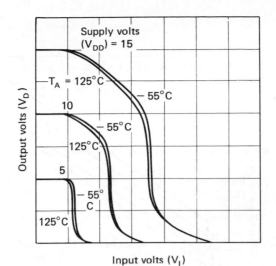

Figure 2-35. Typical COS/MOS transfer characteristics as a function of temperature. *(Courtesy of RCA)*

Ac dissipation characteristics. During the transition from a logic 0 to a logic 1, both devices are momentarily ON. This condition results in a pulse of instantaneous current being drawn from the power supply, the magnitude and duration of which depends upon the following factors:

1. The impedance of the particular devices being used in the inverter circuit.

2. The magnitude of the power supply voltage.

3. The magnitude of the individual device threshold voltages.

4. The input driver rise and fall times.

An additional component of current must also be drawn from the power supply to charge and discharge the internal parasitic capacitances and the load capacitances seen at the output.

The device power dissipation resulting from these current components is a *frequency-dependent parameter*. The more often the circuit switches, the greater the resultant power dissipation; the heavier the capacitive loading, the greater the resultant power dissipation. The power dissipation is *not* duty-cycle dependent. For practical

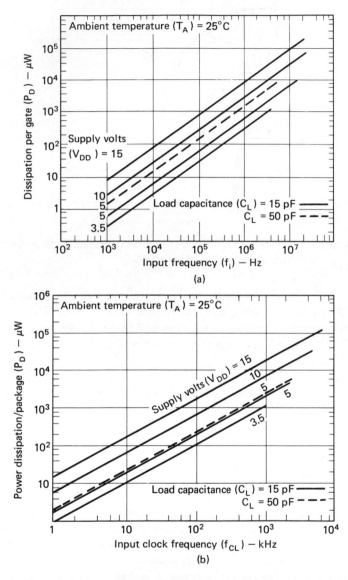

Figure 2-36. Typical power dissipation characteristics: (a) basic gate power dissipation characteristics; (b) MSI device power dissipation characteristics. *(Courtesy of RCA)*

purposes, power dissipation can be considered frequency (repetition-rate) dependent.

Because the COS/MOS product line ranges widely in circuit complexity from device to device, the ac device dissipations vary widely. The effect of capacitive loading on the individual devices also varies. Figure 2-36 shows a family of curves for a typical gate device, and a typical MSI (medium-scale integration) device. These curves illustrate how device power dissipation varies as a function of frequency, supply voltage, and capacitive loading. Note how the MOS devices require less power at higher frequencies.

Ac performance characteristics.   During switching, the capacitances within a given device and the load capacitances external to the circuit are charged and discharged through P- or N-type device conducting channels. As V_{DD} increases, the impedance of the conducting channel decreases accordingly. This lower impedance results in a shorter RC time constant (this nonlinear property of MOS devices can be seen in the curves of Figure 2-32). The result is that the maximum switching frequency of a COS/MOS device increases with increasing supply voltage (see Figure 2-37).

Figure 2-37b shows curves of propagation delay as a function of supply voltage for a gate device. However, the tradeoff or low supply voltage (lower output current to drive a load) is lower speed of operation.

Calculating system power.   The following guidelines have been developed to assist the logic designer in estimating system power for the COS/MOS line. The same general guidelines can be applied to similar MOS lines.

The total system power is equal to the sum of quiescent power and dynamic power. Therefore, system power can be calculated with the following approach:

1. Add all typical package quiescent power dissipations, using published data. Because quiescent power dissipation is equal to the product of quiescent device current multiplied by supply voltage, quiescent power may also be obtained by adding all typical quiescent device currents, and multiplying the sum by the supply voltage (V_{DD}). Quiescent device current is shown in the published COS/MOS data for supply voltages of 5 V and 10 V only.

2. Add all dynamic power dissipations using typical curves of dissipation per package (such as Figure 2-36) as a function of frequency. In a fast-switching system, most of the power dissipation is dynamic, therefore, quiescent power dissipation may be neglected.

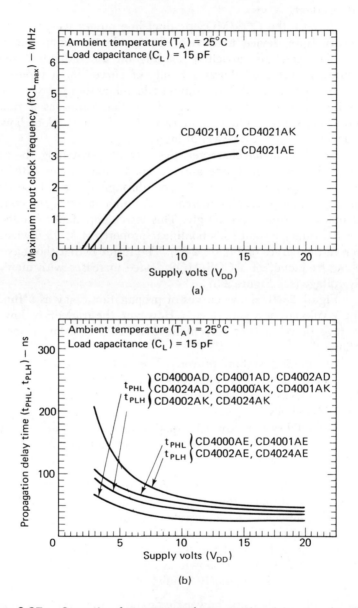

Figure 2-37. Operating frequency and propagation delay as a function of power supply voltage: (a) maximum guaranteed operation frequency as a function of power supply voltage; (b) propagation delay as a function of power supply voltage for the basic gate. *(Courtesy of RCA)*

That is, since the inverters are in a transition state most of the time, the dynamic dissipations govern the total power dissipation.

2–1.13   Texas Instruments MOS/LSI Products

The Texas Instruments line of MOS devices includes shift registers, read-only memories (ROM), programmable logic arrays (PLA), and random access memories (RAM), as well as special purpose devices such as buffers, switches, and custom MOS/LSI devices. These circuits are discussed in their related chapters (switches in Chapter 3, registers in Chapter 4, memories in Chapter 7) and will not be duplicated here.

2–2   SELECTING LOGIC INTEGRATED CIRCUITS

One of the main problems for the logic IC user is to choose the right logic family for a given application. In some cases, price is the all-important factor. In other circumstances, a particular logic function (such as a complete electronic calculator board) may be available in only one logic family. In still other cases, there is a specific design problem (high-speed operation, noise immunity, and so on) that requires a certain logic line.

Before selecting any logic family, the designer should study the characteristics of all available types. At present, there are six basic logic families, three of which (TTL, ECL, and MOS) dominate the field for design of new equipment. Each of these is discussed in Sec. 2–1. It is assumed that the designer will read this material and the discussions showing applications of the various logic families throughout this manual. The designer should also study all available datasheets for logic devices that might suit his requirements. Then, and only then, will the designer be in a good position to choose the right logic IC. However, the designer should keep the following points in mind when reading all of the data.

2–2.1   Availability and Compatibility

TTL is the most available of all logic families. That is, at present, it is possible to obtain the greatest variety of off-the-shelf logic devices in the TTL family. Some designers consider TTL as the "universal" IC logic family, since they can obtain an infinite number of gates (with a variety of input/output combinations), buffers, inverters, counters, registers, arithmetic units, and so on, from more than one manufacturer.

RTL and DTL were, at one time, the next "most available" logic ICs. Today, both RTL and DTL have been replaced by TTL. RTL offers no advantage over TTL. DTL has a slightly higher logic swing (voltage differential from a logic 0 to a logic 1) than TTL (typically 4.5 V for DTL compared to about 3.5 V for RTL). Also, DTL requires slightly less power than TTL, but the operating speeds of DTL are so much slower than TTL (typically TTL is three times faster) that TTL is the preferred family for any present-day application.

HTL is used only where a high logic swing (about 13 V) and high noise immunity is required. Except for this one advantage, HTL is generally of little value to the modern designer. HTL is the slowest and requires the most power of all logic families. In most applications, HTL can be replaced by MOS, since MOS can provide the same logic swing with far less power and at higher speeds. For example, if MOS is operated at with a supply voltage of 15 V, the logic swing can also be almost 15 V.

ECL is used primarily where high speed is essential. ECL is still the fastest of all logic families. The disadvantages of ECL are high power consumption (the highest next to HTL), and a low logic swing (usually less than 1 V, but some ECL will provide nearly 2 V).

MOS is the newest of the commonly-used logic families. The advantages are low power consumption (the lowest of all families), a logic swing equal to any family, and small size. Stated another way, you can get more MOS devices or functions on a given area than any other family. If large scale integration (LSI) is required, MOS is the best choice. TTL and ECL are limited to MSI (medium scale integration).

TTL and DTL are directly compatible with each other. Thus, some designers use only these two families, taking the advantages of both as applicable. Or, they design with TTL only when they must adapt new logic circuits to existing DTL systems.

RTL is the most compatible with linear and analog systems, or any discrete transistor application. This is because RTL is essentially an IC version of conventional solid-state circuits.

Because of their special nature, ECL and HTL are the least compatible with other logic families and with external devices. ECL requires a large supply voltage for a comparatively small logic swing, while HTL produces a very large logic swing which is generally too high for other families. MOS can be made compatible with other families, even though the MOS operating principles are quite different.

Keep in mind that, barring some unusual circumstances, any logic family can be adapted for use with other logic families, or external equipment, by means of *interface circuits*. These interface circuits are discussed in Chapter 8.

2–2.2   Noise Considerations

HTL has the highest noise immunity of any logic family, with the possible exception of MOS. (Or, HTL has the least noise sensitivity, whichever term you prefer.) Typically, signal noise up to about 5 V will not affect HTL. Often, HTL can be used without shielding in noise environments where other families require extensive shielding. MOS devices trigger at about 45 to 50 percent of the supply voltage. If MOS logic is operated at 15 V, the devices will trigger at about 7 V, and will not be affected by lower voltages. Thus, MOS can have higher noise immunity than HTL, if the supply voltage can be kept at 15 V.

RTL has the lowest noise immunity (or is the most noise sensitive). Typically, signal noise in the order of 0.5 V can affect RTL. Considering that RTL operates at logic levels of 1 V, an RTL system is almost always operating near the noise threshold. As a result, RTL is not recommended for noisy environments.

DTL and TTL are about the same in regards to noise immunity or sensitivity. ECL can be operated so that the noise immunity is about equal to that of DTL and TTL. The input of ECL is essentially a differential amplifier. If one base is connected to a fixed bias voltage about half way between the logic 1 and 0 levels, this sets the noise immunity at the level of the bias voltage.

Power supply and ground line noise.   In addition to signal line noise, logic ICs are affected by noise on the power supply and ground lines. Most of this noise can be cured by adequate bypassing as described in Sec. 2–4. However, if there are heavy ground currents due to large power dissipation by the ICs, it may be necessary to use separate ground lines for power supply and logic circuitry.

Noise generation.   In addition to noise immunity or sensitivity, the generation of noise by logic circuits must be considered. Whenever a transistor or diode switches from saturation to cutoff, and vice versa, large current spikes are generated. These spikes appear as noise on the signal lines, as well as the ground and power lines. Since ECL does not operate in the saturation mode, it produces the least amount of noise. Thus, ECL is recommended for use where external circuits are sensitive to noise. On the other hand, TTL produces considerable noise, and is not recommended in similar situations.

Because of its importance in logic design, all of Chapter 9 has been devoted to the subject of noise in logic circuits, particularly noise in IC logic.

114   *Chap. 2   IC Logic Devices*

2–2.3   Propagation Delay and Speed

The speed of an IC logic system is inversely proportional to the propagation delay of the IC elements. That is, ICs with the shortest propagation delay can operate at the highest speed. Since ECL does not saturate, the delay is at a minimum (typically 2–4 ns), and speed is maximum. TTL is the next-to-fastest logic family, with delays of about 10 ns, and can be used in any application except where the extreme high speed of ECL is involved. MOS is slowest of the three currently-popular families (TTL, ECL, and MOS). However, MOS is considerably faster than HTL.

2–2.4   Power Source and Dissipation

MOS requires the least power consumption of all IC logic families, and can be operated over a wide range of power supply voltages (typically 5, 10, or 15 V). CMOS is well suited for battery-operated systems, since little standby power is required. CMOS logic uses power when switching from one state to another, but not during standby (except for some power consumed by leakage).

TTL and ECL generally operate with a 5 V supply, and consume about 15 and 25 mW (per gate), respectively. Power consumption of an MOS gate is generally figured in the μW range, rather than mW.

2–2.5   Fan-out

Fan-out, or the number of load circuits that can be driven by an output, is always of concern to logic designers. Some IC datasheets list fan-out as a simple number. For example, a fan-out of three means that the IC output will drive three outputs or loads. While this system is simple, it may not be accurate. Usually, the term fan-out implies that the output will be applied to inputs of the same logic family, and the same manufacturer. Other datasheets describe fan-out (or load and drive) in terms of input and output current limits. (This system is discussed in Sec. 2–3.)

Aside from these factors, the following typical fan-outs are available from the logic families: RTL 4–5, DTL 5–8, TTL 5–15, HTL 10, ECL 25, MOS 10.

2–3   INTERPRETING LOGIC IC
DATASHEETS

Logic (or digital) IC datasheets are presented in various formats. However, there is a general pattern used by most manufacturers. For example, most logic IC datasheets are divided into four parts. The first part usually provides a logic diagram and/or circuit schematic, plus truth tables, logic equations, general characteristics, and a brief description of the IC. The second part is devoted to test, and shows diagrams of circuits for testing the ICs. These two parts are fairly straightforward and are usually easy to understand.

This is not necessarily true of the remaining two parts (or one large part in some cases), which have such titles as "maximum limits" or "maximum ratings," and "basic characteristics" or "electrical characteristics." These parts are generally in the form of charts, tables, or graphs (or combinations of all three), and often contain the real data needed to design with logic ICs. The terms used by manufacturers are not consistent. Likewise, a manufacturer may use the same term to describe two slightly different characteristics, or use two different terms to describe the same characteristic when different lines or different families are being discussed.

It is impractical to discuss all characteristics found on logic IC datasheets. However, the following notes should help the designer interpret the most critical values.

2–3.1   Electrical Characteristics

It is safe to assume that any electrical characteristic listed on the datasheet has been tested by the manufacturer. If the datasheet also specifies the test conditions under which the values are found, the characteristic can be of immediate value to the designer. For example, if leakage current is measured under worst-case conditions (maximum supply voltage and maximum logic input signal), the leakage current shown on the datasheet can be used for design. However, if the same leakage current is measured with no signal input (inputs grounded or open), the leakage current is of little value. The same is true of such factors as output breakdown voltage and maximum power supply current.

To sum up, if an electrical characteristic is represented as being tested under "typical" (or preferably worst-case) operating conditions, it is safe to take that characteristic as a design value. If the characteristic is measured under no-signal conditions, it is probably included on the datasheet to show the relative merits of the IC. When in doubt, test the IC as described in the datasheet, and use the test results (not the datasheet values) for design.

In the case of quiescent power values for a complementary MOS device, always use the maximum value, rather than the typical value, if both are given. Also, consider clock times, or logic operating speed, when calculating power. In a complementary MOS system, most of the power is dissipated *during transition*. Thus, if transition is slow due to long clock pulses, average power consumption is increased.

2–3.2 Maximum Ratings

Maximum ratings are values that must never be exceeded in any circumstance. They are not typical operating levels. For example, a maximum rating of 15 V for V_{CC} means that if the regulator of the system supply fails, and the V_{CC} source moves up from the normal 8–10 V to 15 V, the IC will probably not be burned out. But never design the system for a normal V_{CC} of 15 V. Allow at least a 10 to 20 percent margin below the maximum, and preferably a 50 percent margin for power supply (voltage and current) limits. Of course, if typical operating levels are given, these can be used even though they are near the maximum ratings.

2–3.3 Drive and Load Characteristics

Generally, the most important characteristics of logic ICs (from the designer's standpoint) are those that apply to the output drive capability, and the input load presented by an IC. This applies to both combinational logic and sequential logic.

No matter which type of logic is involved, the designer must know how many inputs can be driven from one IC output (without amplifiers, buffers, and so on). It is equally important to know what kind of load is presented by the input of an IC on the output of the previous stage (either IC or discrete component). As discussed in Sec. 2–2.5, fan-out is a simple, but not necessarily accurate term to describe drive and load capabilities of an IC.

A more accurate system is where actual input and output currents are given. There are four terms of particular importance. These are:

Output logic 1-state source current I_{OH}

Output logic 0-state sink current I_{OL}

Input forward current I_F

Input reverse current I_R

The main concern is that the datasheet value of I_{OL} must be equal to or greater than the combined I_F value of all gates (or other circuits) connected to an IC output. Likewise, I_{OH} must be equal to or greater than total I_R.

Unfortunately, the same condition exists for datasheet load and drive factors, as exists for other electrical characteristics; the values are not consistent from family to family, and for different manufacturers.

2–3.4   Interfacing Logic ICs

No matter what load and drive characteristics are given on the datasheet, it may be necessary to include some form of interfacing between logic ICs to provide the necessary drive current. This is especially true when the datasheet shows a very close tolerance. For example, assume that an IC with a rated fan-out of 3 is used to drive three gates. If the fan-out rating is "typical" or "average," and the three gates are operating in their "worst case" condition, the IC may not be able to supply (and dissipate) the necessary current. Because of its importance to proper logic design, all of Chapter 8 is devoted to interfacing.

2–4   PRACTICAL CONSIDERATIONS FOR LOGIC ICs

It is assumed that the readers are already familiar with good design practices applicable to all electronic equipment, particularly IC equipment. For example, the readers should understand the basics of selecting IC packages, mounting and connecting ICs, and working with ICs (printed circuit, PC, boards, lead bending, solder techniques, and so on). It is also assumed that the readers are familiar with basic electronic test procedures, and that they are capable of interpreting the test information found on logic IC datasheets. If any of these subjects seem unfamiliar, the reader's attention is invited to the author's *Manual for Integrated Circuit Users* (Reston Publishing Company, Inc., Reston, Virginia, 1973).

2–4.1   Layout of Logic ICs

One area that is of particular importance for any logic design problem is the layout of the ICs. The general rules for logic IC layout are shown in Figure 2-38. The following notes supplement this illustration.

All logic cricuits are subject to noise. Any circuit, discrete or IC, will produce erroneous results if the noise level is high enough. Thus, it is recommended that noise and grounding problems be considered from the very beginning of layout design.

Whenever dc distribution lines run an appreciable distance from the supply to a logic chassis (or a PC board), both lines (positive and negative) should be bypassed to ground with a capacitor, at the point where the wires enter the chassis.

The values for power-line bypass capacitors should be on the order of 1 to 10 μF. If the logic circuits operate at higher speeds

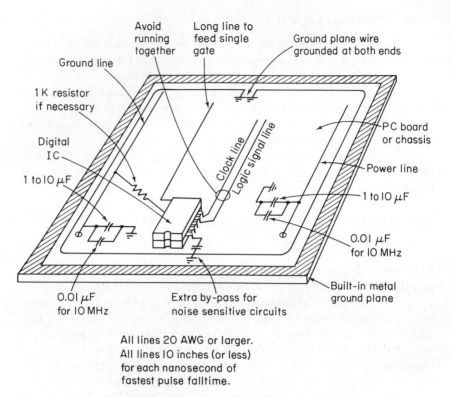

Figure 2-38.   General rules for digital IC layout.

(above about 10 MHz), add a 0.01 μF capacitor in parallel with each 1–10 μF capacitor. Keep in mind that even though the system may operate at low speeds, there are harmonics generated at higher speeds. The high-frequency signals may produce noise on the power lines and interconnecting wiring. A 0.01 μF capacitor should be able to bypass any harmonics present in most logic systems.

If the logic ICs are particularly sensitive to noise, as is the case with TTL, extra bypass capacitors (in addition to those at the power and ground entry points) can be used effectively. The additional power and ground-line capacitors can be mounted at any convenient point on the board or chassis, provided that there is no more than a 7-inch space between any IC and a capacitor (as measured along the power or ground line). Use at least one additional capacitor for each twelve IC packages, and possibly as many as one capacitor for each six ICs.

The dc lines and ground return lines should have large enough cross-sections to minimize noise pickup and dc voltage drop. Unless otherwise recommended by the IC manufacturer, use AWG No. 20 or larger wire for all logic IC power and ground lines.

In general, all leads should be kept as short as possible, both to reduce noise pickup and to minimize the propagation time down the wire. Typically, current logic circuits operate at speeds high enough so that the propagation time down a long wire or cable can be comparable to the delay time through a logic element. This propagation time should be kept in mind during the layout design.

Do not exceed 10 inches of line for each nanosecond of fall time for the fastest logic pulses involved. For example, if the clock pulses (usually the fastest in the system) have a fall time of 3 ns, no logic signal line (either printed-circuit or conventional wire) should exceed 30 inches.

The problem of noise can be minimized if ground planes are used. (That is, if the circuit board has solid metal sides.) Such ground planes surround the active elements on the board with a noise shield. Any logic system that operates at speeds above approximately 30 MHz should have ground planes. If it is not practical to use boards with built-in ground planes, run a wire around the outside edge of the board. Connect both ends of the wire to a common or "equipment" ground.

Do not run any logic signal line near a clock line for more than about 7 inches because of the possibility of *cross-talk* in either direction.

If a logic line must be run a long distance, design the circuits so that the long line feeds a single gate (or other logic element) rather than several gates. External loads to be driven must be kept within the current and voltage limits specified on the IC datasheet. The problems of logic transmission lines are discussed in Chapter 8.

Some logic IC manufacturers specify that a resistor (typically 1 kΩ) be connected between the gate input and the power supply (or ground, depending upon the type of logic), where long lines are involved. Always check the IC datasheet for such notes.

3

Combinational Logic

This chapter is devoted to various forms of combinational logic. The discussion of specific IC logic elements and their applications involves TTL and MOS logic. However, much of the basic information can also be applied to ECL. Before going into specific IC logic devices, we shall cover the basics of decoders and encoders, function generators, parity and comparator networks, as well as data selectors and distributors.

3–1 DECODERS

The basic function of a combinational network is to produce an output (or outputs) only when certain inputs are present. Thus, combinational networks indicate the presence of a given set of inputs by producing the corresponding output. Decoders are classic examples of combinational networks. In effect, all combinational networks are decoders of a sort.

3–1.1 Typical Decoder Circuits

There are many types of decoders in use, and an infinite variety of decoder circuits available in IC form. The following are some examples.

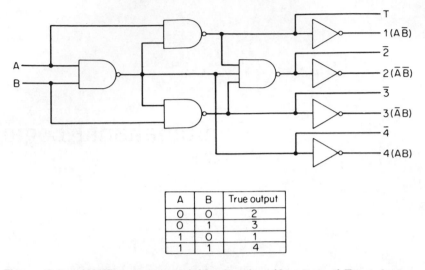

A	B	True output
0	0	2
0	1	3
1	0	1
1	1	4

Figure 3-1. NAND gate two-variable decoder. *(Courtesy of Texas Instruments)*

Variable input decoders. Some decoders produce an output that indicates the state of input variables. For example, if it is assumed that each variable can have only two states (0 or 1), there are four possible combinations for two variables (00, 01, 10, and 11). Thus, a decoder of this type will have two inputs (one for each variable) and four outputs (one for each possible combination). The

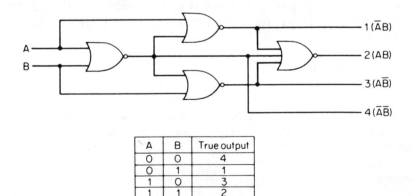

A	B	True output
0	0	4
0	1	1
1	0	3
1	1	2

Figure 3-2. NOR gate two-variable decoder. *(Courtesy of Texas Instruments)*

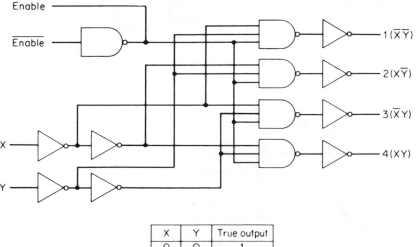

X	Y	True output
0	0	1
0	1	3
1	0	2
1	1	4

Figure 3-3.   Gated two-variable decoder. *(Courtesy of Motorola)*

circuits of Figures 3-1 and 3-2 are NAND and NOR two-variable decoders, respectively. In both circuits, one and only one output will be true for each of the four possible input states. Note that the circuits are similar, except that the NAND circuit (Figure 3-1) requires inverters at the outputs. If a complemented output is acceptable, the inverters can be bypassed.

The circuit of Figure 3-3 is a *gated* two-variable decoder. That is, the outputs are available only when an "enable," "strobe," "gate," or "clock" signal is present. Such "turn-on" signals can be in pulse form, or can be a fixed dc voltage controlled by a switch, whichever is suitable. Note that the enable signal can be either true or complemented.

Similar decoders are available in IC form that provide for three and four input variables.

Even and odd decoders.   With these decoders, the function is to indicate the number of variables that are true (or false) on an "even or odd" basis. For example, an "odd" decoder with three variables produces a true output only when an odd number (1 or 3) of the input variables is true. Such a circuit is shown in Figure 3-4.

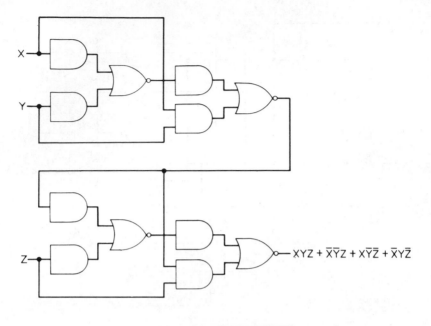

X	Y	Z	Output
O	O	O	O
O	O	1	1
O	1	O	1
O	1	1	O
1	O	O	1
1	O	1	O
1	1	O	O
1	1	1	1

Figure 3-4. AOI gate three-variable odd decoder. *(Courtesy of Texas Instruments)*

Majority and minority decoders. These decoders are used to indicate the majority (or minority) state of variables. For example, if a majority decoder (Figure 3-5) is used with three variables, the single output is true only when two or three of the inputs are true.

The circuits of Figures 3-6 and 3-7 are minority decoders. However, operation of each circuit is somewhat different, as shown by the corresponding truth tables.

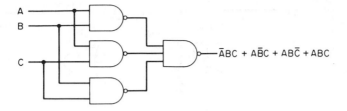

A	B	C	Output
O	O	O	O
O	O	1	O
O	1	O	O
O	1	1	1
1	O	O	O
1	O	1	1
1	1	O	1
1	1	1	1

Figure 3-5. NAND gate majority decoder. *(Courtesy of Motorola)*

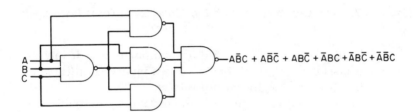

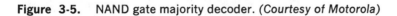

A	B	C	Output
O	O	O	O
O	O	1	1
O	1	O	1
O	1	1	1
1	O	O	1
1	O	1	1
1	1	O	1
1	1	1	O

Figure 3-6. NAND gate minority decoder. *(Courtesy of Cambridge Thermionic Corporation)*

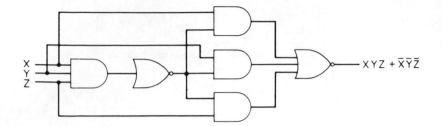

X	Y	Z	Output
0	0	0	1
0	0	1	0
0	1	0	0
0	1	1	0
1	0	0	0
1	0	1	0
1	1	0	0
1	1	1	1

Figure 3-7. AOI gate minority decoder. *(Courtesy of Texas Instruments)*

In the circuit of Figure 3-6, the output is true if any one of the inputs does not agree with the other two inputs. If all three inputs agree, either true or false, the output is false.

In the circuit of Figure 3-7, the output is true only when all three inputs agree, either all true or all false. Note that this circuit uses two AOI gates. The input AOI gate is used as a NAND gate.

Code converter. Still another type of decoder commonly found in IC form is the code converter. Such decoders convert one type of logic code to another. For example, a binary-to-decimal decoder will convert a four-bit binary number into a decimal equivalent.

The circuit of Figure 3-8 is a BCD (binary coded decimal, Sec. 1–3) to decimal decoder. The circuit provides conversion from a four-bit binary coded number to a ten-bit decimal equivalent. The circuit uses all NAND gates.

One and only one output is true when the corresponding inputs are true. For example, the BCD equivalent of decimal 7 is 0111. Thus, the MSB (most significant bit) of the binary input is false, whereas the complement of the MSB input is true. The remaining three bits of the binary inputs are all true (with their complements

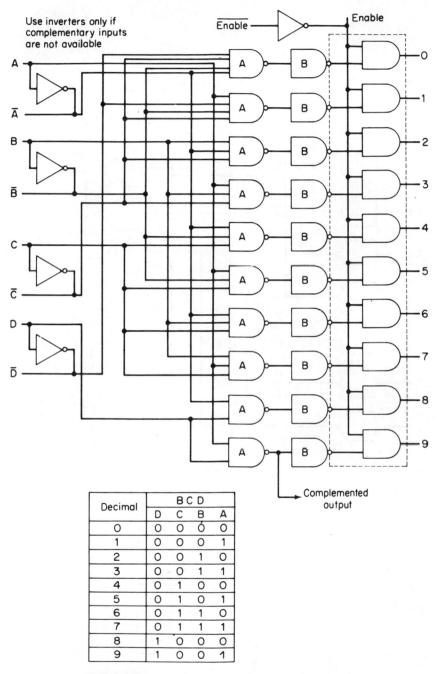

Figure 3-8. BCD to decimal decoder. *(Courtesy of Cambridge Thermionic Corporation)*

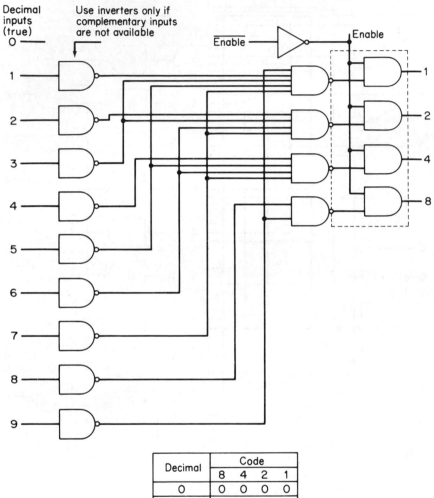

Figure 3-9. Decimal-to-binary encoder. *(Courtesy of Motorola)*

false). Gate 7A of the decoder receives three inputs (lines A, B, and C) and produces a corresponding false output to gate 7B. The output of gate 7B is inverted to true (7B is a NAND gate used as an inverter). All other A gates receive at least one false input, and produce a true output to the corresponding B (or inverter) gate. For example, gate 3A receives two true inputs (lines A and B), and one false input (line $\overline{C}$). Thus gate 3A produces a true output, which is inverted to a false by gate 3B.

A typical line of logic IC code converters would include: binary-to-octal code, 2421 to decimal, 5421 to decimal, excess-3 (XS3) to decimal, and binary to hexadecimal.

3–2   ENCODERS

The term "encoder" is used here to indicate any combinational network that provides the opposite function of a decoder. For example, the decoder of Figure 3-8 converts a four-bit BCD number to a ten-bit decimal equivalent. An encoder, in this context, will convert a ten-bit decimal number to its BCD equivalent. However, both encoders and decoders are forms of combinational networks in that they both produce an output (or outputs) only when certain inputs are present.

The circuit of Figure 3-9 is a decimal-to-BCD encoder. Operation of the circuit is as follows. Assume that the 3 of the decimal input is true, and all other inputs are false. (Only one of the inputs can be true at the same time.) Under these conditions, the output of inverter gate 3 is false, while the output of all other inverter gates is true. Inverter gate 3 is connected to logic gates 1 and 2. Any false input to a NAND gate produces a true output. Thus logic gates 1 and 2 are true. The inputs to logic gates 4 and 8 are all true. All true inputs to a NAND gate produce a false. Thus, logic gates 4 and 8 are false. The binary output of the logic gates is then 0011, or decimal 3.

Note that the 0 decimal input line is not connected. When the decimal input is 0, the 1 through 9 lines are all false, the outputs of the inverter gates are all true, and the outputs of the logic gates are all false, producing a binary 0000, or decimal 0.

3–3   UNIVERSAL FUNCTION GENERATOR

The circuit of Figure 3-10 is a universal function generator. The circuit produces any of the sixteen possible combinations of two variables (and their complements) as a function of the control inputs F1 through F4. That is, the output is dependent upon the presence of

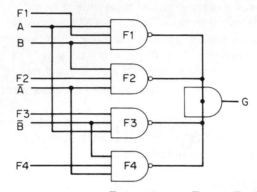

$$G = (F1 + A + B)(F2 + \bar{A} + B)(F3 + A + \bar{B})(F4 + \bar{A} + \bar{B})$$
$$\bar{G} = F1AB + F2\bar{A}B + F3A\bar{B}\ F4\bar{A}\bar{B}$$

F1	F2	F3	F4	G	$\bar{G}$
0	0	0	0	1	0
1	0	0	0	$\bar{A} + \bar{B}$	AB
0	1	0	0	$A + \bar{B}$	$\bar{A}B$
1	1	0	0	$\bar{B}$	B
0	0	1	0	$\bar{A} + B$	$A\bar{B}$
0	1	1	0	$AB + \bar{A}\bar{B}$	$A\bar{B} + \bar{A}B$
1	1	1	0	$\bar{A} + \bar{B}$	$A + B$
0	0	0	1	$A + B$	$\bar{A}\bar{B}$
1	0	0	1	$A\bar{B} + \bar{A}B$	$AB + \bar{A}\bar{B}$
0	1	0	1	A	$\bar{A}$
1	1	0	1	$A\bar{B}$	$\bar{A} + B$
0	0	1	1	B	$\bar{B}$
1	0	1	1	$\bar{A}B$	$A + \bar{B}$
0	1	1	1	AB	$\bar{A} + \bar{B}$
1	1	1	1	0	1

Figure 3-10.   Universal function generator. *(Courtesy of Motorola)*

the inputs, as well as the application of control inputs to the NAND gates. The control signals can be pulses, or a fixed dc voltage controlled by a switch.

The truth table shows the corresponding outputs (true and complement) for each of the possible inputs. For example, if it is assumed that both inputs A and B (and their complements $\bar{A} + \bar{B}$) are present, the output is $\bar{A} + \bar{B}$ when only the $F1$ input is true. This can be understood by the following: With $F1$ true, and $F2$ through $F4$ false, only the A and B inputs will pass. The $F1$ NAND gate will invert the A and B inputs to $\bar{A} + \bar{B}$. Under the same conditions, the complemented output is AB (A and B).

Note that the circuit of Figure 3-10 can be used only with NAND gates capable of being connected in the wired AND configuration. Refer to Sec. 1–19.

3–4   PARITY AND COMPARATOR NETWORKS

In logic systems there are several special codes that have been specifically designed for detecting errors that might occur in digital counting. Any complex logic network that operates on a binary counting system is subject to counting errors due to circuit failure, noise on the transmission line, or some similar occurrence. For example, a defective amplifier can reduce the amplitude of a pulse (representing a binary 1) so that it appears as a binary 0 at the following circuit. Likewise, noise in a circuit could be of equal amplitude to a normal binary 1 pulse. If the noise occurs at a time when a particular circuit is supposed to receive a binary 0, the circuit could react as if a binary 1 were present.

3–4.1   Parity Systems

One method for error detecting is that known as *parity check*. Parity refers to the quality of being equal, and a parity check is actually an equality checking code. The coding consists of introducing additional digits into the binary number. The additional digit is known as a *parity digit* or *parity bit*, and may be either a 0 or a 1. The parity bit is chosen to make the number of all bits in the binary group *even* or *odd*. If a system is chosen in which the bits in the binary number, plus the parity bit, are even, the system is known as *even parity*. If not, the system is referred to as *odd parity*.

For example, the decimal number 7 in binary coded decimal with even parity requires a 1 in the parity bit. Thus, in the total of five bits (four binary bits, plus the one parity bit), there will be an even number of ones. With odd parity, the parity bit is 0 for decimal 7, since there must be an odd number of ones in odd parity. Figure 3-11 illustrates the use of *odd parity check*.

Decimal number	8421 binary number	Parity bit
0	0000	1
1	0001	0
2	0010	0
3	0011	1
4	0100	0
5	0101	1
6	0110	1
7	0111	0
8	1000	0
9	1001	1

Figure 3-11.   Parity bits to be added for odd parity check.

3–4.2   Fundamental Parity Generators and Checkers

Error detection circuits based on the parity system use both parity generators and parity detectors or checkers. The function of a parity generator is to examine the system word (group of binary bits) and calculate the information required for the added parity bit. For example, if there are three 1s in the binary group, and even parity is used, a one must be used for the parity bit to make an even number in the "parity word."

Once the parity bit has been included, the parity word (binary bits, plus parity bit) can be examined after any transmission, or at any point in the system, to determine if a failure or error has occurred. A parity detection circuit or parity checker examines the parity word to see if the desired odd or even parity still exists (say, after passing through several hundred logic gates). If an error has occured, the system control can be informed that the system is not functioning properly (by means of a control panel lamp or alarm).

The fundamental operation required in parity generation and detection circuits, that of comparing inputs to determine the presence of an odd or even number of ones, can be effectively performed by EXCLUSIVE OR logic gates. A basic EXCLUSIVE OR gate performing the function $\overline{A}B + A\overline{B}$ serves to calculate parity over inputs A and B. EXCLUSIVE OR (and EXCLUSIVE NOR) gates can be interconnected to form *parity trees,* and thus perform parity calculations over longer word lengths.

The parity scheme described thus far (known as *simple parity*) detects the presence of a single error in a word (or group of binary bits). If two errors occur, the output does not indicate that an error has occured. Thus, simple parity will detect an odd number of errors, but will fail if an even number of errors occurs.

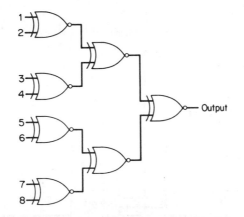

Figure 3-12.   Eight-bit parity tree. *(Courtesy of Motorola)*

3–4.3   Advanced Parity Schemes

Advanced parity schemes have been devised to recognize that an error has occured, and to detect *which bit is in error*. Several extra bits must be added to the system word to make these parity checks. One such scheme is referred to as Hamming parity single-error detection and correction (after R. W. Hamming). When this scheme is used, a four-bit binary word requires that three additional bits (Hamming parity bits) be added.

3–4.4   Simple Parity Checking With Parity Trees

The circuit of Figure 3-12 consists of seven EXCLUSIVE NOR gates connected to form an eight-bit parity tree. Each gate has a 0 logic output if, and only if, one of its two inputs is at a 1 logic level. Thus, the output of the three-stage parity tree is in the 0 state if there is an odd number of ones over the eight inputs.

The circuit of Figure 3-13 is a four-bit parity tree, providing a 0 output state if an odd number of ones exist over the respective inputs.

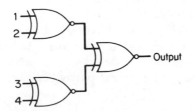

Figure 3-13.   Four-bit parity tree. *(Courtesy of Motorola)*

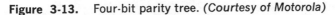

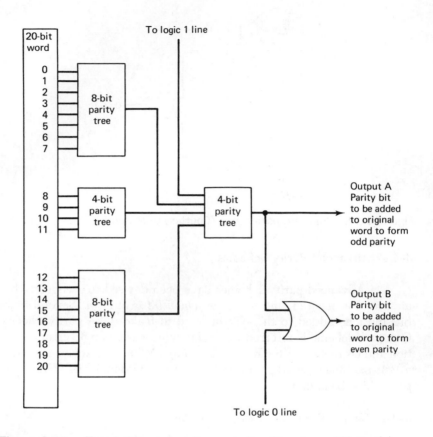

Figure 3-14. Twenty-bit-word parity generator. *(Courtesy of Motorola)*

The circuits of Figures 3-12 and 3-13 form the basic "building blocks" for a twenty-bit-word simple parity generator and detector scheme, shown in Figure 3-14. If a parity word containing odd parity is required (the twenty-bit-word plus the parity bit are to contain an odd number of ones), then the direct output from the parity tree (output *A*) is used as the parity bit. If even parity is required, an inverter (or gate) can be used for inversion, as shown at output *B*.

The twenty-bit-word parity detector or checker is shown in Figure 3-15. The original twenty-bit-word is connected to a two-stage parity tree almost identical to that used for the twenty-bit parity generator. However, for the detection circuit, the output of the tree must be compared with the input parity bit.

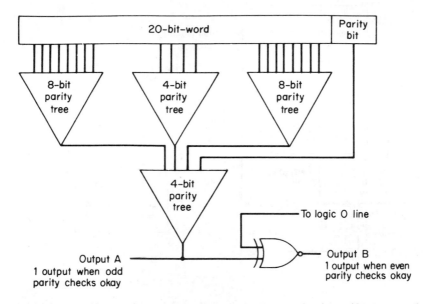

Figure 3-15. Twenty-bit-word parity detector or checker. *(Courtesy of Motorola)*

The parity bit can serve as the input to the second stage of the tree. For odd parity detection, output *A* must be in the 1 state if no error has occurred. An even parity detection system will result in a 1 output (for output *B*) if no error has been introduced.

The circuits of Figures 3-14 and 3-15 can be expanded as necessary to accommodate words of greater length. However, additional parity trees must be used.

3–4.5 Hamming Parity Code Detection and Correction

Advanced parity schemes (often known as *single-error correction*) require that redundant information be added to the word so that the *bit number in error* can be identified. The Hamming parity code is a scheme for adding the required redundant bits of information. In this scheme, the redundant bits are called Hamming parity bits. The number of extra bits that must be added is found from the inequality $k^2 \geqq m + k + 1$, where k is the number of Hamming parity bits and m is the number of information bits.

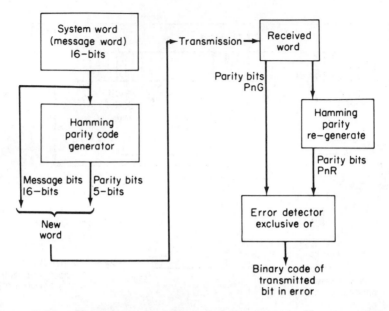

Figure 3-16.   Block diagram of Hamming code system. *(Courtesy of Motorola)*

A block diagram of a Hamming code system is shown in Figure 3-16. This system is for a sixteen-bit message word, and requires five Hamming parity bits. The Hamming code generator calculates the five Hamming parity bits $P1G$, $P2G$, $P4G$, $P8G$, $P16G$ (the letter G indicates the parity bits calculated by the parity generator) from the sixteen message bits, $M1$ through $M16$. The parity bits are then inserted into the message word in a definite order, $M16$, $M15$, $M14$, $M13$, $M12$, $P16G$, $M11$, $M10$, $M9$, $M8$, $M7$, $M6$, $M5$, $P8G$, $M4$, $M3$, $M2$, $P4G$, $M1$, $P2G$, $P1G$. After the parity bits have been inserted in the message bits, the parity word can be processed, where errors are likely to be introduced.

The table of Figure 3-17 indicates which message bits must be examined to calculate the five parity bits. The message bits are the headings for the sixteen rows, with the Hamming parity bits as the column headings. An X is placed in the table to indicate which of the message bits must be examined in order to generate the corresponding parity bit. For example, $P1$ is generated by requiring the combination of $P1$, $M1$, $M2$, $M4$, $M5$, $M7$, $M9$, $M11$, $M12$, $M14$, $M16$ to possess an even parity. Similarly, $P2$ is generated by requiring $M1$,

Parity bits

Message bits	P1	P2	P4	P8	P16
M1	X	X			
M2	X		X		
M3		X	X		
M4	X	X	X		
M5	X			X	
M6		X		X	
M7	X	X		X	
M8			X	X	
M9	X		X	X	
MIO		X	X	X	
MII	X	X	X	X	
MI2	X				X
MI3		X			X
MI4	X	X			X
MI5			X		X
MI6	X		X		X
Total inputs to each parity tree in generator	IO	9	9	7	5

Figure 3-17. Hamming parity generation for sixteen message bits. *(Courtesy of Motorola)*

$M3$, $M4$, and so on, to possess an even parity. Note that the $P1$ calculation requires ten message bits (plus the one parity bit) to be examined.

The Hamming parity detector re-examines the input message exactly as does the generator. However, the parity bits generated at the receiving end of the system, $P1R$, $P2R$, $P4R$, $P8R$, and $P16R$ are compared with the transmitted bits PnG (where n represents 1, 2, 4, 8, and 16 for the example given). Each combination of parity bits ($P1G$ transmitted and $P1R$ received) is compared via an EXCLUSIVE OR circuit. The results of this comparison, $P1E$ through $P16E$, form a binary word that indicates the *bit position* of any single bit in error.

For example, an output code of $P16E = 1$, $P8E = 0$, $P4E = 1$, $P2E = 0$, $P1E = 0$, or 10100, indicates that the twentieth (16 + 4) bit in the transmitted word (or message bit 15) contains an error.

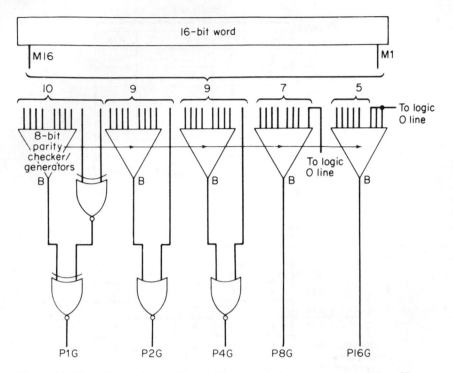

Figure 3-18.   Hamming parity generator for sixteen message bits. *(Courtesy of Motorola)*

The Hamming parity generator (for the sixteen bits shown in the table of Figure 3-17) is shown in Figure 3-18. The eight-bit parity trees (shown in triangle form in Figure 3-18) are equivalent to the circuit of Figure 3-19, with the OR output (output B) used. Note that any unused inputs should be connected to a logic 0 line, as shown. The circuit of Figure 3-19 consists of EXCLUSIVE OR gates and one EXCLUSIVE NOR gate, connected to form an eight-bit parity tree. Note that the EXCLUSIVE NOR gate must also have an OR output (such as found on many ECL gates, described in Sec. 2–1.5). Output A is in the 1 state when an even number of ones exist over the eight inputs. Output B is in the 1 state when an odd number of ones exist over the eight inputs.

The Hamming parity detector is shown in Figure 3-20. The output of the detector circuit is a five-bit binary word that indicates which bit (if any) in the message is in error. If there are no errors, the output is 00000. Note that the generation circuit called for in Figure 3-20 is equivalent to the circuit of Figure 3-18.

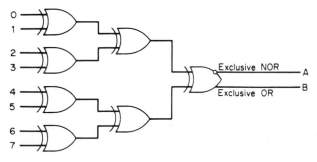

B = O, A = 1 whenever an even number of inputs are true
B = 1, A = O whenever an odd number of inputs are true

Partial Truth Table

0	1	2	3	4	5	6	7	Outputs A	B
0	0	0	0	0	0	0	0	1	0
0	1	1	0	0	1	1	1	0	1
1	1	0	1	1	1	0	1	1	0
1	0	1	1	1	0	1	0	0	1

Figure 3-19.   Eight-bit parity checker/generator. *(Courtesy of Motorola)*

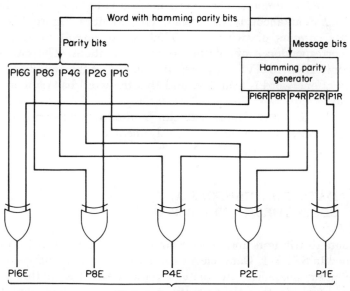

Figure 3-20.   Hamming parity detector for sixteen message bits. *(Courtesy of Motorola)*

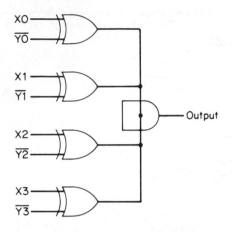

Figure 3-21. Four-bit by two-word equality detector. *(Courtesy of Motorola)*

3–4.6 Comparator Networks

Many logic systems require techniques for comparing two binary or BCD numbers. The fastest method is to compare all bits of each word simultaneously, in parallel. The circuit of Figure 3-21 compares the *equality* of two words. That is, the circuit compares the direct correspondence of bits in like word positions. The circuit requires only EXCLUSIVE OR gates, provided that the complement for each bit of one word is available, and that the word has no more than eight bits.

The circuit of Figure 3-22 can be used in similar applications where it is necessary to detect not only inequality, but also to determine which word is larger.

3–5   DATA DISTRIBUTORS AND SELECTORS (MULTIPLEXERS)

Basic data distributors and selector circuits are similar to the decoders described in Sec. 3–1. Data selectors are also known as *multiplexers* or *multiplex switches*. The two basic circuits described in this section are "universal" in that they can be rearranged to meet almost any logic system needed where data distribution or selection is involved.

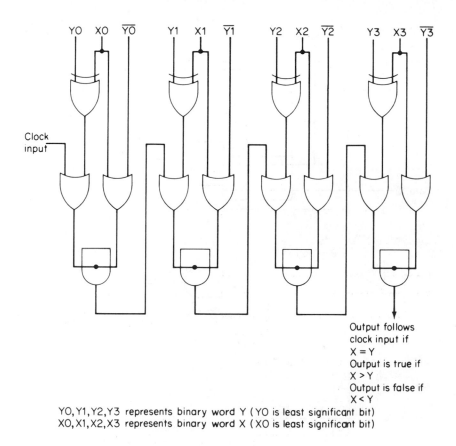

Clock
input

Output follows
clock input if
X = Y
Output is true if
X > Y
Output is false if
X < Y

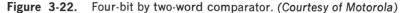

Y0,Y1,Y2,Y3 represents binary word Y (Y0 is least significant bit)
X0,X1,X2,X3 represents binary word X (X0 is least significant bit)

Figure 3-22.   Four-bit by two-word comparator. *(Courtesy of Motorola)*

3–5.1   Data Distributor

The data distributor distributes a single channel of input data to any number of output lines, in accordance with a binary code applied to control lines. Both two-channel and four-channel data distributor circuits are shown in Figure 3-23. Note that both inverters (or gates used as inverters) and AND gates are used. (If the distributor is to be used where the data must be applied to several loads on each output line, the AND gates should be of the power type.)

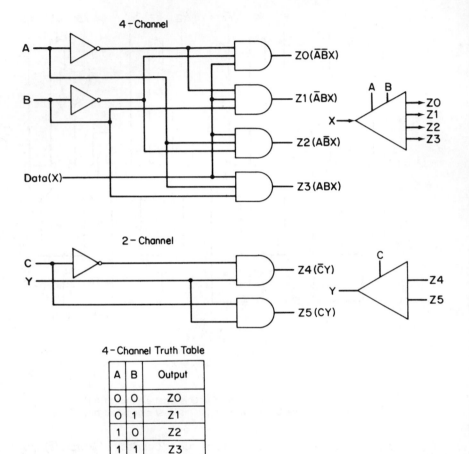

Figure 3-23.   Four- and two-channel data distributor. *(Courtesy of Motorola)*

Information applied to input X is distributed to outputs $Z0$ through $Z3$, in accordance with the binary number applied to control inputs A and B (or the states 0 or 1 of inputs A and B). Information applied to input Y is distributed to outputs $Z4$ and $Z5$, in accordance with the state of control input Y.

The size of the distribution function may be increased by increasing the number of basic distributor circuits (Figure 3-23). For example, a system for distribution of one bit of information to one of sixteen locations is shown in Figure 3-24. Distribution of more than

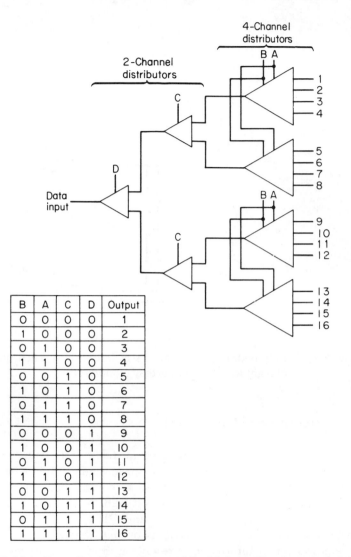

B	A	C	D	Output
0	0	0	0	1
1	0	0	0	2
0	1	0	0	3
1	1	0	0	4
0	0	1	0	5
1	0	1	0	6
0	1	1	0	7
1	1	1	0	8
0	0	0	1	9
1	0	0	1	10
0	1	0	1	11
1	1	0	1	12
0	0	1	1	13
1	0	1	1	14
0	1	1	1	15
1	1	1	1	16

Figure 3-24. Data distribution for one bit of information to one of sixteen
locations. *(Courtesy of Motorola)*

one bit of information is also possible with the basic distributor circuits. For example, a system for distributing N bits of information, with
each bit going to one of eight locations, is shown in Figure 3-25. Information on inputs X through M is distributed in accordance with
control variables A and B as previously described.

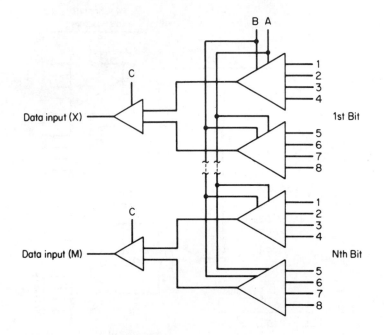

Figure 3-25.   Data distribution for *N* bits of information, each bit to one of eight locations. *(Courtesy of Motorola)*

3–5.2   Data Selector (Multiplexer)

The data selector or multiplex switch selects data on one or more input lines and applies the data to a single output channel, in accordance with a binary code applied to control lines. A four-input data selector circuit is shown in Figure 3-26. Note that inverters (or gates used as inverters), AND gates, and OR gates are used. (If the selector must drive several loads at its output, the OR gate should be of the power type.) Information present on input lines $X0$ through $X3$ is transferred to output $Q0$, in accordance with the state of control inputs A and B, as shown by the equations and truth table.

Data selection from more than four locations (or inputs) can be implemented by using multiple basic data selector circuits (Figure 3-26). An example of expanded selection is given in Figure 3-27. Here, one bit is selected from one of the sixteen inputs, $X0$ through $X16$, and transferred to output $Q0$, in accordance with the truth table.

The basic data selector circuit can also be used to implement a network that provides multiple outputs from multiple inputs. Such

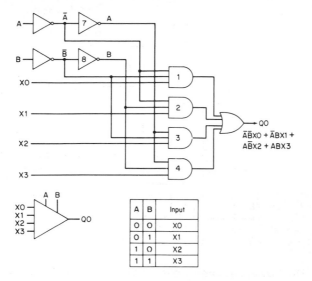

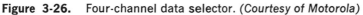

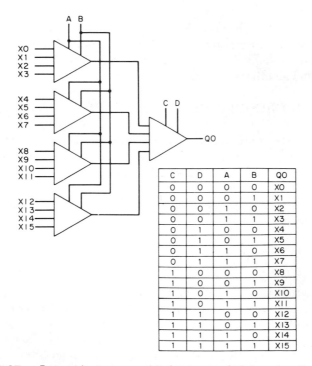

Figure 3-26.   Four-channel data selector. *(Courtesy of Motorola)*

A	B	Input
0	0	X0
0	1	X1
1	0	X2
1	1	X3

C	D	A	B	QO
0	0	0	0	X0
0	0	0	1	X1
0	0	1	0	X2
0	0	1	1	X3
0	1	0	0	X4
0	1	0	1	X5
0	1	1	0	X6
0	1	1	1	X7
1	0	0	0	X8
1	0	0	1	X9
1	0	1	0	X10
1	0	1	1	X11
1	1	0	0	X12
1	1	0	1	X13
1	1	1	0	X14
1	1	1	1	X15

Figure 3-27.   Data selector—one bit from one of sixteen locations. *(Courtesy of Motorola)*

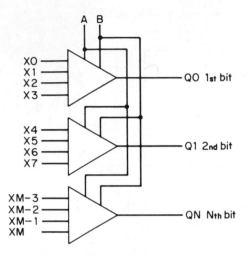

Figure 3-28. Data selector—*N* bits from one of four locations. *(Courtesy of Motorola)*

an *N*-bit network is shown in Figure 3-28. Each output bit is selected from one of its *own group* of four different inputs. For example, with control lines *A* and *B* both at 1, data lines *X*3, *X*7, and *XM* are enabled, and all other data lines are inhibited.

3–6   TTL DECODER/DEMULTIPLEXER

This section describes applications for the Texas Instruments SN54/ 74154 decoder/demultiplexer. The SN54/74154 is a classic TTL MSI device of the 54/74 series. (The designation 54 applies to military, whereas 74 applies to industrial IC logic devices. Similar devices, with characteristics equivalent to the 54/74 series, are available from other manufacturers.)

3–6.1   Logic Description

Figure 3-29 shows the logic diagram for the SN54/74154, and Figure 3-30 shows the truth table and pin connection diagram. As shown, the device is housed in a 24-pin IC. Power is applied to pins 12 and 24. The remaining pins are for active signals. The dc and switching characteristics of the device are available on the data sheet, and will not be repeated here. Instead, we will concentrate on logic design applications. However, in summary of the device characteristics, the SN54/74154 is compatible with any standard TTL system

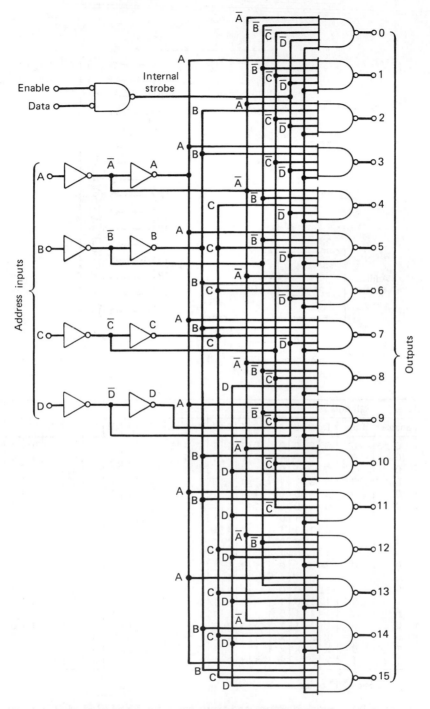

Figure 3-29. Logic diagram from SN54/74154. *(Courtesy of Texas Instruments)*

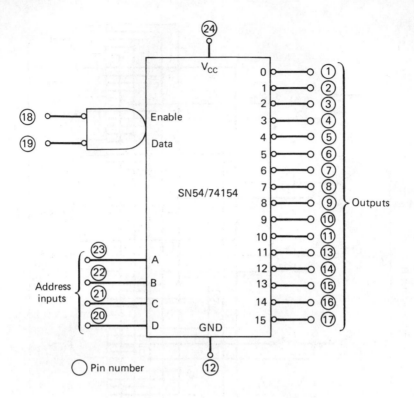

Inputs						Outputs															
Enable	Data	D	C	B	A	0	1	2	3	4	5	6	7	8	9	10	11	12	13	14	15
0	0	0	0	0	0	0	1	1	1	1	1	1	1	1	1	1	1	1	1	1	1
0	0	0	0	0	1	1	0	1	1	1	1	1	1	1	1	1	1	1	1	1	1
0	0	0	0	1	0	1	1	0	1	1	1	1	1	1	1	1	1	1	1	1	1
0	0	0	0	1	1	1	1	1	0	1	1	1	1	1	1	1	1	1	1	1	1
0	0	0	1	0	0	1	1	1	1	0	1	1	1	1	1	1	1	1	1	1	1
0	0	0	1	0	1	1	1	1	1	1	0	1	1	1	1	1	1	1	1	1	1
0	0	0	1	1	0	1	1	1	1	1	1	0	1	1	1	1	1	1	1	1	1
0	0	0	1	1	1	1	1	1	1	1	1	1	0	1	1	1	1	1	1	1	1
0	0	1	0	0	0	1	1	1	1	1	1	1	1	0	1	1	1	1	1	1	1
0	0	1	0	0	1	1	1	1	1	1	1	1	1	1	0	1	1	1	1	1	1
0	0	1	0	1	0	1	1	1	1	1	1	1	1	1	1	0	1	1	1	1	1
0	0	1	0	1	1	1	1	1	1	1	1	1	1	1	1	1	0	1	1	1	1
0	0	1	1	0	0	1	1	1	1	1	1	1	1	1	1	1	1	0	1	1	1
0	0	1	1	0	1	1	1	1	1	1	1	1	1	1	1	1	1	1	0	1	1
0	0	1	1	1	0	1	1	1	1	1	1	1	1	1	1	1	1	1	1	0	1
0	0	1	1	1	1	1	1	1	1	1	1	1	1	1	1	1	1	1	1	1	0
0	1	X	X	X	X	1	1	1	1	1	1	1	1	1	1	1	1	1	1	1	1
1	0	X	X	X	X	1	1	1	1	1	1	1	1	1	1	1	1	1	1	1	1
1	1	X	X	X	X	1	1	1	1	1	1	1	1	1	1	1	1	1	1	1	1

X = Logical "1" or logical "0"

Figure 3-30. Block diagram and truth table for SN54/74154. *(Courtesy of Texas Instruments)*

148

(with a power supply of about 5 V, a typical dissipation of 170 mW, and a maximum dissipation of less than 300 mW). The longest switching or propagation delay is 37 ns (to a logic 1 level at output from A, B, C, or D inputs through three logic levels).

As shown in Figure 3-29, four binary-coded address inputs A, B, C, and D are decoded internally to address only one of sixteen mutually-exclusive outputs. Outputs with decimal names 0, 1, 2, 3 . . . 15 correspond to the binary-weighted input codes at the address inputs for $A = 2^0$, $B = 2^1$, $C = 2^2$, and $D = 2^3$. For example, if $A = D = 1$ and $B = C = 0$, gate 9 is addressed.

From the truth table of Figure 3-30 it can be seen that all sixteen outputs will be at 1 unless both enable and data inputs are at 0. When enable and data inputs are 0, the device operates on a mutually exclusive sixteen-line NAND-gate decoder of the four input address lines. (The addressed output gate will be at 0, with all other outputs at 1.)

The device can also be used as a one-line to sixteen-line demultiplexer by setting enable at 0, and connecting binary information to the data input. This information is routed, unchanged, to the output selected by the address inputs. For example, if $A = D = 1$, $B = C = 0$, and enable = 0, then output 9 = data. All other outputs are at 1.

If enable = 1 when the address inputs are changed, the possibility of decoding spikes and/or more than one output being momentarily 0 is eliminated. Since the *internal strobe* line is 0, address information is locked out, causing all outputs to be 1. After switching transients have subsided, the enable input can be returned to 0, thus returning control to the address inputs and data input.

Due to the symmetry of the positive AND gate with inverted inputs (positive NOR gate) used, enable and data functions are interchangeable. The names "enable" and "data" given these functions are arbitrary. Likewise, the four address inputs may be interchanged and the device will still produce the correct output when the output pins are relabeled. This may be convenient for printed circuit board layouts.

3–6.2 Decoder

The use of the SN74154 as a four-line-to-sixteen-line decoder is shown in Figure 3-31. Since enable and data are at 0, the addressed output is 0. This type of circuit is often referred to as a 1-of-16 decoder.

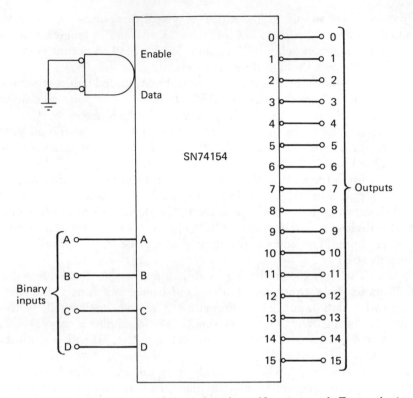

Figure 3-31. Four-line-to-16-line decoder. (*Courtesy of Texas Instruments*)

Figure 3-32 shows four SN74154 decoders used to decode six variables to one of sixty-four lines. Both enable and data inputs are used on the four devices to determine which device is activated. Since enable and data must be 0 to activate the device, proper decoding of the signals at these inputs is necessary to activate only one device at a time.

This method can be extended to any number of variables with additional SN74154 devices. For example, an eight-line-to-256-line (1-of-256) decoder can be constructed with seventeen SN74154 devices. As shown in Figure 3-33, one SN74154 decoder controls the data inputs of sixteen others. By controlling the enable line of device-16, all 256 outputs can be disabled simultaneously.

By using a SN7493 binary counter to drive the address inputs, a *clock time generator* can be constructed. (Binary counters are discussed in Chapter 4.) To get positive pulses from a particular

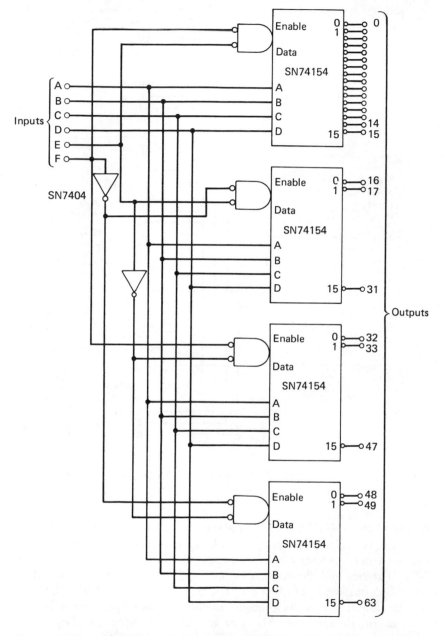

Figure 3-32. Six-line-to-sixty-four-line decoder with inhibit capability. *(Courtesy of Texas Instruments)*

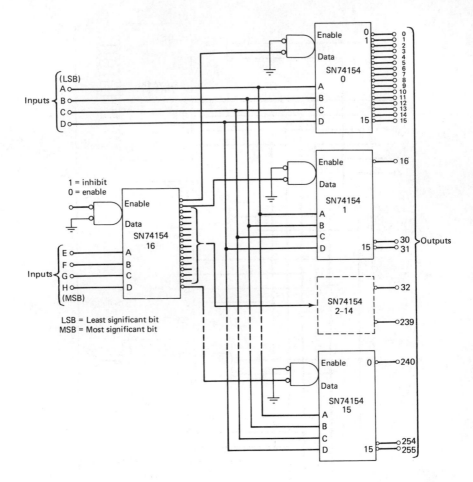

Figure 3-33. Eight-line-to-256-line decoder with inhibit capability. *(Courtesy of Texas Instruments)*

output of the SN74154, an inverter must be inserted as shown in Figure 3-34. By using cross-coupled NAND gate latches, pulses can be extended over several clock pulse periods such as shown in Figure 3-34. (Latches and flip-flops are discussed in Chapter 4.) The latch sets when output 9 goes low and resets when output 14 goes low. If the data input is used as a strobe, decoding spikes (pulse overshoots) may be eliminated from the output. This is especially important when an asynchronous counter (Chapter 4) is used on the address inputs.

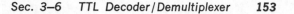

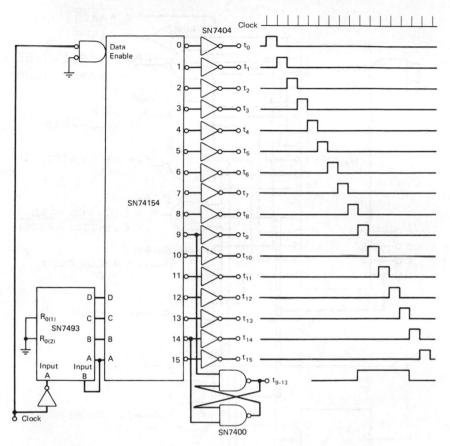

Figure 3-34. Clock time generator. *(Courtesy of Texas Instruments)*

3–6.3 Minterm Generator

When operated as a decoder, the SN74154 can function as a low active-output minterm generator (function generator) producing fifteen possible minterms from four variables. The desired minterms can be summed by a positive NAND (negative NOR) gate as shown for functions F_1, F_2, F_3, F_4, and F_6 in Figure 3-35.

Note that the circuit of Figure 3-35 requires the use of an SN7442, which is a BCD-to-decimal decoder (or converter) such as described in Sec. 3–1.1. When the two devices are used as shown in Figure 3-35, any combination of the seven input variables A, B, C, D, E, F, and G can be obtained as an AND function by programming with only a two input NOR gate (F_9).

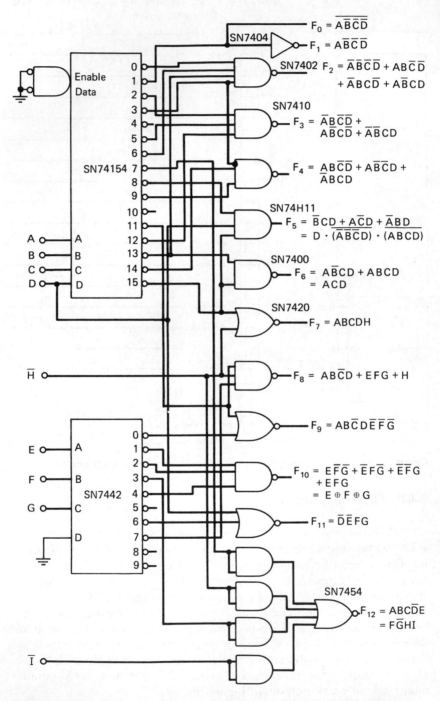

Figure 3-35. Minterm generator. *(Courtesy of Texas Instruments)*

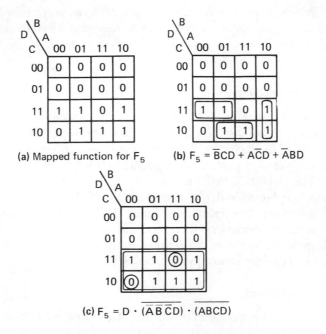

(a) Mapped function for F_5

(b) $F_5 = \overline{B}CD + A\overline{C}D + \overline{A}BD$

(c) $F_5 = D \cdot \overline{(\overline{A}\,\overline{B}\,\overline{C}D)} \cdot \overline{(ABCD)}$

Figure 3-36.   Mapping for function F_5 of minterm generator. *(Courtesy of Texas Instruments)*

To illustrate the method used to obtain function F_5, refer to the maps in Figure 3-36. The function is mapped in Fig. 3-36b. As shown, the ones are grouped to get the following minimal expression:

$$F_5 = \overline{B}CD + A\overline{C}D + \overline{A}BD$$

As shown in Figure 3-36c, another method of grouping ones is to encircle the group whose D is 1, and inhibit those individual cells or blocks within the group containing 0 (that is, inhibit $\overline{A}\overline{B}\overline{C}D$ and $ABCD$).

$$F = D \cdot \overline{(\overline{A}\overline{B}\overline{C}D)} \cdot \overline{(ABCD)}$$

$$F = \overline{B}CD + A\overline{C}D + \overline{A}BD = F_5$$

Note that F shown in Figure 3-36c is identical to F_5 when simplified. Since the inverted output exists for all sixteen possible combinations of inputs, the latter method lends itself quite readily to use with the SN75154. A single three-input AND gate (such as the SN74H11) or a NAND gate and an inverter will decode the complete expression from the SN74154.

Figure 3-35 is only one example of how very complex logic operations can be implemented with a few packages by using the SN74154 in combination with other MSI circuits (such as the SN74150 series of data selectors, discussed in Sec. 3–7).

3–6.4 Code Decoders

As discussed in Sec. 3–1.1, there are many special-purpose ICs designed to convert (or decode) from one logic code to another (such as BCD to decimal, XS3 to decimal, and so on. Each special-purpose decoder accepts only one specific four-bit code.

The SN74154, with strobe capability, can decode (or convert) *any four-bit code* if the appropriate outputs are selected in the desired sequence. For example, Figure 3-37a shows the output selection sequence necessary to completely decode the four-bit Gray code. Figure 3-37b shows an output selection matrix by which the SN74154 can decode several typical codes.

3–6.5 Demultiplexer

A typical application of the SN74154 used as a demultiplexer is shown in Figure 3-38. Parallel input information is converted to

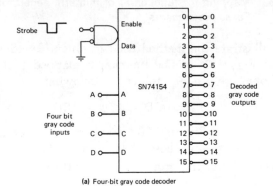

(a) Four-bit gray code decoder

Decimal digit	SN74154 output selection			
	8421	Excess 3	2421	Excess 3 gray
0	0	3	0	2
1	1	4	1	6
2	2	5	2	7
3	3	6	3	5
4	4	7	4	4
5	5	8	11	12
6	6	9	12	13
7	7	10	13	15
8	8	11	14	14
9	9	12	15	10

(b) Output selection for decoding four-bit decimal codes

Figure 3-37. Using an SN54/74154 to decode any four-bit codes by selecting appropriate outputs. (*Courtesy of Texas Instruments*)

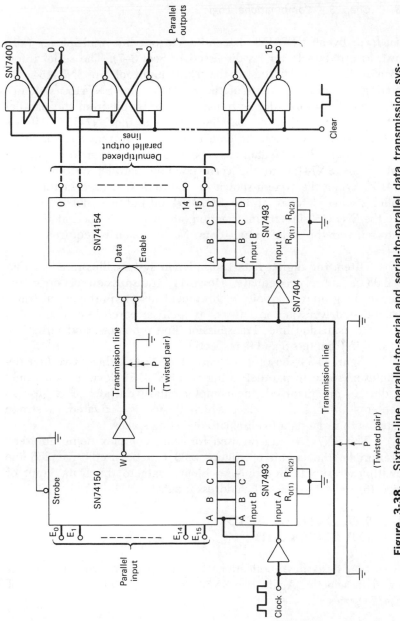

Figure 3-38. Sixteen-line parallel-to-serial and serial-to-parallel data transmission system. *(Courtesy of Texas Instruments)*

157

serial form by an SN74150 sixteen-line to one-line multiplexer (also known as a data selector, as discussed in Sec. 3–7). The serial information is then transmitted to the Data input of an SN74154. By operating the address inputs of the SN74154 and SN74150 synchronously, parallel information is transferred bit-by-bit from the parallel inputs of the SN74150 to the parallel outputs of the SN74154. Latches may then be used to store the data in parallel form, as shown.

Multiple bits of data may be transmitted from one parallel input of the SN74150 to the corresponding parallel output of the SN74154. When the system shown in Figure 3-38 is used in this way, parallel storage latches at the SN74154 output are not necessary. Since the SN74150 inverts information, an inverter is used at the output to reinvert the serial output data. No inversion is required at the SN74154.

Often, the digital data transmission system illustrated in Figure 3-38 is entirely adequate. However, transmission of error-free data over long lines in a noisy environment may require special transmission-line drivers and receivers, as well as careful selection of a suitable transmission line. Transmission line problems, and noise, are discussed in Chapters 8 and 9, respectively.

Figure 3-39 shows a method of getting a thirty-two-line demultiplexer (serial to parallel) using two SN74154 devices. Since only two devices are involved, the complementary outputs of a flip-flop (Chapter 4) are used to select which device is activated. A strobe is gated with serial data to eliminate decoding spikes.

The SN74154 can be used for more complex demultiplexers. For example, the circuit of Figure 3-33 can be used in a 256-line demultiplexer simply by applying binary data to the Data input of device-16, and sequencing the address inputs.

3–7 TTL DATA SELECTOR (MULTIPLEXER)

This section describes applications for the Texas Instruments SN54/74150 data selector. Again, the SN54/74150 is a TTL MSI device of the 54/74 series.

3–7.1 Logic Description

Figure 3-40 shows the logic diagram for the SN54/74150. As shown, the device is basically an AOI gate with sixteen OR branches. Each AND gate has its individual data input (E_0 through

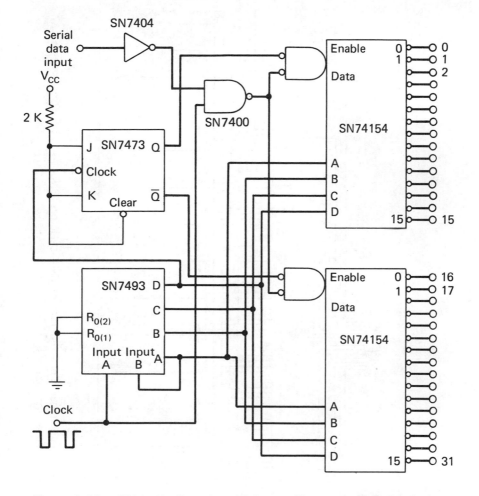

Figure 3-39.   Thirty-two-line demultiplexer. *(Courtesy of Texas Instruments)*

E_{15}). Any one of these inputs can be selected by applying the appropriate binary-coded address to the data-select terminals (A, B, C, D). A strobe signal (S) can be applied concurrently with the address signal. A 0 at the strobe input enables the data present at the selected data-input to be coupled through to the output. A one at the strobe input inhibits data transmission.

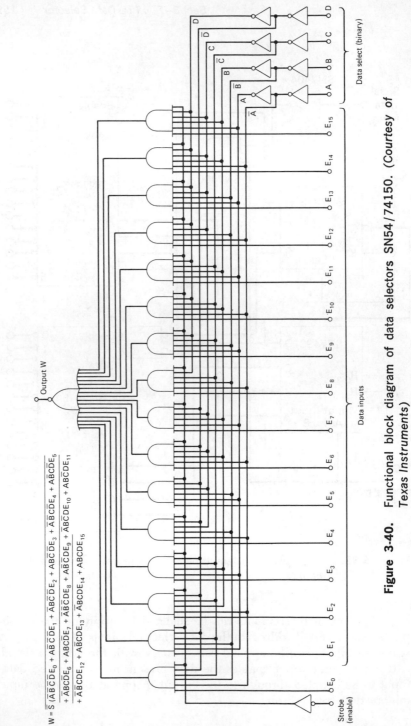

$$W = \overline{S}\ \overline{(A\overline{B}\overline{C}\overline{D}E_0 + A\overline{B}\overline{C}\overline{D}E_1 + \overline{A}B\overline{C}\overline{D}E_2 + AB\overline{C}\overline{D}E_3 + \overline{A}\overline{B}C\overline{D}E_4 + A\overline{B}C\overline{D}E_5}$$
$$\overline{+\ \overline{A}BC\overline{D}E_6 + ABC\overline{D}E_7 + \overline{A}\overline{B}\overline{C}DE_8 + A\overline{B}\overline{C}DE_9 + \overline{A}B\overline{C}DE_{10} + AB\overline{C}DE_{11}}$$
$$\overline{+\ \overline{A}\overline{B}CDE_{12} + A\overline{B}CDE_{13} + \overline{A}BCDE_{14} + ABCDE_{15}}$$

Figure 3-40. Functional block diagram of data selectors SN54/74150. (Courtesy of *Texas Instruments*)

3–7.2   Random Data Selection

Data selectors can select at random one source out of a multitude of information sources, and couple the output of this source through a single information channel or input. Data input can be selected by applying the appropriate binary-coded address to the data-select inputs (A, B, C, or D).

The number of data inputs can be increased by using additional data selectors. A system using two data selectors is shown in Figure 3-41. When a strobed system is required, the control network of Figure 3-41a should be used with the network of Figure

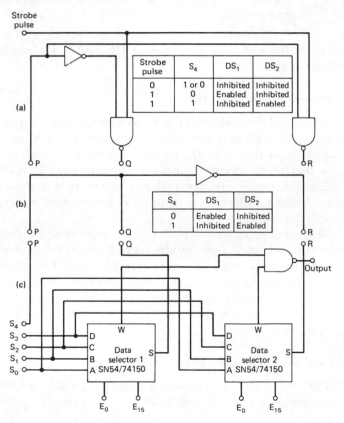

Control circuit (b) + circuit (c) gives a non-strobed system
Control circuit (a) + circuit (c) gives a strobed system

Figure 3-41.   Block diagram of data selectors being used to select one out of thirty-two information sources. *(Courtesy of Texas Instruments)*

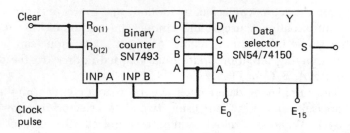

Figure 3-42.   Block diagram using a data selector to sequentially select one out of sixteen information sources. *(Courtesy of Texas Instruments)*

3-41c. Use of the networks of Figure 3-41b and 3-41c produces a nonstrobed system.

3–7.3   Sequential Data Selection

Sequential data selection can be performed with data selectors if the data-select address is taken from the outputs of a binary counter (Chapter 4), as shown in Figure 3-42. Operation of such a system resembles that of an electromechanical stepping switch. With each clock pulse, the binary counter switches to the next state, causing the data selector to select in sequence the information sources connected to its data inputs.

The number of data-inputs can be expanded by cascading data selectors and using appropriate control networks. A sequential data selection system with two data selectors, and two alternative control networks is shown in Figure 3-43. Note that two counters are used. Counter I supplies the binary-coded address. Counter II (a single flip-flop) followed by a decoding network, sequentially enables the data selectors.

3–7.4   Parallel-to-serial Conversion

Data selectors may be used to serialize parallel information. The circuits of Figures 3-42 and 3-43 may be used for parallel-to-serial conversion.

The circuit of Figure 3-42 is capable of serializing one word of sixteen bits, or m words of $16/m$ bits.

The circuit of Figure 3-43 is capable of serializing one word of thirty-two bits, or m words of $32/m$ bits.

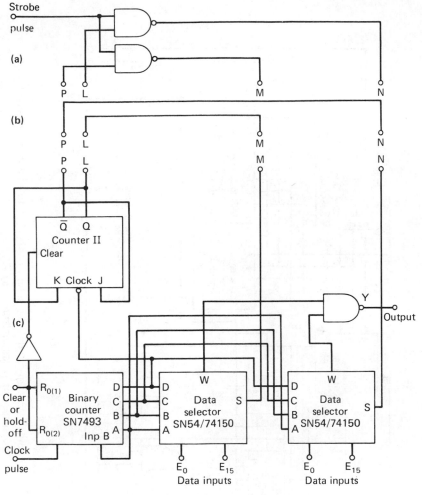

Control circuit (a) + circuit (c) gives a strobed system
Control circuit (b) + circuit (c) gives a non-strobed system

Figure 3-43. Block diagram of data selector used to sequentially select one out of thirty-two information sources. *(Courtesy of Texas Instruments)*

3–7.5   Multiplexing to One Line

The operations described thus far are for multiplexing to one line. The multiplexing capabilities of the selectors in Figures 3-42 and 3-43 can be enlarged by using shift registers (Chapter 4) as information sources. Thus, with a single data selector, 16 words can be serially

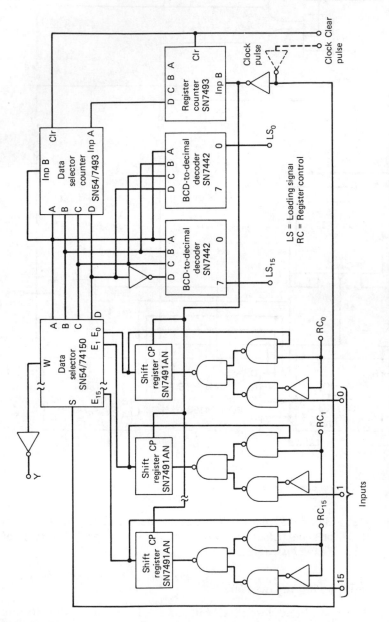

Figure 3-44. Block diagram of data selectors connected to multiplex a single output. *(Courtesy of Texas Instruments)*

multiplexed onto one line by either shifting out or circulating (if the data must be preserved) the contents of the shift registers, as shown in Figure 3-44. The word length is determined by the number of stages of the shift registers. Very long words may be stored in adjacent shift registers. However, the word capacity of the system will then be reduced.

A low-going loading signal (LS) indicates the shift register that is selected. The mode of operation of the shift register is controlled by appropriate signals on the Register Control (RC) terminal. The signals are:

$$RC = 0 \text{ for shift-out/serial-load}$$

$$RC = 1 \text{ for data circulation}$$

A complete cycle of either data shift-out/serial-loading or data circulation is performed during each count cycle of the register (R) counter. Consequently, the R-counter must have as many states as the shift registers have stages. Loading of a shift register may take place during the cycle in which it is selected. In this case, LS and RC terminals with the same number must be connected. If the loading must occur in the following selection cycle, the RC terminals must be connected to the LS terminals with the next higher number. Any loading sequence may be obtained by connecting the RC terminals to the appropriate LS signals.

The data selector is strobed by the incoming clock-pulse, while the shift registers and counters are operated by the inverted incoming clock-pulse. This is done to inhibit the data selector during the transition periods of the counters, shift registers, and data inputs.

3–7.6   Multiplexing to Multiple Lines

In general, multiplexing means transmitting a large number of information units over a smaller number of channels or lines. Consequently, the number of outgoing lines of a multiplexing system is not necessarily restricted to one. Multiplexing to multiple lines is necessary, for instance when a multitude of words must be transferred, a whole word at a time. For multiplexing n-bit words, n data selectors are necessary.

A system for multiplexing up to sixteen words of n bits onto n parallel lines is shown in Figure 3-45. This system can be used either for random word selection, or with a binary counter for sequential word selection.

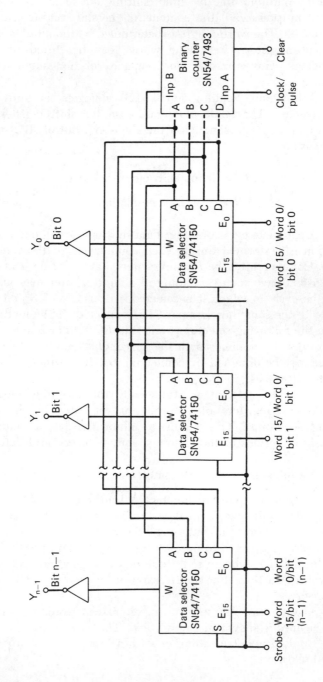

Figure 3-45. Block diagram of data selectors used to multiplex up to sixteen words of *n* bits onto *n* parallel lines. (*Courtesy of Texas Instruments*)

3-7.7 Data Selectors as Character Generators

Data selectors, when used with a binary counter for sequential selection, can operate as character generators. The characters to be generated may be either fixed (wired-in), or manually changeable (switches). Further, the characters may be controlled and/or determined by a logic system.

Figure 3-46 shows how an SN74151 data selector is used as a character generator with manually changeable characters. The

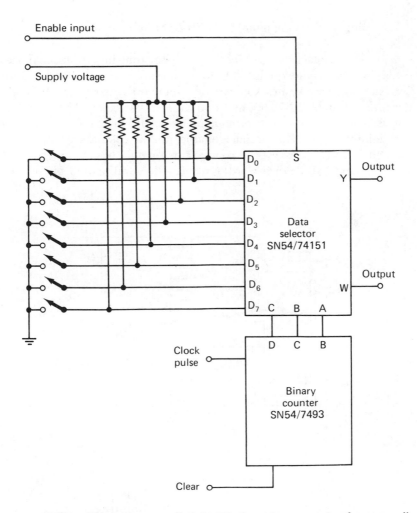

Figure 3-46. Block diagram of eight-bit character generator for manually changeable characters. *(Courtesy of Texas Instruments)*

SN74151 is similar to the SN54/74150 shown in Figure 3-40, except that the SN74151 has eight data inputs, three data select lines, and two outputs (Y is the inverted form of output W).

Almost any data selector circuit, with sequential selection, can be used as a character generator. The circuits shown in Figures 3-42, 3-43, and 3-44 can be used as character generators with each character appearing in serial form at the output. The circuit of Figure 3-45 is capable of generating characters in parallel form, or multiple characters in serial form.

3–7.8 Binary Word Comparison With Data Selectors

Data selectors may be used with four-line-to-decimal decoders to determine quality (or to compare) binary words. Figure 3-47 shows a comparator for two four-bit words using two BCD-to-decimal decoders (SN7442), and one SN74150 selector. The decoders are used as three-line binary-to-octal decoders with input D as enable/inhibit input. The decoder enable/inhibit inputs are controlled through NAND gates by the comparator enable-input and the most significant bit (MSB) of a word. (In fact, the combination of the two

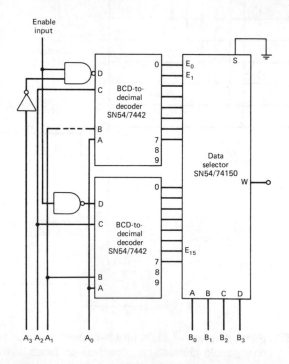

Figure 3-47. Data selector used as a four-bit word comparator. *(Courtesy of Texas Instruments)*

decoders form a four-line binary-to-hexadecimal decoder; see Figure 1-4 for hexadecimal code.) A logic one on the enable input of the Figure 3-47 circuit enables the comparator function, whereas a zero on the enable input inhibits the comparator.

Output signals in the enabled and inhibited states are as follows:

Enable input $= 1$ (comparator enabled)

$W = 1$ if word $A =$ word B

$W = 0$ if word $A \neq$ word B

Enable input $= 0$ (comparator inhibited)

$W = 0$ regardless of whether words A and B

are equal or not

Consequently, output W can be used as the compare-output as well as the enable-output.

The comparator of Figure 3-47 may be cascaded to compare longer words, as shown in Figure 3-48. If a 1 output signal is required to indicate equality, an additional inverter (dashed in Figure 3-48) is necessary.

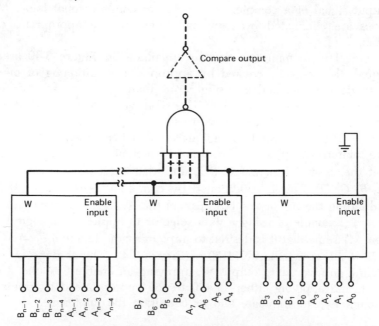

Figure 3-48.   Block diagram of cascaded comparators used to compare two words of n bits. (*Courtesy of Texas Instruments*)

3–7.9 Implementing Logic Functions With Data Selectors

Almost any logic function can be implemented with data selectors. The simplest implementation is obtained when the logic function can be written either as a true or an inverted sum of products of the data selector's logic variables. This is because the characteristic logic expression of a data selector is an inverted sum of products.

To implement a given logic function, it is necessary to use a data selector which is able to satisfy all minterms of the function, directly or through conditioning. For example, the logic function of the SN74150 is shown in Figure 3-40. If all of the data inputs are at 1, and the strobe S is at 0, the data selector is enabled, and the logic expression for the SN74150 is equal to the function shown in Figure 3-40. If an inverter is used at the ouput W, the logic function equals that shown in Figure 3-40, but in complemented form (the overbar shown in Figure 3-40 is removed).

To implement a specific function made up of the variables shown in Figure 3-40, it is necessary to condition the desired minterms, and eliminate unused minterms. A minterm is conditioned by applying the appropriate logic signal to the corresponding data input. Such a conditioning signal can vary from a simple one, to the output signal of a complex gate array of combinational logic. Minterms are eliminated by applying a zero to the appropriate data input.

If more minterms than are available in Figure 3-40 are required, they may be created by appropriate conditioning of one or more data inputs and/or using more than one data selector connected as shown in Figure 3-49. Use of n data selectors with m data inputs provide a total of mn minterms.

The term "conditioning" as applied here, means to combine two minterms to produce the desired function. For example, assume that the desired function is $\bar{B}\bar{D}$, and that there are four variables (A, B, C, D). This is conditioned by combining $\bar{A}\bar{B}\bar{C}\bar{D}$ and $A\bar{B}C\bar{D}$, and letting the $\bar{A}\bar{C}$ of one term cancel the AC of the other term.

Example of using a data selector to implement a logic function. Assume that it is desired to implement the function $F = A\bar{C} + \bar{A}CD + \bar{B}\bar{D}$ using the SN74150 of Figure 3-40. This is done by connecting an inverter to output W, applying an enable 0 to the strobe input, and applying either 1 or 0 to the appropriate data inputs E_0 through E_{15} (to eliminate and/or condition the minterms).

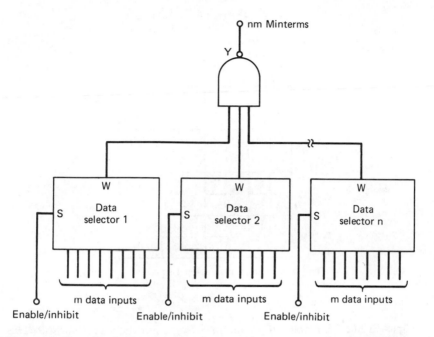

Figure 3-49. Block diagram of minterms expanded by cascading data selectors. *(Courtesy of Texas Instruments)*

Several approaches are possible to determine which minterms are to be eliminated or conditioned for a given logic function. One such method uses Karnaugh maps described in Chapter 1. When a logic expression of a data selector is represented by a Karnaugh map, each cell (or block) of the map represents a minterm, and the numerical value assigned to each cell corresponds with the number of the data input which governs the minterm represented by the cell.

For example, Figure 3-50 shows a four-variable map representing the SN74150. (Note that the map format is somewhat different from those of Chapter 1, but the results are the same.) The upper right-hand cell represents data input E_2, and the minterm $\overline{A}BC\overline{D}$; the lower left-hand cell represents data input E_8, and the minterm $A\overline{B}\overline{C}D$.

By comparing the function $F = AC + ACD + BD$ with the function of the SN74150 (Figure 3-40), it will be seen that the minterms controlled by E_2, E_6, E_7, E_8, E_{10}, E_{13}, and E_{15} are not needed. These data inputs are connected to a zero. All remaining minterms are

Logic function: $F = \overline{A}C + \overline{A}CD + \overline{B}\overline{D}$
Data selector: SN54/74150

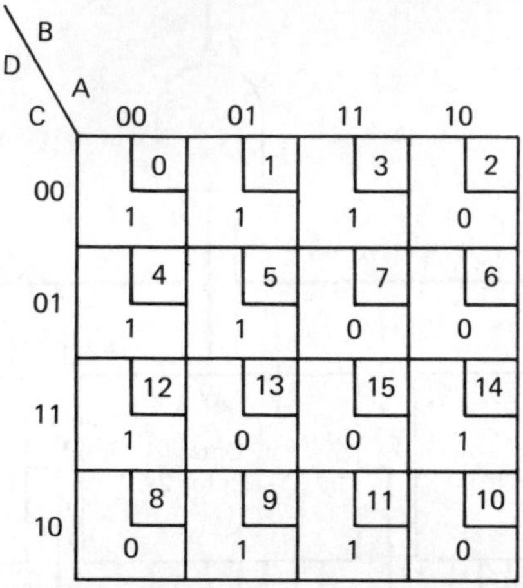

The conditioning signals to be used are
$$E_0 = E_1 = E_3 = E_4 = E_5 = E_9 = E_{11} = E_{12} = E_{14} = 1$$
$$E_2 = E_6 = E_7 = E_8 = E_{10} = E_{13} = E_{15} = 0$$

Figure 3-50.   Example of mapping a logic function of the SN74150. *(Courtesy of Texas Instruments)*

needed for the desired function, so the remaining data inputs E_0, E_1, E_3, E_4, E_5, E_9, E_{11}, E_{12}, and E_{14} must be at logic 1. With these connections, and an inverter at output W, the desired function is produced when an enable 0 is applied to the strobe.

4

Sequential Logic

This chapter is devoted to various forms of sequential logic. The discussion of specific IC logic elements and their applications involves TTL, ECL, and MOS logic. Before going into specific IC logic devices, we shall cover the basics of flip-flops, counters, and registers.

4–1 FLIP-FLOPS AND LATCHES

As discussed in Chapter 1, the basic flip-flop (FF) or latch is essentially a multivibrator (MV). Most logic MVs and FFs are implemented by means of logic gates fabricated as part of the IC. However, it is possible to make up FFs and MVs using basic gates.

4–1.1 Logic FF Forms

This section describes the various logic FFs in common use. Some of these simpler forms first appeared as discrete component circuits. However, most of the forms are the result of packaging logic elements as ICs. Although there are many variations of FF logic circuits, there are only three basic types: *RS*, *JK*, and master-slave.

The following descriptions are provided for the student who is not familiar with the history and development of the logic FF. This information should also be of interest to the designer, since it forms

NOR gate FF

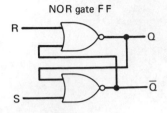

R	S	Q^{M+1}
Lo	Lo	Q^N
Lo	Hi	Hi
Hi	Lo	Lo
Hi	Hi	*

Positive logic equations

$$Q = \overline{R} \cdot \overline{\overline{Q}} = \overline{(R + \overline{Q})}$$
$$\overline{Q} = \overline{S} \cdot \overline{Q} = \overline{(S + Q)}$$

Q is defined as the 1 side
Q^N = present state
Q^{N+1} = next state after applying
input signals
Hi is more positive than Lo
* = both outputs go to Lo state
when Hi is applied to both inputs

NAND gate FF

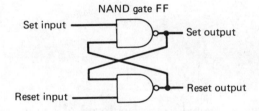

Set input ── ── Set output

Reset input ── ── Reset output

Conditions					
Inputs				Outputs	
Old		New			
S	R	S	R	S	R
—	—	0	0	1	1
—	—	0	1	1	0
—	—	1	0	0	1
0	1	1	1	1	0
1	0	1	1	0	1
0	0	1	1	*	*

Blanks means that the
condition has no effect
on the output
* = the output might
be 1 or 0

Figure 4-1. Basic *RS* FF. *(Courtesy of Motorola)*

the basis for all sequential circuits (counters, registers, and so on) described in this manual.

RS flip-flop. Figure 4-1 shows the basic *RS* FF circuit, together with truth tables. The same basic circuit can also be called a latch. Note that either NAND or NOR gates can be used. The circuits are formed by cross-coupling the two gates. The NOR gate *RS* FF is set to one of its stable states by applying a high logic level on one of the inputs. (The inputs to NOR gate FFs are normally low.)

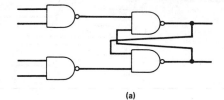

(a)

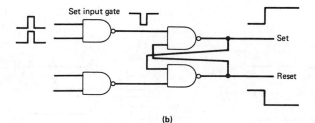

(b)

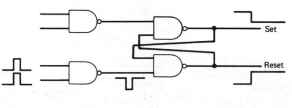

(c)

Figure 4-2.   Gated RS FF.

High logic levels on both inputs produce simultaneous low logic levels at both outputs. However, if both high levels are removed from the inputs simultaneously, the state of the *output cannot be predicted.*

Note that the terms SET and RESET refer to the two states of the FF. When the FF is in the SET state, the SET output is 1, and the RESET output is 0. The opposite occurs in the RESET state.

In the case of the NAND gate RS FF, as long as both inputs are not 0 at the same time, the FF state is that of the input *that was 0 last.* This is shown in the truth table for the NAND gate FF. If two zeros are put in simultaneously, the FF goes into a condition of being neither SET nor RESET (both inputs = 1). The designer must realize this possibility when using NAND gate FFs. With any NAND gate circuit, *if any input is 0, the output is 1; only if all inputs are 1 is the output 0.*

Gated RS flip-flop.   An improved version of the RS FF is the gated RS FF shown in Figure 4-2. The improvement over the basic

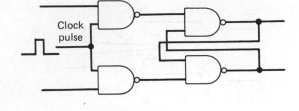

(a)

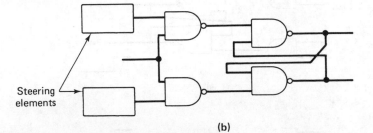

(b)

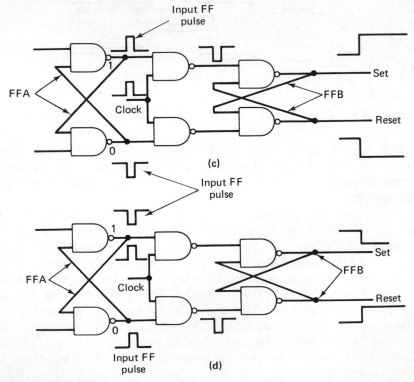

(c)

(d)

Figure 4-3.   Steered or clocked *RS* FF.

RS is the addition of two input gates. In order to set the gated FF, it is necessary to have two positive pulses appearing simultaneously at the inputs to the SET gate, as shown in Figure 4-2b. The reset condition requires two pulses at the RESET gate (Figure 4-2c).

Steered or clocked *RS* flip-flop.   Sometimes one of the pulses used with the SET gate or the RESET gate is a *timing* or *clock* pulse. Often, a single timing or clock pulse can be applied to both the SET and RESET gates. In that case, one input from the SET gate and one input from the RESET gate are tied together and appear as a single input, as shown in Figure 4-3.

This type of FF is called a *clocked* or *triggered* FF by some logic designers, since it requires both a clock pulse and a data input to produce a change in output. Other designers prefer to call the circuit a *steered* FF, since it is directed by another element. This element is shown as two blocks in Figure 4-3b. The element could be a computer data bus, or simply an input from another source.

Quite often, the steering element is another FF, the outputs of which are connected to the inputs of the steered FF, as shown in Figures 4-3c and 4-3d. With this circuit, if FF *A* is in the SET state, the logic 1 output of the FF is up, and the logic 0 is down, as shown in Figure 4-3c. When a positive timing or clock pulse appears at the common input, the output of the SET gate of FF *A* goes down, and FF *B* goes to the SET state. If FF *A* is in the RESET state, the logic 1 output of FF *A* is down, and the logic 0 output is up. When a positive timing pulse appears at the common input, the output of the RESET gate of FF *B* goes down, and FF *B* goes into the RESET state. Thus, FF *A* is in a position to control or steer FF *B*.

Self-steered or toggle *RS* flip-flop.   It is often necessary to change the state of FFs by means of a single trigger (or clock or timing) pulse. This effect can be produced by self-steering as shown in Figure 4-4. Such an arrangement is often referred to as the toggle or *T*-type FF, where the 1 output is connected to the RESET gate input, and the 0 output is connected to the SET gate.

This self-steering approach to toggle operation has one major drawback. As soon as the FF changes states, the conditions at the inputs of the SET and RESET gates reverse, and, if the timing pulse is still present (this is very possible, since the timing pulse duration is usually longer than the delay of FF), it is quite possible that a second pulse will appear at the output of either the SET gate or the RESET gate. This changes the state of the FF once more. The condition could repeat itself several times until the timing pulse is no longer present. Under these circumstances, it is very difficult to predict the end result. That is, the FF can end up being SET or RESET.

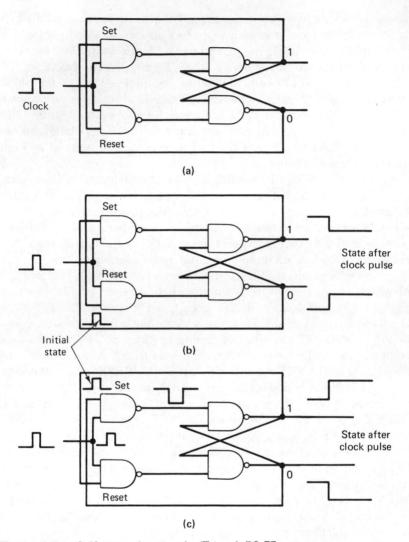

Figure 4-4.   Self-steered or toggle (T-type) *RS* FF.

Cascaded or classic *JK* flip-flop.   In order to solve the problem of the basic self-steered FF (that is, the problem of not knowing
the end result of SET or RESET) two gated *RS* FSs can be cascaded
as shown in Figure 4-5. The first FF steers the second FF, and, likewise, the second FF steers the first FF. If FF *A* is SET, then FF *B* is
SET. If FF *B* is RESET, then FF *A* will be RESET.

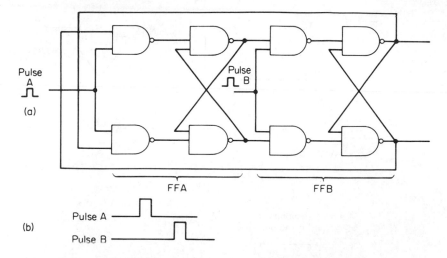

Figure 4-5.   Cascaded or *JK* FF. *(Courtesy of ITT Semiconductors)*

For this scheme to work successfully, the common inputs of the two FFs are separated. Thus, two timing or clock inputs are required. Also, one pulse must be delayed from the other, as shown in Figure 4-5b. For example, if FF *B* is SET, FF *A* will be RESET when pulse *A* arrives. Then if FF *A* is RESET, FF *B* will be RESET when pulse *B* arrives. Since there is no feedback action taking place after the first change for each FF, there are no ambiguous situations, and the end result is predictable.

This approach was used in early logic design, and is sometimes used (in modified form) today. Early designs of FFs with this capability were known as *JK* FFs, in contrast to *RS* FFs. However, today, the term *JK* is generally applied to any FF made up of cascaded elements in which the outputs are connected back to the input. Most current *JK* FFs require only one clock input, together with a SET and RESET input, to change states. That is, they can be *preset* to either state. The term *JK* is also applied to the master-slave FF which operates as follows.

Master-slave flip-flop.   The master-slave configuration is used extensively for FFs in IC form. The original discrete component *JK* FF consisted of an *RS* FF, plus two *RC* (resistor-capacitor) networks for gating. Since it is very difficult to produce a capacitor in IC form, a master-slave configuration is used as shown in Figure 4-6. This circuit is similar to the cascaded FF of Figure 4-5, with the following

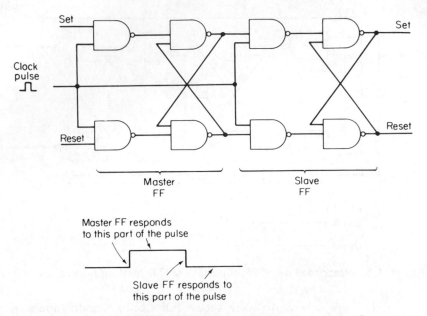

Figure 4-6.   Master-slave FF. *(Courtesy of ITT Semiconductors)*

two major differences.

First, the outputs of the slave FF (Figure 4-6) are not tied back (internally, within the IC package) to the inputs of the master FF. This is done for greater flexibility. For example, if logic designers want to use only the master or slave portion of the IC, they may do so. In other circuits, both master and slave are used, but the designer must tie the slave outputs back to the master.

The second major difference is that the common inputs of both master and slave FFs are tied together. This eliminates the need for two clock or timing pulses. If it is assumed that positive logic is used, the master FF is arranged so that the timing inputs of the SET and RESET gates respond to a positive pulse, while the slave FF responds to negative pulses. The master FF responds to the positive part of a single pulse, and the slave FF responds to the negative part of the same pulse. That is, the master is triggered by the positive-going edge, and the slave by the negative-going edge of the same pulse.

Double-rail flip-flop. As discussed, the basic *RS* FF (or simple latch circuit) has many drawbacks. One is that with two zeros at inputs, the *RS* is neither SET nor RESET. Another disadvantage is that the *RS* outputs will change in direct response to the inputs, instead of only with clock pulses. These two problems are solved by

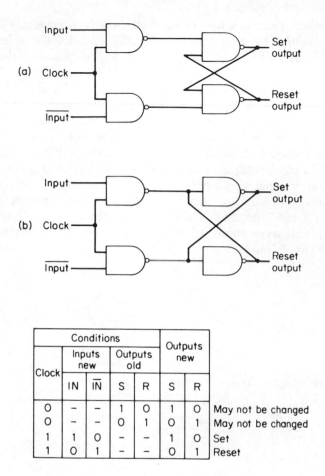

Conditions						Outputs new	
Clock	Inputs new		Outputs old			new	
	IN	$\overline{\text{IN}}$	S	R	S	R	
0	–	–	1	0	1	0	May not be changed
0	–	–	0	1	0	1	May not be changed
1	1	0	–	–	1	0	Set
1	0	1	–	–	0	1	Reset

Figure 4-7.   Double-rail FF. *(Courtesy of Honeywell)*

use of the double-rail FF, shown in Figure 4-7. As shown, the circuit is essentially a basic *RS* or latch, with two gates added, similar to the clocked FF of Figure 4-3. However, in Figure 4-7, two circuits are shown, together with a complete truth table (common to both circuits).

Note that the clock line in either circuit could be labeled "strobe," "trigger," "timer," "gate," or some similar term. No matter what term is used by the designer (or manufacturer in the case of FFs in IC form), when the clock line carries a zero, the output of the FF cannot be changed by an input at either SET or RESET. (Both the circuits and truth tables of Figure 4-7 assume that the SET and RESET inputs are always complementary.)

When the clock line carries a 1 (which is generally a pulse short in duration by comparison to the SET and RESET pulses), the FF is SET or RESET depending on the logical value of the input lines.

The circuits of Figure 4-7 are called double-rail FFs since they provide for two inputs (in addition to the clock input). These circuits can be used for *jam transfer of data;* that is, transfer of data into the FF with no regard for the previous state of the FF. The circuit of Figure 4-7b is faster than the circuit of Figure 4-7a. The Figure 4-7b circuit output uses two single-input NAND gates instead of the conventional two-input NAND gates of Figure 4-7a. Single-input gates (which are essentially inverters) introduce less delay than two-input gates. Thus, the overall FF circuit introduces less delay, or is faster; whichever term you prefer. However, the truth table of Figure 4-7 applies to both circuits.

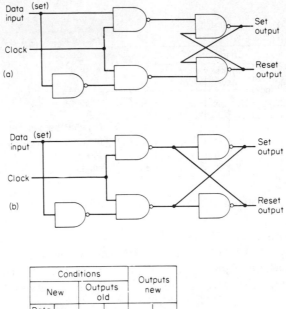

Conditions				Outputs new	
New		Outputs old			
Data input	Clock	S	R	S	R
–	0	1	0	1	0	May not be changed
–	0	0	1	0	1	May not be changed
1	1	–	–	1	0	Set
0	1	–	–	0	1	Reset

Figure 4-8.   *D*-type (single-rail) FF. *(Courtesy of Honeywell)*

D-type flip-flop (single-rail FF). The *D*-type FF requires only one data input (a SET input) in addition to the clock input. *D*-type FFs are thus referred to as *single-rail*. Two *D*-type FF circuits are shown in Figure 4-8, along with the truth table. Again, the circuit of Figure 4-8b is faster than that of Figure 4-8a, since the circuit of Figure 4-8b uses inverters at the output instead of conventional two-input gates. With either circuit, a single data line is connected directly to the SET gate input. An inverter is provided between the data input line and the RESET input. This ensures that the SET and RESET levels cannot be high (or at 1) at the same time.

   D-type FFs are especially suited to counters and registers, as described in Secs. 4–2 and 4–3. Note that these *D*-type circuits trigger on the leading edge (or positive-going edge, if positive logic is assumed) of the clock pulse. Once the clock line goes high, all input changes will affect the FF. Thus, the FF is locked to the input. The data input is not unlocked until the clock input trailing edge (negative-going voltage) falls below the threshold of the input gates.

   Double-rank flip-flop. The double-rank FF shown in Figure 4-9 is similar to the cascaded FF (Figure 4-5) and the master-slave FF

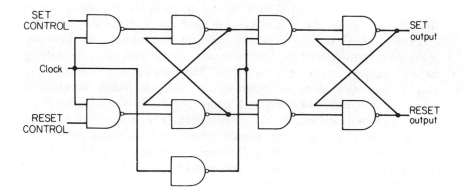

Clock	Inputs S R	Master output	Slave output
1	1 1	Same as input	Previous state of slave
1	1 0	Same as input	Previous state of slave
1	0 1	Same as input	Previous state of slave
1	0 0	Previous state of master	Previous state of slave
0	Any	Previous state of master	Same as master

Figure 4-9. Double-rank FF. *(Courtesy of Honeywell)*

(Figure 4-6). The term "double-rank" is used because the FF is essentially two complete toggle-type FFs connected together.

The circuit of Figure 4-9 is often called a *JK* FF, even though it is not truly *JK* wired (outputs returned to inputs). The circuit operates on a principle usually referred to as *pulse dodging*. This means that half of the storage cycle occurs on the 0-to-1 transition of the clock line, and the other half occurs on the 1-to-0 transition.

When the clock line is at 0, the output of the master FF is unchangeable, while the output of the slave FF is determined by the state of the master. Thus, when the clock is at 0, the output of the slave (and thus the output of the complete FF circuit) is fixed to whatever the input last was. On the other hand, when the clock is at 1, the slave FF is locked and the master responds to the input. Such an FF cannot be set into oscillation.

JK master-slave FF (with preset).   The circuit of Figure 4-10 is a dc triggered (capable of being preset by dc voltages), master-slave *JK* FF. The circuit may be set or reset asynchronously (that is, may be preset, without regard to the clock or any other input) with the P_J and P_K inputs. The circuit may also be switched synchronously by using the *J* and *K* inputs, together with a clock pulse.

When the circuit is switched asynchronously, it behaves as an *RS* FF (or simple latch), and can be used for jam transfer of data. With the asynchronous (or preset) function, the master and slave are coupled together, and the outputs are set immediately. Presetting should be performed with the clock line low (at 0). However, in applications such as ripple counting (Sec. 4–2), in which the state of the clock line cannot be predicted, presetting can be done by lowering the *J* input while raising the P_J, or lowering the K input while raising P_K. Both the *J* and *K* inputs should remain stable during clock time (that is, when the clock is at 1). If the preset inputs P_J and P_K are not used, they should be tied down (connected to a fixed logic level, usually 0).

When switched synchronously, the rising clock pulse cuts the slave off from the master (through the transfer gate). As the clock line rises still higher, it allows the logic at the *J* and *K* inputs to be set into the master. Then, when the clock returns to its low level, the state of the master is transferred to the slave, which, in turn, sets the output levels.

The thresholds of the transfer gate and the master FF gates must be separated by sufficient voltage to guarantee that input and transfer cannot occur simultaneously. That is, the transfer gate must have a much lower voltage threshold than the master FF gates. This

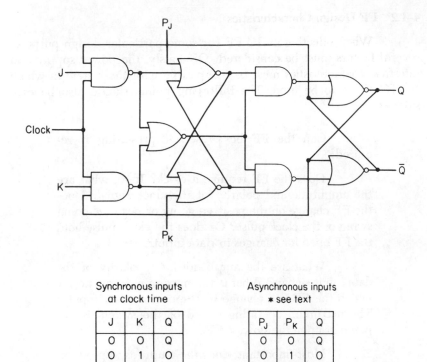

Synchronous inputs at clock time		
J	K	Q
0	0	Q
0	1	0
1	0	1
1	1	$\bar{Q}$

Asynchronous inputs ∗ see text		
P_J	P_K	Q
0	0	Q
0	1	0
1	0	1
1	1	∗

Figure 4-10.   *JK master-slave FF (with preset). (Courtesy of Cambridge Thermionic Corporation)*

will prevent the circuit from switching back and forth between states (or oscillating) during the rise of the clock pulse. Such undesired switching between states is often known as a "race" condition or "critical race." When there is no undesired switching of states in the presence of simultaneous input, the FF is said to be "race free."

The synchronous inputs strictly follow the "conventional" definitions of a *JK* FF. This is shown on the truth table. If both *J* and *K* inputs are at 0, the *Q* output will remain in the previous state. If both *J* and *K* are at 1, the *Q* output will change states ($\bar{Q}$).

The preset inputs are also shown on the truth table. Note that the preset (P_J and P_K) truth table is identical to the synchronous input (*J* and *K*), with one exception. When both P_J and P_K inputs are up (at 1), there is no immediate effect on the outputs. The P_J or P_K *input that falls last* will control the *final state* of the FF.

4–1.2 FF Design Characteristics

When selecting an IC FF for some particular design purpose, several factors must be considered. Obviously, the power supply and interface characteristics must be compatible with the system in which the IC FF is to be used. The following points should also be considered.

Can the FF be preset, if presetting is required?

Does the FF require a clock? If so, what are the amplitude and polarity of the clock signal? Does the FF change on the positive-going or negative-going swing of the clock pulse? Or does the clock pulse hold the FF open for changes in data input?

What are the amplitude and polarity of the data input signals? What is the maximum frequency at which the FF will change in response to data inputs? The maximum operating speed of the system is dependent upon the slowest FF.

Most important, does the truth (as appears on the datasheet) match the system requirements?

Usually, the critical factor is response of the FF to *simultaneous inputs,* particularly where the inputs are of the same logic (1 or 0). For example, in the FF of Figure 4-10, the Q output remains in the previous state with J and K at 0, but changes states (no matter what the previous state, and whether desired or not) when J and K are at 1, simultaneously with the clock. Assume that J and K are supposed to be at 1 and 0, respectively, at clock time. Then assume that, because of some undesired delay in the circuits ahead of the FF, K moves to 1, even momentarily. The FF could possibly switch states.

4–2 COUNTERS

There are three basic types of counters available in IC form: serial or ripple counters, synchronous counters, and shift counters. The basic difference among them is in the method of counting. Counters can be made to count either serially or in parallel.

Serial counters, also known as *ripple counters,* use the output of a counting element (generally a flip-flop) to drive the input of the following counting elements. In ripple counters, the FFs operate in a *toggle* mode (that is, they change with each clock pulse), with the output of each FF driving the clock input of the following stage. Ripple counters operating in this manner are able to operate at higher input frequencies than most other types of counters, require little or no gating and few interconnections, and present only one clock input to the input line. A limitation of all ripple counters is that a change of state may be required to ripple through the entire length of the counter.

A *synchronous* or *clocked counter* is one in which the next state depends on the present state through gating, and all state changes occur simultaneously with a clock pulse. Since all FFs change simultaneously, the output (or count) can be taken from synchronous counters in *parallel* form (as opposed to serial form for ripple counters).

Shift counters are a specialized form of clocked counter. The name is derived from the fact that the operation is similar to a shift register (Sec. 4–3). In general, the shift counter produces outputs that may easily be decoded, and normally requires no gating between stages (thus permitting high-speed operation).

4–2.1   Ripple or Serial Counters

Generally speaking, ripple counters of all types require less gating to perform a function, as compared to synchronous counters. In ripple counters, the output of one FF is used as the clock input to another FF, whereas synchronous counters use the same clock input to drive all FFs. With the ripple system, each FF changes on each 1-to-0 (or 0-to-1) transition of the previous stage. The counter is implemented by connecting the clock input of each stage to the Q output of the previous stage. (See Figure 4-10 for typical Q output and clock input connections.)

The propagation delays through the counting elements are added together. In long high-speed counters, the output of the last stage can be considerably out of phase with the input of the first stage. For example, consider a counter operating at 20 MHz using 10 counting elements, each with a 10 ns propagation delay. The time required for an input to propagate (ripple) to the output is 100 ns, yet there are only 50 ns between input pulses. This is not a problem unless phasing is required (as is the case with some decoders). Ripple counters are very useful for straight frequency division, and are excellent for high-speed counting that must be read out after the count is complete.

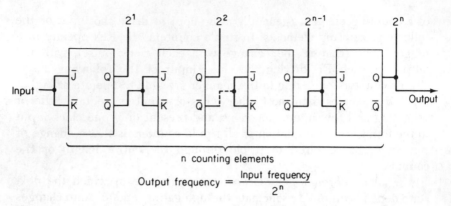

$$\text{Output frequency} = \frac{\text{Input frequency}}{2^n}$$

Figure 4-11.   Basic binary counter. *(Courtesy of Motorola)*

Ripple counters fall into two groups: straight binary and feed-back. Straight binary ripple counters divide the input by 2^n, where n is the number of counting elements. Straight binary counters count in a binary code, with the *first* counting element (FF) containing the least significant bit (LSB).

Feedback counters also count in a binary code, but a number of higher-value states are eliminated by the feedback. The basic feedback counter divides the input frequency by $2^{x-1} + 1$, where x is the number of counting elements. A division by any arbitrary number can be done by using various combinations of the two groups (straight and feedback).

Straight binary ripple counters. A straight binary ripple counter can be implemented with *JK* FFs by connecting a *J* and *K* input together on each FF, and driving this "clock" with the *Q* output of the previous counting element (FF), as shown in Figure 4-11. When Q changes from a one to a zero, the $\overline{Q}$ output undergoes a positive transition, and this changes the state of the following stage. Counters of moduli, (number of different states), 2, 4, 8, and so on; that is, counters that will divide by 2, 4, 8, and so on, can be implemented as binary ripple counters.

An alternate method of implementing the basic ripple counter is shown in Figure 4-12. Here, the clock input of the LSB FF receives the count line input, and the *Q* output of each FF is connected to the clock input of the following FF.

Feedback ripple counters. The output of a feedback ripple counter is connected to the input to *reduce* the counter's modulus from binary power of the number of elements in the counter. In a feedback counter, the 2^{x-1}th count causes the output FF to go to a 1 state, and

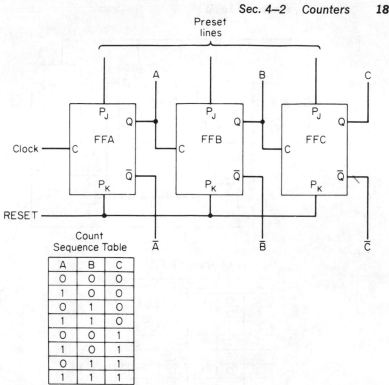

Figure 4-12. Basic binary counter with preset and reset. *(Courtesy of Cambridge Thermionic Corporation)*

the rest of the FFs to go to 0. The output 1 is also fed back, and inhibits the first element, while the next count resets the output elements, as shown in Figure 4-13. This restores the counter to contain all zeros.

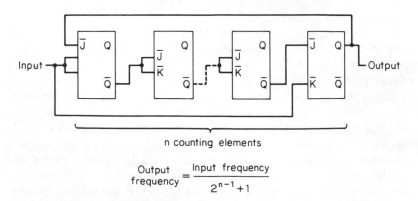

$$\frac{\text{Output}}{\text{frequency}} = \frac{\text{Input frequency}}{2^{n-1}+1}$$

Figure 4-13. Basic feedback ripple counter. *(Courtesy of Motorola)*

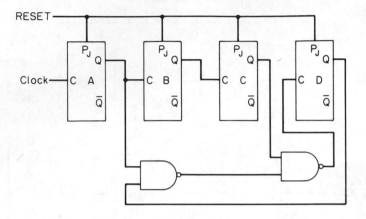

Count Sequence Table

A	B	C	D	Decimal
O	O	O	O	O
1	O	O	O	1
O	1	O	O	2
1	1	O	O	3
O	O	1	O	4
1	O	1	O	5
O	1	1	O	6
1	1	1	O	7
O	O	O	1	8
1	O	O	1	9

Figure 4-14.   Feedback ripple counter with BCD format count. *(Courtesy of Cambridge Thermionic Corporation)*

Feedback ripple counters produce counts of 3, 5, 9, 17, 33, and so on.

For proper operation of feedback counters, the period of the input pulses must be large enough to allow the signal to ripple through *all of the FFs within the feedback loop.* A basic feedback counter circuit is shown in Figure 4-14. As shown by the count sequence table, the feedback provides for a count in the BCD format.

Combination ripple counter. If the output of a straight binary ripple counter is used to drive the input of a feedback ripple counter, a frequency division of $2^n(2^{x-1} + 1)$ is obtained, where n is the number of counting elements in the straight binary counter and x is the number of counting elements in the feedback counter. The basic combination is shown in Figure 4-15. When the scheme is used,

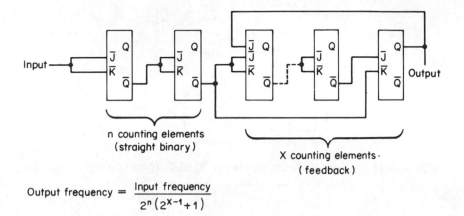

$$\text{Output frequency} = \frac{\text{Input frequency}}{2^n(2^{X-1}+1)}$$

Figure 4-15. Basic combination ripple counter. *(Courtesy of Motorola)*

counts of 6, 10, 12, 18, 20, 24, 34, 36, 40, and so on, are obtainable. If a symmetrical output is desired, one stage of straight binary counter can be moved to the output of the feedback counter. However, the counter will no longer count with a binary code.

Divide-by counters. The counters discussed thus far are "divide-by" counters. That is, the counters divide by a specific number of counts. In a straight binary counter, the number of counts is a power of two. Specifically, if the number of FFs is N, the number of counts is 2^N. If means are provided to prevent the counter from entering some of the possible states, then the length of the sequence can be limited to any desired number (N).

There are many means of accomplishing this result. One way is to form the counter to go from the state representing $N-1$ back to the beginning of its normal sequence (all zeros) on the Nth clock pulse. Another way is to recognize the state representing $N-1$, to inhibit stages normally changing to 1, and to force stages with 1 to go to 0. If one is not interested in all states up to the number N, *cascading* of lesser divide-by counters can be used to form the higher-numbered divide-by circuits. The circuits of Figures 4-16 and 4-17 use this approach. By cascading these circuits (a divide-by-3 and a divide-by-7 counter), the input count is divided by the product of the two counters, or 21. By cascading similar circuits, a wide variety of counts can readily be formed with the intended goal of having an output for a predetermined number of input events (or clock pulses).

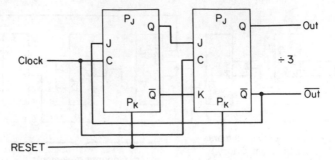

Figure 4-16.   Divide-by-3 counter with RESET. *(Courtesy of Cambridge Thermionic Corporation)*

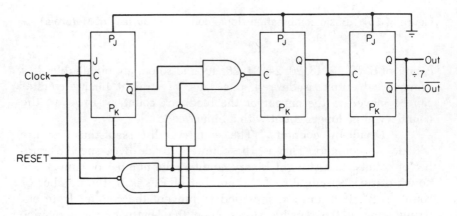

Figure 4-17.   Divide-by-7 counter with RESET. *(Courtesy of Cambridge Thermionic Corporation)*

BCD ripple counter with decoder. Many of the counters available in IC form also include auxiliary circuits such as decoders, drivers, and so on. Figure 4-18 is a typical example, in that the circuit is a BCD counter (from 0 to 9), but it also provides both a decoded output (indicating the count in decimal form), and the outputs (true and complementary) of each FF stage.

Note that two NAND gates are required for the counter portion of the circuit (in addition to the FFs). Also note that *JK* FFs are used, and that each FF has a RESET function tied to a common RESET line. This permits all the stages to be RESET (or CLEARED) to zero simultaneously.

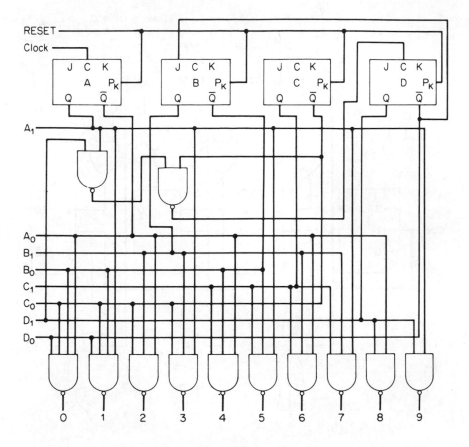

Figure 4-18.   BCD ripple counter with decoder. *(Courtesy of Cambridge Thermionic Corporation)*

4–2.2 Synchronous Counters

Synchronous counters can be designed to count up, count down, or do both, in which case they are called *bidirectional* counters. The bidirectional (also known as up-down) counter is actually an up-counter superimposed on a down-counter. The two counters use the same FFs, but have independent gating logic. An up or down signal (usually a fixed dc voltage) must be provided to instruct the counter circuit in which direction to count by activating the proper gating circuitry.

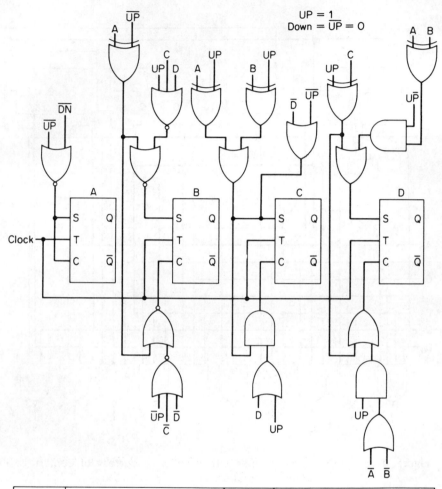

Figure 4-19. Bidirectional XS3 counter. (Courtesy of Motorola)

	Up								Down							
Decimal	n				n + 1				Decimal	n				n + 1			
	D	C	B	A	D	C	B	A		D	C	B	A	D	C	B	A
0	0	0	1	1	0	1	0	0	9	1	1	0	0	1	0	1	1
1	0	1	0	0	0	1	0	1	8	1	0	1	1	1	0	1	0
2	0	1	0	1	0	1	1	0	7	1	0	1	0	1	0	0	1
3	0	1	1	0	0	1	1	1	6	1	0	0	1	1	0	0	0
4	0	1	1	1	1	0	0	0	5	1	0	0	0	0	1	1	1
5	1	0	0	0	1	0	0	1	4	0	1	1	1	0	1	1	0
6	1	0	0	1	1	0	1	0	3	0	1	1	0	0	1	0	1
7	1	0	1	0	1	0	1	1	2	0	1	0	1	0	1	0	0
8	1	0	1	1	1	1	0	0	1	0	1	0	0	0	0	1	1
9	1	1	0	0	0	0	1	1	0	0	0	1	1	1	1	0	0

The following examples cover both bidirectional and single-directional counters, as they appear in IC form. The bidirectional counter has an advantage in that it can perform two separate functions, and thus becomes a "universal" counter. The single-directional counter has the advantage of requiring fewer components (and thus should be smaller, use less power, cost less, and so on).

XS3 bidirectional counter. The circuit of Figure 4-19 is a bidirectional clocked counter using the XS3 code. As shown in Figure 1-4, the XS3 (excess 3) code is similar to the BCD code, except that the BCD value of three is added to each digit before decoding. The XS3 code is self-complementing, which means that if all the zeros in a number are changed to ones, and ones to zeros, the nine's complement of the number is obtained. This ability to obtain the nine's complement is useful in reducing the required hardware in subtraction operations. Another advantage is that all decimal representations have at least one 1 in the XS3 coding. This makes it possible to distinguish zero information from no information.

In the circuit of Figure 4-19, note that both complementary and true up-down signals are required. Also note that the circuit is shown in simplified form. That is, the gate inputs are labeled as to destination, but the circuit wiring is not shown. The truth table shows the states of each FF during the count, for both up and down counts. The n column represents the present (or existing) state, while the $n + 1$ column represents the state at the next bit or clock pulse time.

BCD single-directional counter. The circuit of Figure 4-20 is a single-directional clocked counter that counts up in BCD form. The circuit is designed with an optional gate on the K input of the first FF. The purpose of this gate is to provide an optional zero in the count that can be used in certain instrumentation applications. The optional zero in this case is the binary number 15. By designing a decoder/display system to follow the counter such that the binary number 15 does not produce a display output, one can insure that unnecessary zeros will not be displayed. As an example, a frequency meter capable of 10 MHz display with eight-place readout can be made to read 455 kHz as 455,000 Hz, rather than 00,455,000 Hz.

4–2.3 Shift Counters

Generally, IC shift counters are found in two forms: decade shift counters, and ring shift counters.

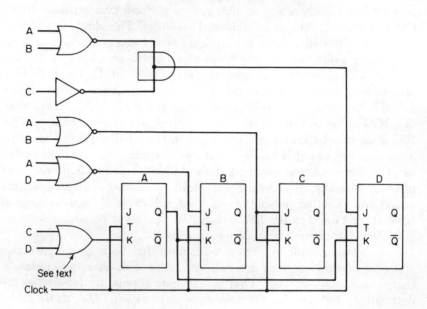

Counter Sequence

Bit weight	A	B	C	D
0	0	0	0	0
1	1	0	0	0
2	0	1	0	0
3	1	1	0	0
4	0	0	1	0
5	1	0	1	0
6	0	1	1	0
7	1	1	1	0
8	0	0	0	1
9	1	0	0	1

Decoding Logic

0	$\bar{A}$	$\bar{B}$	$\bar{C}$	$\bar{D}$
1	A	$\bar{B}$	$\bar{C}$	$\bar{D}$
2	$\bar{A}$	B	$\bar{C}$	
3	A	B	$\bar{C}$	
4	$\bar{A}$	$\bar{B}$	C	
5	A	$\bar{B}$	C	
6	$\bar{A}$	B	C	
7	A	B	C	
8	$\bar{A}$	D		
9	A	D		

Unused State Analysis

Without zero suppression	
Unused state	Next state
A B C D →	A B C D
0 1 0 1 →	1 1 0 1
1 1 0 1 →	0 0 1 0
0 0 1 1 →	1 0 1 1
1 0 1 1 →	0 0 1 0
0 1 1 1 →	1 1 1 1
1 1 1 1 →	0 0 0 0

With zero suppression	
Unused state	Next state
A B C D →	A B C D
0 1 0 1 →	1 1 0 1
1 1 0 1 →	0 0 1 0
0 0 1 1 →	1 0 1 1
1 0 1 1 →	1 0 1 0
0 1 1 1 →	1 1 1 1
1 1 1 1 →	1 0 0 0

Figure 4-20. BCD single-directional counter. *(Courtesy of Motorola)*

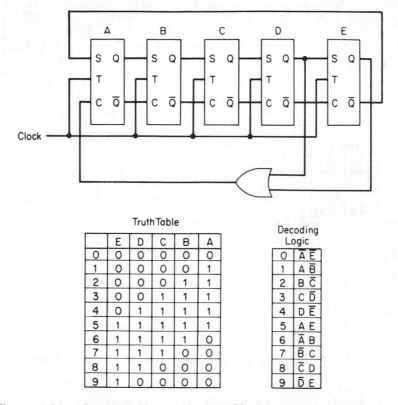

Truth Table

	E	D	C	B	A
0	0	0	0	0	0
1	0	0	0	0	1
2	0	0	0	1	1
3	0	0	1	1	1
4	0	1	1	1	1
5	1	1	1	1	1
6	1	1	1	1	0
7	1	1	1	0	0
8	1	1	0	0	0
9	1	0	0	0	0

Decoding Logic

0	$\overline{A}\,\overline{E}$
1	$A\,\overline{B}$
2	$B\,\overline{C}$
3	$C\,\overline{D}$
4	$D\,\overline{E}$
5	$A\,E$
6	$\overline{A}\,B$
7	$\overline{B}\,C$
8	$\overline{C}\,D$
9	$\overline{D}\,E$

Figure 4-21. Decode shift counter with OR gate error correction. *(Courtesy of Motorola)*

Decade shift counters. The counter shown in Figure 4-21 is sometimes called a switch-tail ring counter, a Johnson counter, or simply a shift counter. Flip-flops A through E provide a decade output, but require decoding as shown by the output decoding table. For example, if you want a count of 8, you monitor the simultaneous condition of the FF D Q output and the FF C $\overline{Q}$ output. FF A represents the least significant bit, and E is the most significant bit. The output of the E stage is fed back into the A stage as shown. Note that the circuit uses feedback (through an OR gate) for error correction. This presets the counter to zero before the counting can begin.

Ring shift counters. In some applications, it is more important to eliminate (or minimize) the decoding logic that follows a counter. In such applications, a ring counter can be used. Ring counters use one FF for each count. That is, a ten-count device (or decade

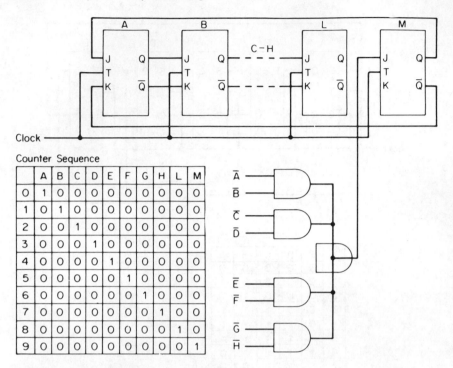

Figure 4-22. Decode ring counter with error correction. *(Courtesy of Motorola)*

counter) requires ten FFs. Only one FF is true at any time. Thus, no decoding is required.

The circuit of Figure 4-22 is a typical ring counter with error correction. The *K* input of the last FF (FF *M*) should be returned to a high or logic 1 level at all times. Also note that no output decoding table is provided on the diagram (as is the case for a decade shift counter) since a ring counter needs no decoding. For example, in the circuit of Figure 4-22, if you want a count of 8, you monitor the eighth FF (FF *L*).

4–3   SHIFT REGISTERS AND SHIFT ELEMENTS

The term "register" can be applied to any logic circuit that stores information on a temporary basis. Permanent or long-term storage is generally done with magnetic cores, tapes, and drums. The term "register" is usually applied to a logic circuit consisting of FFs and gates that will store binary or other coded information.

A *storage register* is an example of such a logic circuit. The various counter circuits discussed in Sec. 4–2 are, in effect, a form of storage register. Counters accept information in serial form (the clock or count input) and parallel form (by presetting the FFs to a given count), and hold this information (the count) as long as power is applied, and provided that no other information (serial or parallel) is added. With proper gating, the information can be read in or read out.

In many logic circuits, particularly computers, it is necessary to manipulate the data held in registers. For example, registers are used with binary adders and subtractors (Chapter 6) to hold and transfer data. Registers can also be used to multiply and divide binary numbers. Both multiplication and division are a form of "shifting." For example, when the binary number 0111 (decimal 7) is shifted one place to the left, it becomes 1110 (decimal 14). Thus, a one-place left shift for a binary number is the same as multiplication by two.

A *shift register* is a circuit for storing and shifting (or manipulating) a number of binary or decimal digits (rather than the simple storage function of a storage register). In addition to their use in arithmetic operation, shift registers are used for such functions as conversion between parallel and serial data.

4–3.1   Storage Registers

Figure 4-23 shows the circuit of a typical binary storage register available in IC form. All FFs are *JK* type with preset (P_J) and preclear (P_K) inputs, in addition to the clock or toggle input. Serial information is entered into the register by means of the clock input. Parallel information is entered through the gates and $P_J P_K$ inputs. The circuits are, in effect, ripple counters with gates for parallel entry of data.

The gates are arranged so that the P_J and P_K inputs receive opposite states when information is applied to the parallel data lines, and the STROBE line is enabled. For example, assume that the P_J input must be low (and the P_K input high) for the Q input to be high. Under these conditions, the STROBE line is lowered, and the data line is raised, when the Q output is to be high (or at 1).

Now assume that a one is to be entered into the A FF. This

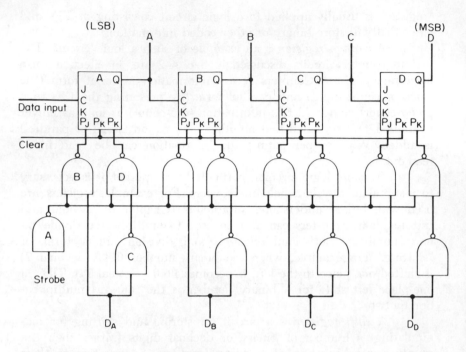

Figure 4-23.   Binary storage register. *(Courtesy of Cambridge Therm-ionic Corporation)*

will preset or "store" a binary 0001 into the register, since FF A is the least significant bit (with the FF D as the most significant bit). The D_A data input line is then set at 1, with the STROBE line at 0. The one appears at the input of gate B, along with a one from the STROBE line (inverted from a zero by gate A). The two ones produce a 0 output from gate B. This 0 output is applied to the P_J input, and sets the Q output of FF A to 1.

At the same time, the 1 input at the D_A line is inverted by gate C, and appears as a zero at the input of gate D, along with a one from the STROBE line. The 1 and 0 inputs produce a 1 output from gate D, which is applied to the P_K input. This assures that the Q output of FF A is 1 (with the $\bar{Q}$ output at 0). All of the remaining FFs receive opposite states; P_J input at 1, and P_K at 0. The Q outputs of FFs B, C, and D go to 0, and a binary 0001 results.

All the data is cleared when the CLEAR line is enabled (lowered). That is, all of the Q outputs move to 0, without regard to

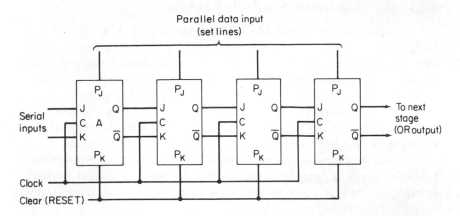

Figure 4-24.   Basic shift register. *(Courtesy of Cambridge Thermionic Corporation)*

their existing state. Data can be taken from the Q or $\overline{Q}$ outputs, as desired. Serial information is usually taken from the Q or $\overline{Q}$ outputs of FF D.

4–3.2   Basic Shift Registers

A basic or simplified shift register is shown in Figure 4-24. (Note that the circuit is similar to the shift counters described in Sec. 4–2.3.) Circuits such as the one shown in Figure 4-24 shift their contents *one position to the right* for each occurrence of the clock pulse. That is, each stage switches states once for each clock pulse.

Data may enter the registers in serial or parallel form from other registers or counters. The RESET line common to all stages will preset all stages to 0. All SET lines are independently brought out to allow parallel transfer of data into the register. Both Q and $\overline{Q}$ of all stages are also brought to permit parallel transfer of data out of the register. Thus, data may be shifted serially in and out of the register, or may be transferred in parallel form in and out of the register. When shifting in serial form, the most significant bit (FF D) is transferred out first. To clear the register automatically after all information is shifted out, the J and K inputs of the first stage (FF A) are tied to 0 and 1, respectively.

4–3.3 Shift Registers With Multiple Functions

Shift registers available in IC form usually have more versatile features than those of the basic register shown in Figure 4-24. For example, a typical IC shift register is capable of shifting data stored in the register one position to the right (as with the basic register), or one position to the left, with each clock pulse.

When shifting information out of the register, it may be desired to replace the original information back into the register. In IC registers, this is usually done by an *end-around shift* feature.

For high-speed systems, parallel transfer of data is essential. For this reason, IC registers usually provide both serial and parallel transfer, as well as combinations such as serial-to-parallel, and vice versa.

For arithmetic operations, such as subtraction and division (Chapter 6) a complementation feature is also desirable. IC registers often include a feature whereby a given set of commands cause the FFs to set up in a toggle mode, and will complement with the next clock pulse (all FFs with a zero will go to a one, and vice versa).

There are many means by which to accomplish these functions in IC registers. The methods will not be discussed here. However, the logic designer should be aware that registers with the features are available in IC form.

4–3.4 MOS Shift Registers

Although MOS shift registers perform the same functions as TTL and other registers, the MOS registers operate on different principles. FFs, as such, are not used in MOS registers. The following descriptions apply to the Texas Instruments MOS/LSI line. However, similar principles are used by other manufacturers of MOS registers.

Basic configuration. MOS shift registers can be supplied in the following configurations: serial-in/serial-out, parallel-in/serial-out, and serial-in/parallel-out. The *serial-in/serial-out* configuration is by far the most popular. A MOS register is able to store N bits, each on a *basic cell* consisting of two MOS inverters and timing devices, as shown in Figure 4-25.

Static versus dynamic. Dynamic shift registers use two independent inverters (not cross-coupled). The information is stored temporarily on a capacitor inherent to the MOS device (the gate capacitance). The device cannot be operated below a certain clock frequency or the data storage will be lost.

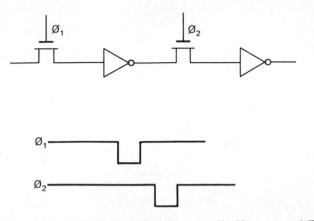

Figure 4-25. Basic MOS/LSI shift register cell. *(Courtesy of Texas Instruments)*

A *static shift register* (Figure 4-26) operates in the same way as a dynamic shift register, as long as the frequency is high. The two inverters used in a static shift register are the static type (unclocked load). When the frequency falls below a certain level, a third phase is generated internally, and this signal is used to close a feedback loop between the output of the second inverter and the input of the first inverter. Comparing the two, the dynamic shift registers are faster and use less power than static shift registers. However, dynamic shift registers are not as flexible to use in a system.

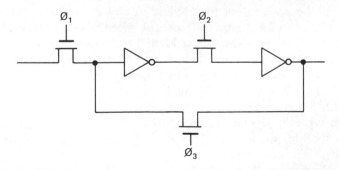

Figure 4-26. Basic MOS-LSI static shift register. *(Courtesy of Texas Instruments)*

Static shift registers. As shown in Figure 4-27, a static shift register uses two static MOS inverters (Sec. 2–1.12 and Figure 2-33).

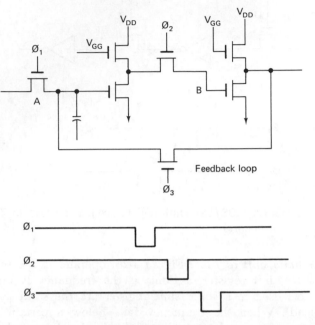

Figure 4-27. Basic cell of MOS/LSI static shift register. *(Courtesy of Texas Instruments)*

Three phases (or clocks) are necessary for operation. The third-phase clock is always *generated internally,* and is used to time the feedback loop. The second-clock phase is often generated internally.

In the basic cell of Figure 4-27, A and B are storage elements. The device operates dynamically except when phase 3 is on. Phase 3 is present only when phase 1 is at logic 0 and phase 2 is at logic 1 for more than 10 microseconds. This condition must be maintained for long-time data storage. As shown in Figure 4-28, phase 3 is actually a delayed phase 2, generated by an inverter. The load devices associated with the static shift register remain on at all times. Static shift registers typically operate in the 0-to-2.5 MHz clock range, are extremely flexible, and can hold data indefinitely (as long as power is supplied).

Dynamic shift registers. Dynamic shift registers use either two or four phases (or clocks). These phases can be generated on the chip or supplied externally. Two-phase shift registers can be classified as *ratio* or *ratioless* circuits.

The two-phase ratio-type shift register (Figure 4-29) consists of two simple dynamic inverters and timing devices. When phase 1 is at a logic 1 (low), the capacitance C_1 charges at the inverse of the data input. Information is transferred out when phase 2 goes to 1.

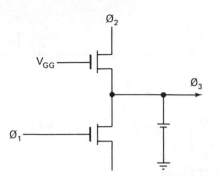

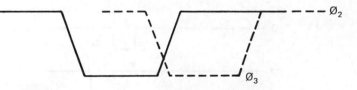

Figure 4-28.   Generation of a Ø3 in MOS/LSI register. *(Courtesy of Texas Instruments)*

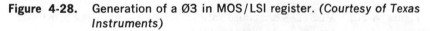

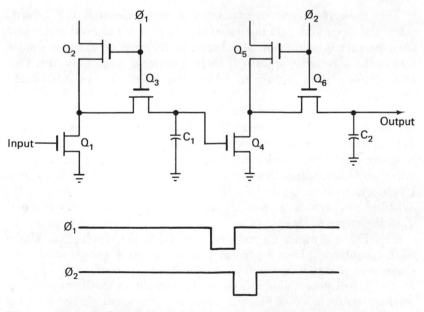

Figure 4-29.   Basic cell for MOS/LSI dynamic shift register. *(Courtesy of Texas Instruments)*

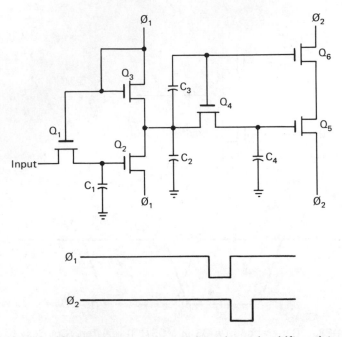

Figure 4-30.   MOS/LSI two-phase ratioless dynamic shift register. *(Courtesy of Texas Instruments)*

In a ratio-type circuit, current flows through the inverter when the clock and data input are at a logic 1 simultaneously. There must be certain minimum ratio between the size of the two transistors in the inverters (typically 5 to 1), requiring more chip area than in a ratioless shift register in which the MOS devices are usually identical in size.

The *two-phase ratioless dynamic shift register* (Figure 4-30) has been designed to decrease the power dissipation and chip area. This ratioless register uses identical transistors throughout, and can thus work at higher clock rates because the precharging paths are of lower impedance than those in the circuit. When phase 1 goes to 1, C_2 charges to 1 via Q_3, and C_1 charges to the data input level via Q_1. When phase 1 returns to zero, Q_2 turns on if the input level was a one. This discharges C_2.

For a 0 input, Q_2 stays off and C_2 is not discharged. Under these conditions, phase 2 goes to a one and turns on Q_4 so that C_2 shares any charge it has with C_4. Capacitor C_3 is used to compensate for the loss of potential across C_2 by introducing a small extra charge on the negative edge of phase 2. However, the small charge does not introduce enough energy to destroy a logic 0 on C_2. When phase 2

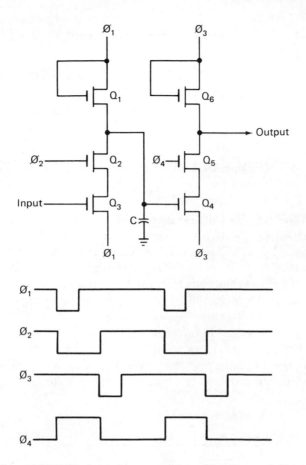

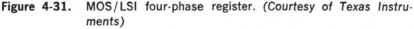

Figure 4-31. MOS/LSI four-phase register. *(Courtesy of Texas Instruments)*

returns to a zero, the charge on C_4 transfers the data-input level to the output.

Four-phase shift registers (Figure 4-31) are used for very high density circuits operated at very high speeds. In the basic four-phase dynamic shift register, C is precharged via Q_1 during phase 1. After phase 1, phase 2 holds Q_2 on, so C takes a level which is the inverse of the input. The process is repeated by the slave section $Q_4 - Q_6$ so that the input level is transferred to the output after phase 3 and during phase 4. The four-phase register uses similar transistors throughout, giving high package density. Power dissipation is low, speed can be high, but a relatively complex clock drive (four phases) is required.

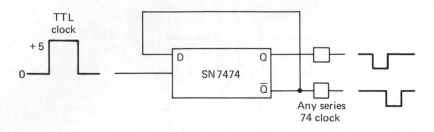

Figure 4-32.   Clocking MOS/LSI from TTL. *(Courtesy of Texas Instruments)*

Clocking the shift registers.   Most of the current shift registers designed by Texas Instruments feature total TTL compatibility. This is done by including a clock driver (driven directly by TTL circuits) on the chip. In older MOS shift registers, the designer must generate the clocks, and shift their level (from TTL to MOS).

When the MOS shift register does not have direct TTL compatibility, interfacing can be accomplished with a D-type FF, as shown in Figure 4-32. This arrangement is favorable because Q and $\overline{Q}$ will be of opposite polarities when the TTL clock is stopped, allowing static storage in static registers. Two FFs can be used when four clock phases are required (such as with the register of Figure 4-31).

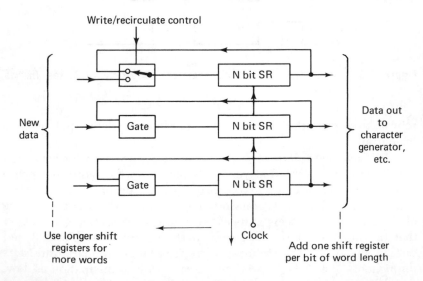

Figure 4-33.   MOS/LSI shift register used as refresh memory. *(Courtesy of Texas Instruments)*

The FFs must be capable of shifting the level to that suitable for the MOS register. The subject of interfacing MOS logic with other logic forms is discussed in Chapter 8.

Typical applications.   Typical applications for MOS registers are discussed in Sec. 4–7. The following is a brief description of some typical memories using basic MOS shift registers.

An N-bit shift register can be used as a *refresh memory* by returning outputs to inputs as in Figure 4-33. The rate (in seconds) at which a particular bit of information is available at the output is determined by the expression: N/clock frequency.

This arrangement is particularly useful for renewing fading displays such as CRT character-generator systems. New information is written in via a two-way input gate circuit as shown. This gate circuit is incorporated on current Texas Instruments shift registers.

By adding an address counter and comparator, the refresh memory becomes a *scratch-pad memory*, as shown in Figure 4-34.

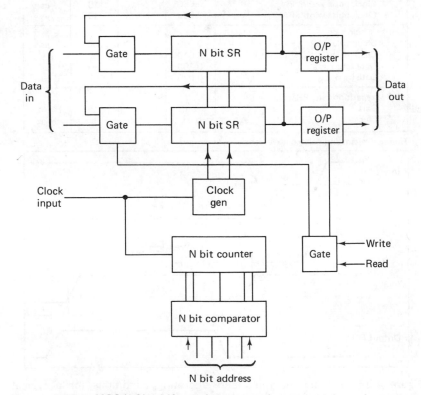

Figure 4-34.   MOS/LSI shift registers used as scratch pad memory. *(Courtesy of Texas Instruments)*

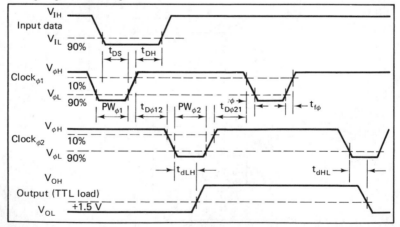

Recommended operating conditions

Parameter	Min	Nom	Max	Units
Operating voltage				
Substrate supply, V_{SS}	+4.75	+5	+5.25	V
Gate supply, V_{GG}	−13	−12	−11	V
Logic levels				
Input high level, V_{IH}	$V_{SS} - 1.6$		V_{SS}	V
Input low level, V_{IL}	$V_{SS} - 18$		$V_{SS} - 4.2$	V
Clock voltage levels				
Clock HIGH level, $V_{\phi H}$	$V_{SS} - 1$		V_{SS}	V
Clock LOW level, $V_{\phi L}$	$V_{SS} - 19$		$V_{SS} - 16$	V
Pulse timing				
Clock pulse transition, $t_{r\phi}$, $t_{f\phi}$			1	μs
Clock pulse width, $PW_{\phi 1}$	0.080		10	μs
Clock pulse width, $PW_{\phi 2}$	0.080		10	μs
Pulse spacing				
Clock delay, $t_{d\phi 12}$	0.010		50	μs
Clock delay, $t_{d\phi 21}$	0.010		50	μs
Data setup, t_{DS}	0.100			μs
Data hold, t_{DH}	0.020			μs
Pulse repetition rate, PRR				
Data	0		5	MHz
Clock	0.01		5	MHz

Timing diagram and voltage waveforms

Figure 4-35.   Timing diagrams and operating conditions for a typical MOS/LSI twelve-bit dynamic shift register. *(Courtesy of Texas Instruments)*

Information can be written into and read out of *any point* specified by the input address code. An output register is necessary to store the required output data and to provide a one-bit delay so that the Read address is the same as the Write address (because there is a one-bit delay between output and input).

4–3.5   Register Design Characteristics

When selecting an IC register for some particular design purpose, several factors must be considered. Obviously, the power supply and interface characteristics must be compatible with the system in which the IC register is to be used. The following points should also be considered.

Keep in mind that interface not only includes logic levels, but pulse timing and spacing. For example, Figure 4-35 shows the timing diagrams and recommended operating conditions for a Texas Instruments MOS/LSI 512-bit dynamic shift register. A quick study of this information shows that data is transferred into the register when the ϕ_1 clock is low (nominally -12 V). The data must set up at least 100 ns *before* the clock ϕ_1 goes high ($+5$ V) and held steady at least 20 ns after ϕ_1 reaches this state. Also, output delay time is defined as the time required for the output to reach the TTL change-over threshold after the ϕ_2 clock reaches 90 percent of its low voltage. This time is shorter than 100 ns.

Often, characteristics such as those shown in Figure 4-35 are critical to operation of registers (and counters) when used in a system. For this reason, the datasheets must always be consulted.

4–4   UNIVERSAL COUNTER CIRCUIT

IC logic manufacturers often provide *universal counter* or *programmable counter* devices. Such units permit the designer to implement a counter circuit of any given count. This eliminates the need to combine two off-the-shelf counters to provide some count not readily available (such as a noninteger division). Section 4–5 describes two programmable or universal counters available in IC form. This section discusses how universal counters can be made up using basic FFs and gates.

The block diagram of Figure 4-36 shows the basic universal counter circuit. The circuit is a four-bit counter consisting of four clocked *JK* FFs and internal feedback. Inputs and outputs may be connected to count any number from 2 through 12, except 7 and 11. Counting to 7 or 11 requires external gating. The four FFs may also

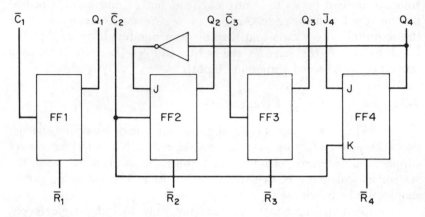

Figure 4-36. Universal counter circuit block diagram. *(Courtesy of Motorola)*

be used independently as divide-by-2 and divide-by-5, or divide-by-2 and 6, or two divide-by-2 and a divide-by-3 counters. Individual inputs are available at each FF directly to RESET the FF, or to hold a stage at zero for special counting functions. The RESET inputs may be tied together to form a common RESET.

Count	Input	Output	External Connections
2	$\bar{C}1$	Q1	None
3	$\bar{C}2$	Q4	Q2 To $\bar{J}4$
4	$\bar{C}1$	Q3	Q1 To $\bar{C}3$
5	$\bar{C}2$	Q4	Q2 To $\bar{C}3$, Q3 To $\bar{J}4$
6	$\bar{C}1$	Q4	Q1 To $\bar{C}2$, Q2 To $\bar{J}4$
8	$\bar{C}1$	Q3	Q1 To $\bar{C}2$, Q2 To $\bar{C}3$
9	$\bar{C}2$	Q4	Q2 To $\bar{C}3$, Q3 To $\bar{C}1$, Q1 To $\bar{J}4$
10	$\bar{C}1$	Q4	Q1 To $\bar{C}2$, Q2 To $\bar{C}3$, Q3 To $\bar{J}4$
12	$\bar{C}3$	Q4	Q3 To $\bar{C}1$, Q1 To $\bar{C}2$, Q2 To $\bar{J}4$

Figure 4-37. Inputs, outputs, and external connections for universal counter circuit. *(Courtesy of Motorola)*

Divide-by-2

C̄1	Q1
0	0
1	1

Divide-by-5

C̄2	Q2	Q3	Q4
0	0	0	0
1	1	0	0
2	0	1	0
3	1	1	0
4	0	0	1

Divide-by-10

C̄1	Q1	Q2	Q3	Q4
0	0	0	0	0
1	1	0	0	0
2	0	1	0	0
3	1	1	0	0
4	0	0	1	0
5	1	0	1	0
6	0	1	1	0
7	1	1	1	0
8	0	0	0	1
9	1	0	0	1

Divide-by-3

C̄2	Q2	Q4
0	0	0
1	1	0
2	0	1

Divide-by-6

C̄1	Q2	Q3	Q4
0	0	0	0
1	1	0	0
2	0	1	0
3	1	1	0
4	0	0	1
5	1	0	1

Divide-by-4

C̄1	Q1	Q3
0	0	0
1	1	0
2	0	1
3	1	1

Divide-by-11

C̄3	Q3	Q1	Q2	Q4
0	0	0	0	0
1	1	0	0	0
2	0	1	0	0
3	1	1	0	0
4	0	0	1	0
5	1	0	1	0
6	0	1	1	0
7	1	1	1	0
8	0	0	0	1
9	1	0	0	1
10	0	1	0	1
11	1	1	0	1

Divide-by-9

C̄2	Q2	Q3	Q1	Q4
0	0	0	0	0
1	1	0	0	0
2	0	1	0	0
3	1	1	0	0
4	0	0	1	0
5	1	0	1	0
6	0	1	1	0
7	1	1	1	0
8	0	0	0	1

Divide-by-8

C̄1	Q1	Q2	Q3
0	0	0	0
1	1	0	0
2	0	1	0
3	1	1	0
4	0	0	1
5	1	0	1
6	0	1	1
7	1	1	1

Figure 4-38.   Counting sequence tables for universal counter circuit. *(Courtesy of Motorola)*

The inputs and outputs shown in Figure 4-36 must be connected externally for the count function selected. The table of Figure 4-37 shows the inputs and external connections necessary to get the various counts. The binary count sequence and FF outputs for each count are shown in Figure 4-38.

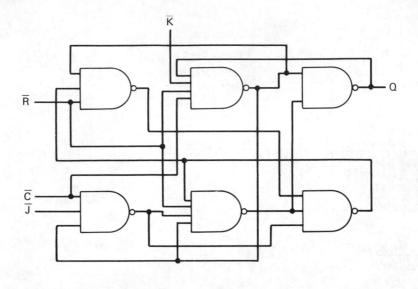

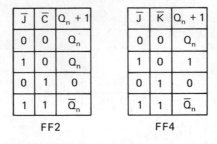

J̄	C̄	Q_n + 1
0	0	Q_n
1	0	Q_n
0	1	0
1	1	Q̄_n

FF2

J̄	K̄	Q_n + 1
0	0	Q_n
1	0	1
0	1	0
1	1	Q̄_n

FF4

Q_n = Existing state
$Q_n + 1$ = State after clock

Figure 4-39. FF circuit for universal counter. *(Courtesy of Motorola)*

Each of the FFs is composed of six NAND gates, as shown in Figure 4-39. The gates are interconnected to operate as a clocked *JK* FF with a direct RESET ($\overline{R}$). FFs 1 and 3 do not use the *JK* inputs, and thus operate in the toggle mode. A positive pulse applied to the clock input causes the FF output to change state on the negative-going clock edge. FFs 2 and 4 operate as shown in the truth table of Figure 4-39. The $\overline{C}$ input of FF 2 and the $\overline{J}$ and $\overline{K}$ inputs of FF 4 refer to *dynamic logic swings* in the truth table, with a logic 1 being the negative edge. The $\overline{J}$ input of FF 2 refers to a static logic level with a 1 being a positive voltage.

The direct RESET inputs are normally high during clocked operation of the FFs. A low on a RESET input causes the associated

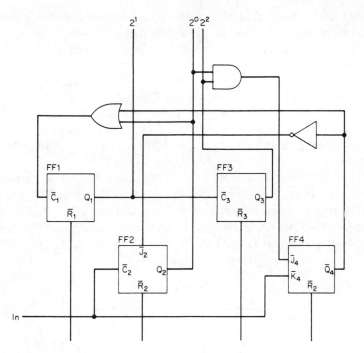

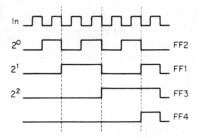

Figure 4-40. Divide-by-7 connections and timing diagram for universal counter. *(Courtesy of Motorola)*

FF to RESET (Q output goes low), regardless of the signals present on the clocked inputs ($\bar{J}$, $\bar{K}$, or $\bar{C}$).

4-4.1 Dividing by 7 and 11

Dividing by 7 and 11 with the circuit requires external gating. The technique shown in Figure 4-40 for divide-by-7 uses all four FFs and the internal feedback. $\bar{J}$ of FF 4 is gated with Q2Q3;

an AND gate is used. This results in a positive output on FF 4 during the count of six, which holds FF 2 low between counts of six and zero. The timing diagram for the divide-by-7 function is also shown in Figure 4-40.

There is no danger of a "race" condition occurring between counts five and six at the input of the OR gate because of the combined delays of the AND gate and FF 4. For example, if we assume a gate delay of 9 ns and an FF delay of 16 ns, the total delay time of 25 ns is safely greater than the typical 10-ns negative pulse required to toggle FF 1.

A divide-by-11 counter using the basic circuit is shown in Figure 4-41. Here, the F_1 function is gated to the $\overline{R}$ of FF 1 to inhibit the toggle action between counts ten and zero. The F_2 function is gated to the $\overline{C}$ input of FF 3 to complete the counting from ten to zero. The input signal is used in the gating to eliminate any chance of a race condition occurring.

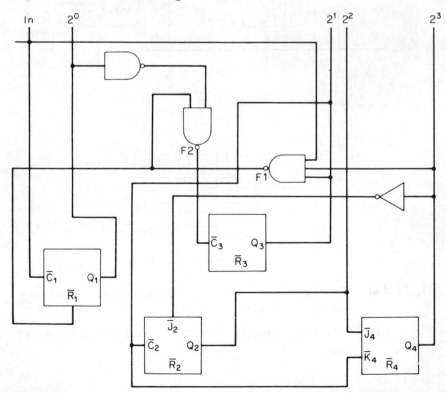

Figure 4-41. Divide-by-11 connections for universal counter. *(Courtesy of Motorola)*

4–4.2 Noninteger Division

In many logic system designs, a convenient clock frequency is not an integral multiple of the required outputs. In some cases, the clock may be raised to a common denominator of the needed outputs, but this has limitations due to maximum circuit speeds. In many cases, intermediate stages of the circuit may be used to give noninteger division. The table of Figure 4-42 shows the possible divisors that may be generated from the basic circuit. The numbers indicated by an asterisk are divisors generated directly at the counter outputs; the other numbers require extra gating (to produce an output when the desired count occurs). The numbers on the left side of Figure 4-42 refer to the count setting of the circuit. The numbers along the top are the number of *output pulses* needed for each count sequence.

Divide by	Possible divisions									
	2	3	4	5	6	7	8	9	10	11
3	$1-\frac{1}{2}$									
4		$1-\frac{1}{3}$								
5	$2-\frac{1}{2}^*$	$1-\frac{2}{3}$	$1-\frac{1}{4}$							
6				$1-\frac{1}{5}$						
7	$3-\frac{1}{2}^*$	$2-\frac{1}{3}^*$	$1-\frac{3}{4}$	$1-\frac{2}{5}$	$1-\frac{1}{6}$					
8		$2-\frac{2}{3}$		$1-\frac{3}{5}$		$1-\frac{1}{7}$				
9	$4-\frac{1}{2}^*$		$2^*-\frac{1}{4}$	$1-\frac{4}{5}$		$1-\frac{2}{7}$	$1-\frac{1}{8}$			
10		$3-\frac{1}{3}$				$1-\frac{3}{7}$		$1-\frac{1}{9}$		
11	$5-\frac{1}{2}$	$3-\frac{2}{3}^*$	$2-\frac{3}{4}$	$2^*-\frac{1}{5}$	$1-\frac{5}{6}$	$1-\frac{4}{7}$	$1-\frac{3}{8}$	$1-\frac{2}{9}$	$1-\frac{1}{10}$	
12				$2-\frac{2}{5}$		$1-\frac{5}{7}$				$1-\frac{1}{11}$

* See text

Figure 4-42. Possible noninteger divisors that may be generated from the basic universal counter. *(Courtesy of Motorola)*

The diagram and waveforms of Figure 4-43 show use of the table with a divide-by-3½ circuit. The counter is set for a divide-by-10, and the $A\overline{B}$ function (one NAND gate for $\overline{B}$, and one AND gate for A) gives the required three output pulses.

When divisors less than 2 are used, it is necessary to use the clock signal in the gating logic. The counter outputs enable the clock for the required number of pulses for each count sequence. It should

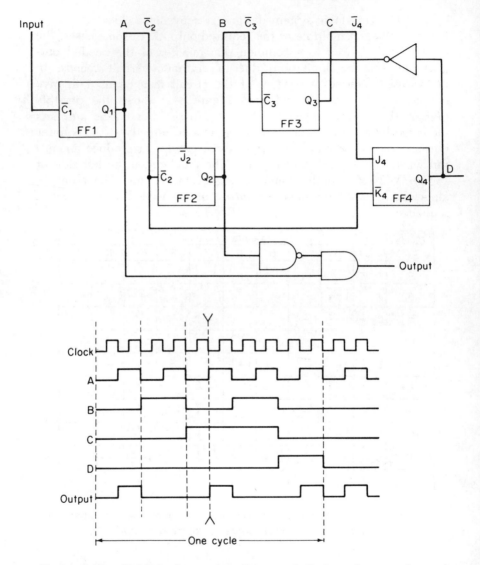

Figure 4-43   Divide-by-3½ connections and timing diagram for universal counter. *(Courtesy of Motorola)*

be noted that although the output pattern is repetitive, the individual output pulses are *not equally spaced* when this technique is used for noninteger division.

4–5   USING PROGRAMMABLE OR UNIVERSAL COUNTERS

This section describes operation of two MECL 10,000 universal counters (the MC10136 and MC10137) and their use in high-speed (100 MHz) programmable counter functions. (Note the MECL is the Motorola Semiconductor Products designation for ECL.)

4–5.1   Counter Operation

Figure 4-44 shows the logic configuration for each counter, and the function select table applicable to both counters. Operation of both counters is similar. Three control lines (S1, S2, and $\overline{C}_{in}$) determine the operational mode of the counter. Lines S1 and S2 control one of four operations: preset (program), increment (count up), decrement (count down), or hold (stop count).

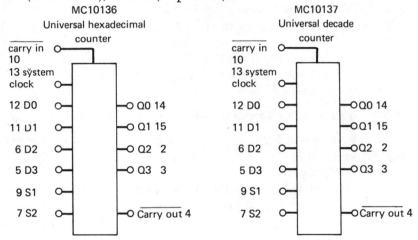

Function select table

S1	S2	Operating mode
0	0	Preset (program)
0	1	Increment (count up)
1	0	Decrement (count down)
1	1	Hold (stop count)

Figure 4-44.   MECL 10,000 universal counter logic diagrams and function select table. *(Courtesy of Motorola)*

In the *preset mode* a clock pulse is necessary to load the counter. When the S1 and S2 select lines are both in the LOW state, the information present on the data inputs ($D0$, $D1$, $D2$, $D3$) will be entered into the counter on the positive transition of the clock.

The system clock is defined as positive going, and the counters change state only on the rising edge of the clock signal. Due to the master-slave construction of the FFs, any other data or control input may change at any time other than during the positive transition of the clock (by observing proper set-up and hold times, as specified on the datasheet).

The $\overline{C}_{in}$ line overrides the clock when the counter is in either the increment or decrement modes. This input allows several devices to be cascaded into a fully synchronous multistage counter. The $\overline{C}_{out}$ ouput goes LOW on the terminal state of the counter, whether in the increment or decrement modes. With both counters, the $\overline{C}_{out}$ output goes LOW on the zero state when operating in the decrement mode. The increment mode causes the $\overline{C}_{out}$ to go LOW at the count of 15 for the MC10136, and at the count of 9 for the MC10137.

$\overline{C}_{out}$ is obtained by ORing $\overline{C}_{in}$ with the outputs of the counter. In this manner, the carry is rippled through the counter for multistage applications.

4–5.2   Programmable Counters Using No External Gating

Both counters may be used in a programmable counter without external gating. Figure 4-45 shows a technique in which $\overline{C}_{out}$ is used to control the mode.

The counter is normally in the decrement mode and counts down from the number N preset into the device. On reaching the "zero" state, the $\overline{C}_{out}$ goes LOW and allows the next clock pulse to reload the number N into the counter. The divide modulus M then is equal to $N + 1$.

The clock signal is tied common to the $\overline{C}_{in}$ and S2 control lines to prevent a latch-up state when reloading the counter. In the program mode $\overline{C}_{out}$ is forced LOW and $\overline{C}_{in}$ is disabled. $\overline{C}_{out}$, fed back to the S1 line, would latch up the counter unless the clock is tied to the other control lines.

For the MC10136, M may vary from 1 to 16; for the MC10137, M may vary from 1 to 10. Maximum toggle frequency in both cases is over 50 MHz. Figure 4-45 shows typical waveforms at 50 MHz.

If a larger divide modulus is required, two or more devices may be used in a larger counter, at a sacrifice in maximum operating frequency. Figure 4-46 shows a two-stage configuration with maximum frequency typically 35 MHz. The divide modulus is extended to 256 with the MC10136 and to 100 with the MC10137.

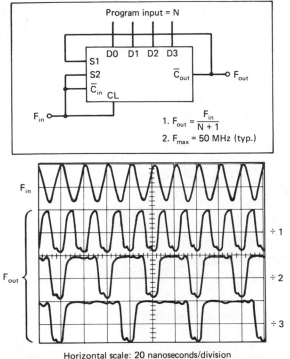

Horizontal scale: 20 nanoseconds/division
Vertical scale: 500 millivolts/division

Figure 4-45.   Programmable counter using no external gating, divide modulus $M = N + 1$. *(Courtesy of Motorola)*

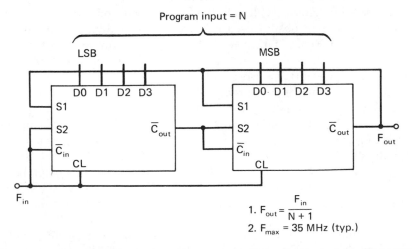

Figure 4-46.   Two-stage programmable counter, divide modulus M is 256 maximum with MC10136. *(Courtesy of Motorola)*

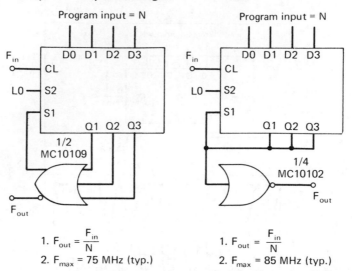

1. $F_{out} = \dfrac{F_{in}}{N}$

2. $F_{max} = 75$ MHz (typ.)

1. $F_{out} = \dfrac{F_{in}}{N}$

2. $F_{max} = 85$ MHz (typ.)

Figure 4-47. Programmable counter using external decoding; either a gate or a wired-OR may be used. *(Courtesy of Motorola)*

4–5.3 Higher Frequency Counters

Other techniques may be used to produce higher frequency programmable counters. External decoding and pulse "gobbling" allow higher performance at the cost of an increased package count. (Pulse "gobbling" is a technique by which an external FF is used to hold a pulse, thus acting as an auxiliary counter.)

One such technique is used in the counter of Figure 4-47. A gate is used to externally decode the preset condition for the counter. This can decrease delay time by 2 to 3 ns. The divide modulus for the counter of Figure 4-47 is equal to the program input N, $(M = N)$. The preset condition is decoded *one clock pulse before* the zero state of the counter. Thus, the clock pulse necessary for preset is included in the programmed input number N. For the MC10136, M may vary from 2 to 15, and from 2 to 9 for the MC10137.

A wired OR may be used in place of the OR gate. Maximum operating frequency can be extended to 85 MHz with this technique. A gate should still be used, however, to buffer the signal out (F_{out}).

In Figure 4-48, F_{out} and F_{in} waveshapes are shown for both circuit configurations. Notice that the frequency of the wired OR is displayed at 85 MHz, as opposed to the 75 MHz shown for the gate version. Both are dividing by 3.

Another higher-frequency counting technique uses pulse "gobbling" as shown in Figure 4-49. In addition to externally decoding the preset condition, an FF is used to "gobble" a pulse and provide an even shorter preset delay time than the external decoding version.

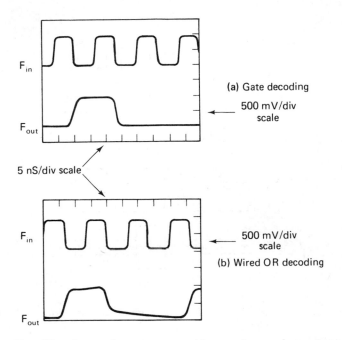

F_{in}

F_{out}

(a) Gate decoding

500 mV/div
scale

5 nS/div scale

F_{in}

500 mV/div
scale

(b) Wired OR decoding

F_{out}

Figure 4-48.   Waveforms for programmable counters using external decoding. *(Courtesy of Motorola)*

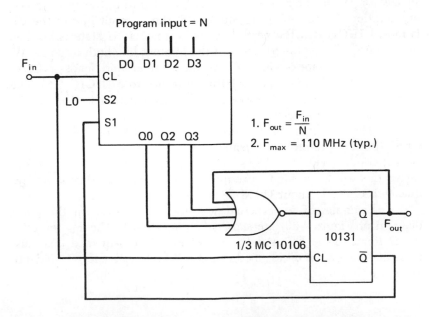

Program input = N

F_{in}

D0 D1 D2 D3
CL

L0 — S2

S1

Q0 Q2 Q3

1. $F_{out} = \dfrac{F_{in}}{N}$

2. $F_{max} = 110$ MHz (typ.)

1/3 MC 10106

D   Q

10131

F_{out}

CL $\overline{Q}$

Figure 4-49.   Programmable counter using external decoding and pulse
gobbling, divide modulus $M = N$. *(Courtesy of Motorola)*

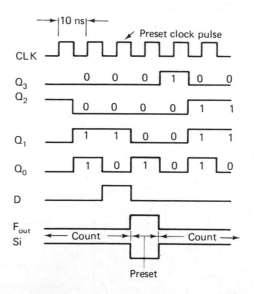

Figure 4-50. Pulse diagram for pulse gobbling technique. (*Courtesy of Motorola*)

The pulse diagram of Figure 4-50 shows the sequence of signals for this counter. The S1 line is HIGH during the count phase of operation. On reaching the count of 2, the D input line to the FF is forced HIGH. On the next clock pulse the HIGH state is clocked into the FF, causing the S1 line and the D input line both to go LOW. The succeeding clock pulse presets the counter and loads a LOW back into the FF, causing the S1 line to return to a HIGH state. The counter is then ready to proceed in the decrement count mode. In Figure 4-50, the number 8 is loaded into the counter.

The advantage of this technique is that the decode delay and set-up time for presetting the counter do not have to occur within one clock period. These two times occur within separate clock periods. Again this allows a higher frequency of operation. Maximum frequency is typically about 110 MHz.

Examples of some actual waveforms (idealized in the pulse diagram of Figure 4-50) are shown in Figure 4-51. The F_{in} (or CL), the D input to the FF, and F_{out} are shown. The sequence is as discussed, although the divide modulus is 3. Two frequencies, 33 and 110 MHz, are shown.

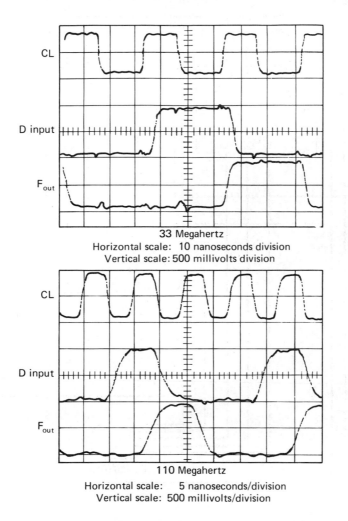

Figure 4-51. Waveforms for programmable counter using pulse gobbling.

The divide modulus M is similar to the preceding design. That is, $M = N$. For larger moduli counters, two counter may be cascaded using the pulse gobbling technique (Figure 4-52). The divide modulus may be extended to 255 with two MC10136s. The maximum frequency is about 80 MHz for such a configuration.

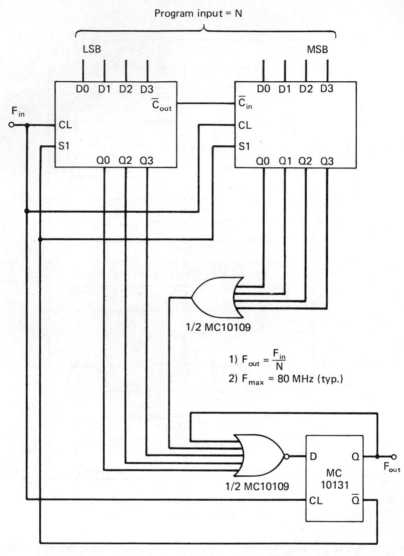

Figure 4-52. Two-stage pulse gobbler counter, divide modulus *M* is 255 maximum with MC10136. *(Courtesy of Motorola)*

4–6 USING SHIFT REGISTERS AS PULSE DELAY NETWORKS

It is possible to use high-speed FFs to form a shift register that can act as a digital incremental delay. Such a register may be clocked with a frequency division counter (Sec. 4–2) to accomplish any desired

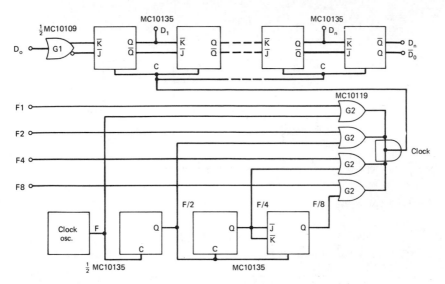

Figure 4-53.   Delay register diagram. *(Courtesy of Motorola)*

delay with increments typically as small as 7.5 ns. As a typical application, such a shift register may be used for timing basic computer decisions, or as an adjustable digital delay line for pulses.

Figure 4-53 is the logic schematic for an N bit shift register which includes the necessary logic for electronically controlling the delay of the register in powers of two. The FFs used are Motorola MC10135 high-speed MECL devices. The MC10135 is a dual, master-slave, dc coupled $\overline{J}\overline{K}$ FF with separate asynchronous set and reset inputs and a common clock input. Both FFs in the package have complementary Q and $\overline{Q}$ outputs. FF specifications are: typical toggle frequency of 140 MHz, typical propagation delay time of 3 ns. Operating voltage is -5.2 V with negligible degradations in operation for $\pm$ 10 percent variations in supply voltage. Figure 4-54 shows an FF logic block connected as a dual stage shift register. The $\overline{J}\overline{K}$ and RS truth tables are also shown. For shift register operation, the $\overline{J}$ and $\overline{K}$ inputs are always complements.

4–6.1   Delay Register Operation

In the circuit of Figure 4-53, incoming information to be delayed is considered to be asynchronous with the internal clock oscillator, and random in nature. The input data (D_0) might appear as shown in Figure 4-55, which illustrates typical waveforms that would result from D_0 and a 50-MHz clock. G1 is used to split the input into

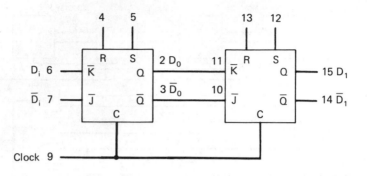

V_{CC1} = Pin 1
V_{CC2} = Pin 16
V_{EE} = Pin 8

$\bar{J}$	$\bar{K}$	Q_{n+1}
1	1	Q_n
0	1	1
1	0	0
1	1	$\bar{Q}_n$

R	S	Q
0	0	Q_n
0	1	1
1	0	0
1	1	ND

ND = not defined

Figure 4-54.   Two-stage MC10135 shift register with truth tables. *(Courtesy of Motorola)*

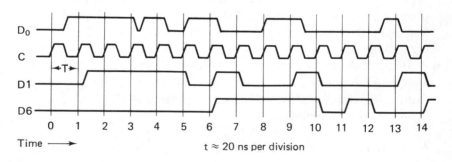

DO = data into first $\bar{J}$ – $\bar{K}$ flip-flop   D1 = delay data out of first $\bar{J}$ – $\bar{K}$ flip-flop
C = clock to shift register   D6 = delayed data out of sixth $\bar{J}$ – $\bar{K}$ flip-flop

Figure 4-55.   Typical delay register waveforms for a 50 MHz clock frequency. *(Courtesy of Motorola)*

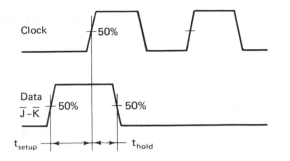

Figure 4-56.   Setup and hold time definitions. *(Courtesy of Motorola)*

Data (D_0) and Data NOT $(\overline{D}_0)$ since the $J\overline{K}$ FF requires complementary data (dual-rail logic). The OR and NOR outputs of G1 have essentially the same propagation delay so output skew (outputs occurring at different times) is no problem.

Several timing condition possibilities that may occur between D_0 and the input clock are shown in Figure 4-55. For example, D_0 is low between time marks 3 and 4. This low data input is not picked up by the first FF due to no positive clocking edge being present during this low data pulse. For timing purposes, the worst case numbers for the MC10135 are: t_{pd}(max) clock to output (or time delay of the pulse from start of the clock to output) = 4.5 ns, setup time = 2.5 ns, and hold time = 1.5 ns. Figure 4-56 illustrates the definitions of these parameters. Setup time is defined as the minimum time prior to the positive-going clock that any data to the $J\overline{K}$ inputs must be settled. The hold time is the minimum time after the positive-going clock that the data must be held steady.

The input levels $F1$, $F2$, $F4$, and $F8$ of Figure 4-53 allow the selection of the clock frequency $f/2^M$, where M is the number of FFs used to divide the basic frequency of the clock oscillator. Inputs $F1 - F8$ are negative logic inputs, and only one input should be low at a time. The low input enables the desired clock frequency. If all inputs go high, the register stores the data that was being shifted in the register.

The gate for G1 is the MC10109 and G2 is the MC10119, which are standard logic gates in the MC10,000 series. These devices have delays of about 2 and 3 ns respectively. Five dual FFs may be driven from the MC10119 with a rise time of 3 ns and fall time of 4 ns. If lead lengths are kept short, and low inductance printed circuit wiring is used, the register of Figure 4-53 will typically shift at clock rates of 140 MHz.

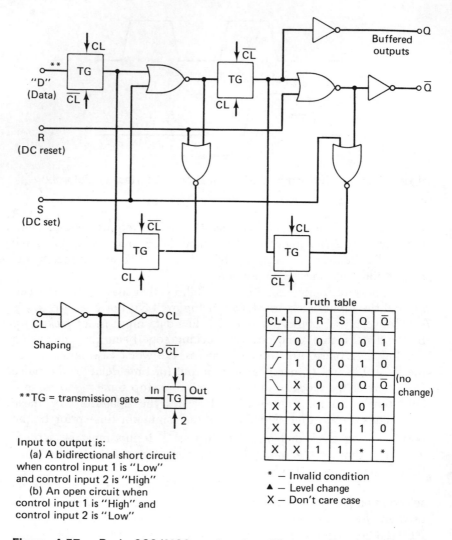

Truth table

CL▲	D	R	S	Q	Q̄	
⌡	0	0	0	0	1	
⌡	1	0	0	1	0	
⌉	X	0	0	Q	Q̄	(no change)
X	X	1	0	0	1	
X	X	0	1	1	0	
X	X	1	1	*	*	

* — Invalid condition
▲ — Level change
X — Don't care case

Figure 4-57. Basic COS/MOS master-slave FF stage. *(Courtesy of RCA)*

The data delay through the first FF is a minimum of the setup time plus the delay time of the FF, or a maximum of the clock period plus the delay time and the setup time of the FF. With a worst case delay time of 4.5 ns, a setup time of 2.5 ns and operation at a clock frequency of 50 MHz, the MC10135 yields a minimum delay of 7 ns or a maximum of 27 ns.

4–6.2   Alternate FFs

Two other dual FFs may also be used to improve the delay register performance. Both the MC10131 and MC10231 devices are D-type master-slave FFs. The D-type is a single rail as opposed to the JK which is dual rail (Sec. 4–1). This feature eliminates the requirement for G1 which allows a reduction in component count. Typically, the MC10131 will operate at 150 MHz, allowing clock periods as small as 6.7 ns. The MC10231 is an even higher speed MECL 10,000 device which allows clock periods as small as 5 ns.

Two MSI devices will provide four bits of delay per package at high speeds. These are the MC10141 and the MC1694 four-bit shift registers. Typically, the MC10141 is capable of shift rates in excess of 100 MHz. The MC1694 device will shift up to 300 MHz, and is fully compatible with the MECL 10,000 logic series.

4–7   TYPICAL COUNTER AND REGISTER APPLICATIONS

This section describes typical applications for one counter and one register in the RCA COS/MOS line.

4–7.1   Basic COS/MOS Master-slave FF Stage

Figure 4-57 shows the basic static master-slave FF used in the COS/MOS line. The logic level preset at the D (or data) input is transferred to the Q output during the positive-going transition of the clock pulse. Direct-current SET and RESET are done by a high level at the respective input. Either or both SET and RESET can be omitted. Output lead isolation at the Q and $\overline{Q}$ outputs is obtained by use of inverters. The inverters also provide additional drive, noise immunity, and operating speed. The size of the MOS devices in the output inverters is tailored to meet the desired drive and sink current requirements of the particular IC counter or register.

4–7.2   Decade Counter/Divider

Figure 4-58 shows the logic diagram of a CD4017A decade counter which uses FFs similar to those of Figure 4-57. The CD4017A is an IC counter which includes ten decoded decimal outputs in the package. A five-stage Johnson counter (Sec. 4–2.3) is used to implement the decade counter. CLOCK, RESET, INHIBIT, and C_{OUT} (carry out) signals are provided.

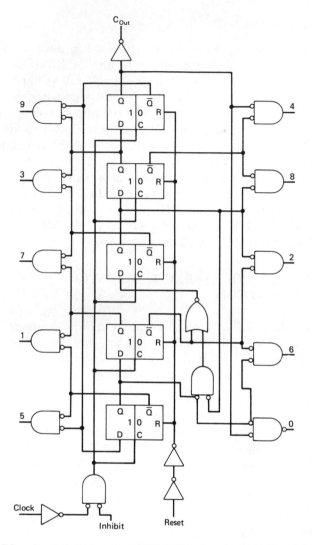

Figure 4-58.   Logic diagram of CD4017A decade counter. *(Courtesy of RCA)*

      The decade counter advances one count at the positive clock signal transition, provided the INHIBIT signal is low. Counter advance by means of the clock line is inhibited when the INHIBIT signal is high. A high RESET signal clears the decade counter to its zero count. The Johnson counter configuration permits high-speed operation, two-input decimal decode gating, and spike-free decoded output. Antilock gating is provided to permit only the proper counting sequence.

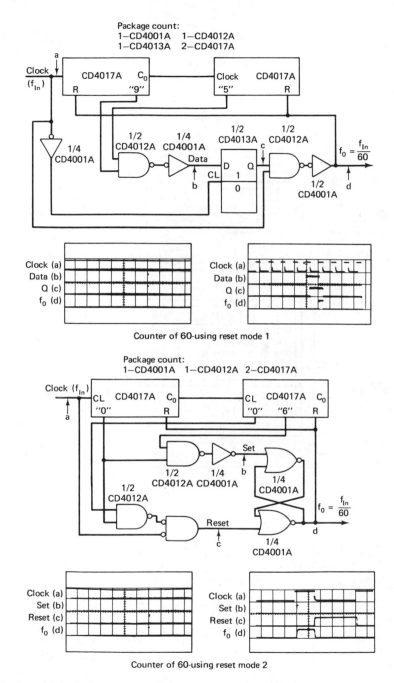

Figure 4-59. CD4017A divide-by-*N* counter configurations. *(Courtesy of RCA)*

233

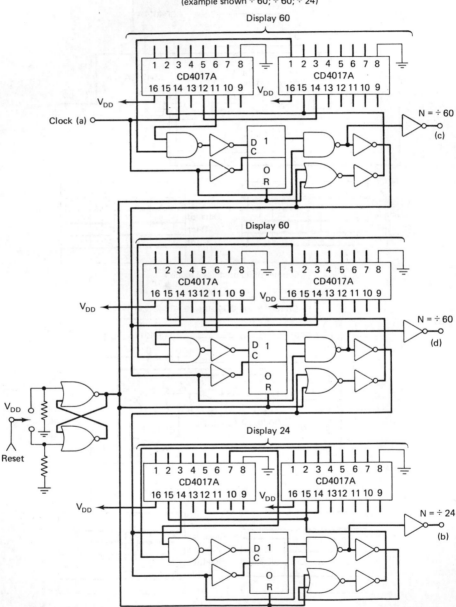

Figure 4-60. Divide-by-60 and divide-by-24 configuration. *(Courtesy of RCA)*

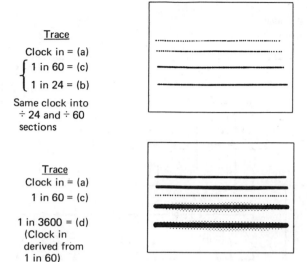

Trace

Clock in = (a)

$\left\{\begin{array}{l} \text{1 in 60 = (c)} \\ \text{1 in 24 = (b)} \end{array}\right.$

Same clock into
÷ 24 and ÷ 60
sections

Trace
Clock in = (a)
1 in 60 = (c)

1 in 3600 = (d)
(Clock in
derived from
1 in 60)

Package count

6—CD4017A
2—CD4011A
2—CD4001A
2—CD4013A
2—CD4009A

Figure 4-60.   (Continued)

The ten decimal outputs are normally low, and go high only at their respective decoded decimal time slot. Each decimal output remains high for one full clock cycle. The carry-out signal completes one cycle for every ten clock input cycles, and is used to clock the following decade directly (in multiple decade applications).

Figures 4-59 and 4-60 show use of the CD4017A to obtain various divide-by-N counter configurations. Figure 4-59 illustrates two reset methods when using the CD4017A for counting to the number 60. Figure 4-60 shows an example of six CD4017As used in a divide-by-60, divide-by-60, and divide-by-24 circuit. Note that the CR4017A permits easy decimal display of each divide-by-N section because each IC package contains its own decoded decimal outputs.

4–7.3 Static Shift Register

Figure 4-61 shows the logic diagram of the CD4014A, an 8-stage synchronous parallel-input/serial-ouput register. A clock input and a single serial data input, along with individual parallel "jam" inputs to each register stage, and a common parallel/serial control

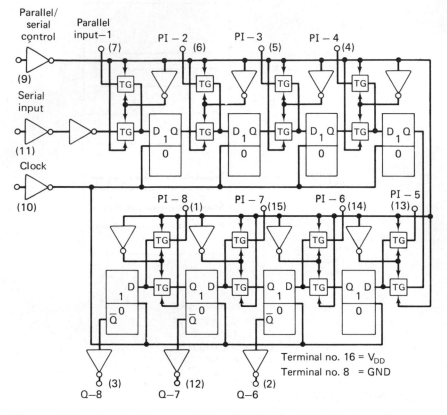

Figure 4-61.   Logic diagram of the CD4014A eight-stage synchronous parallel-input / serial-output register. *(Courtesy of RCA)*

signal are provided. Q outputs from the 6th, 7th, and 8th stages are available. All register FFs are similar to that shown in Figure 4-57, except that extra transmission gates permit parallel or serial entry.

Parallel or serial entry is made into the register synchronous with the positive clock transition, and under control of the parallel/serial input. When the parallel/serial input is low, information is serially shifted into the eight-stage register synchronous with the positive clock transition. When the parallel/serial input is high, information is jammed into the eight-stage register by way of the parallel input lines and synchronous with the positive clock transition. Register expansion with multiple CD40114A packages is possible.

Figure 4-62 shows use of the CD4014A in an eight-stage synchronous parallel-input/serial-output register application. In this configuration, the CD4013A (an FF of the COS/MOS line) allows parallel transfer to be made into the CD4014A register every eight

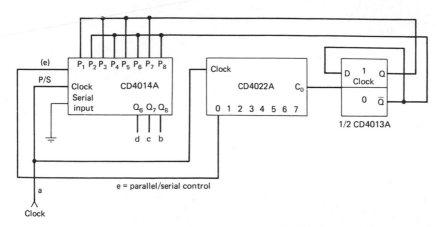

Figure 4-62. CD4014 eight-stage synchronous parallel-input/serial-output register application. *(Courtesy of RCA)*

clock pulses. Use of the divide-by-2 outputs of the CD4013A as parallel inputs to alternate CD4014A stages permits changeover from a 10101010 to a 01010101 (or complement) parallel input pattern every eight pulses. Note that the CD4022A is a four-stage divide-by-8 Johnson counter of the COS/MOS line, and is shown in Figure 4-63.

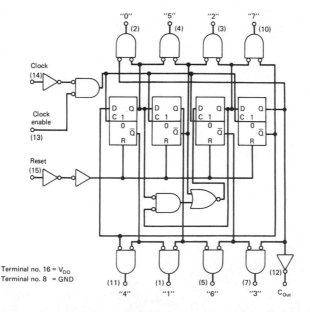

Figure 4-63. Logic diagram of the CD4022A divide-by-8 counter. *(Courtesy of RCA)*

5

Digital-to-Analog
and Analog-to-Digital
Conversion

This chapter is devoted to the various digital-to-analog (D/A) and analog-to-digital (A/D) techniques found in current logic design. In this chapter we shall concentrate on explanations of the basic principles and techniques of D/A and A/D converters. By studying this information, the designers may more intelligently decide which type of conversion is best suited for their systems. The chapter also includes specific design examples of state-of-the-art D/A and A/D conversion.

5–1 CONVERSION BETWEEN ANALOG AND DIGITAL INFORMATION

There are several methods for converting voltage (or current) into digital form. One method is to convert the voltage into a frequency or series of pulses. Then the pulses are converted into BCD form, and the BCD information is converted into a decade readout.

It is also possible to convert voltage directly into BCD form by means of an A/D converter, and to convert BCD data back into voltage by means of a D/A converter. Before going into the operation of these conversion circuits, let us discuss the signal formats for BCD data, as well as the *four-bit system.*

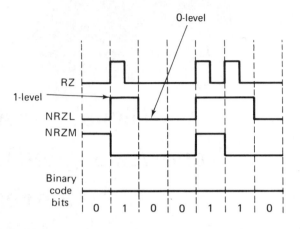

Figure 5-1. Typical BCD signal formats.

5–1.1   Typical BCD Signal Formats

Although there are many ways in which pulses can be used to represent the 1 and 0 digits of a BCD code, there are only three ways in common use. These are the NRZL (nonreturn-to-zero-level), the NRZM (nonreturn-to-zero-mark), and the RZ (return-to-zero) formats. Figure 5-1 shows the relation of the three formats.

In NRZL, a one is one signal level, and a zero is another signal level. These levels can be 5 V, 10 V, 0 V, −5 V, or any other selected values, provided that the one and zero levels are entirely different.

In RZ, a 1 bit is represented by a pulse of some definite width (usually a one-half bit width) that returns to the zero signal level, while the 0 bit is represented by a zero-level signal.

In NRZM, the level of the pulse has no meaning. A 1 is represented by a *change in level*.

5–1.2   The Four-bit System

A four-bit system is one that is capable of handling four information bits. Any fractional part (in sixteenths) can be stated by using only four binary digits. The number 15 is represented by 1111 and zero is represented by 0000. Any number between 0 and 15 requires only four binary digits. For example, the number 1000 is 8, the number 0111 is 7, and so on. Although not all logic codes use the four-bit system, it is common and does provide a high degree of accuracy for conversion between analog and digital data. Most conversion systems use the four-bit code, or a multiple of the basic code (8 bits, 16 bits, and so on).

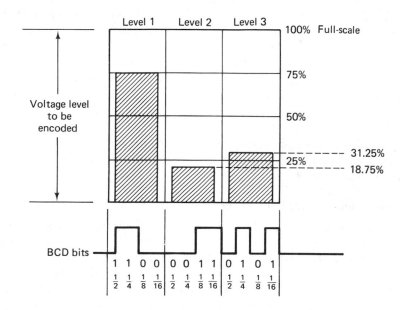

Figure 5-2. Relationship among three voltage levels to be encoded and the corresponding BCD code (using the four-bit system).

In practice, a four-bit A/D converter (also called a *binary encoder* by some logic designers) samples the voltage level to be converted and compares the voltage to ½ scale, ¼ scale, ⅛ scale, and ¹⁄₁₆ scale (in that order) of some given full-scale voltage. The A/D converter (or encoder) then produces four bits, in sequence, with the decision (or comparison) made on the most significant (or ½ scale) first.

Figure 5-2 shows the relation among three voltage levels to be converted (or encoded) and the corresponding binary code (in NRZL form). As shown, each of the three voltage levels is divided into four equal time increments. The first time increment is used to represent the ½-scale bit, the second increment is used for the ¼-scale bit, the third increment for the ⅛-scale bit, and the fourth increment for the ¹⁄₁₆-scale bit.

In level 1, the first two time increments are at a binary 1, with the second two increments at 0. This produces a 1100, or decimal 12. Twelve is three-fourths of 16. Thus level 1 is 75 percent of full scale. For example, if full scale is 100 V, level 1 is at 75 V.

In level 2, the first two increments are at 0, while the second two increments are at 1. This is represented as 0011, or 3. Thus, level 2 is three-sixteenths of full scale (or 18.75 V).

This can be expressed in another way. In the first or half-scale increment, the converter produces a 0 because the voltage (18.75) is *less than* half-scale (50). The same is true of the second, or quarter-scale increment (18.75 V is less than 25 V). In the third or one-eighth scale increment, the converter produces a 1, as it does in the fourth or one-sixteenth scale increment, because the voltage being compared is *greater than* one-eighth of full scale (18.75 is greater than 12.5) and greater than one-sixteenth of full scale (18.75 is greater than 6.25).

Thus, the half and quarter increments are at 0 or off, while one-eighth and one-sixteenth scale increments are on. Also $\frac{1}{8} + \frac{1}{16} = \frac{3}{16}$, or 18.75 percent.

5–1.3   Analog-to-digital Conversion

One of the most common methods of direct A/D conversion involves the use of a converter that operates on a sequence of *half-split, trial-and-error steps*. This produces code bits in serial form.

The heart of a converter is the *conversion ladder* such as that shown in Figure 5-3. The ladder provides a means of implementing a four-bit binary coding system and produces an output that is equivalent to the switch positions. The switches can be moved to either a 1 or a 0 position, which corresponds to a four-place binary number. The output voltage describes a percentage of the full-scale reference voltage, depending on the binary switch position. For example, if all switches are in the 0 position, there is no output voltage. This produces a binary 0000, represented by 0 V.

If switch A is at 1, and the remaining switches are at 0, this produces a binary 1000 (decimal 8). Since the total in a four-bit system is 16, 8 represents one-half of full scale. Thus the output voltage is one-half of the full-scale reference voltage. This is accomplished as follows.

The 2-, 4- and 8-Ω switch resistors and the 8-Ω output resistor are connected in parallel. This produces a value of 1-Ω across points X and Y. The reference voltage is applied across the 1-Ω switch resistor (across points Z and X) and the 1-Ω combination of resistors (across points X and Y). In effect, this is the same as two 1-Ω resistors in series. Since the full-scale reference voltage is applied across both resistors in series, and the output is measured across only one of the resistors, the output voltage is one-half the reference voltage.

In a practical converter, the same basic ladder is used to supply a comparison voltage to a *comparison circuit* which compares the voltage to be converted against the binary-coded voltage from the ladder. The resultant output of the comparison circuit is a binary code representing the voltage to be converted.

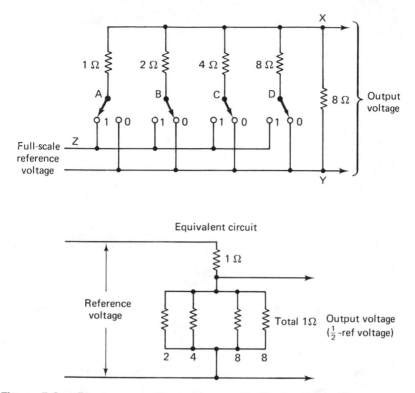

Figure 5-3.   Binary conversion ladder used in the four-bit system.

The mechanical switches shown in Figure 5-3 are replaced by electronic switches, usually FFs. When the FF is "on," the corresponding ladder resistor is connected to the reference voltage. When "off," the resistor is disconnected from the reference voltage. The switches are triggered by four pulses (representing each of the four binary bits) from the system clock. An enable pulse is used to turn the comparison circuit on and off, so that as each switch is operated, a comparison can be made of the four bits.

Simplified A/D operating sequence. Figure 5-4 is a simplified block diagram of an A/D converter. Here, the reference voltage is applied to the ladder through the electronic switches. The ladder output (comparison voltage) is controlled by switch positions, which, in turn, are controlled by pulses from the clock, as shown.

The following paragraphs outline the sequence of events necessary to produce a series of four binary bits that describe the input voltage as a percentage of full scale (in one-sixteenth increments). Assume that the input voltage is three-fourths of full scale (or 75 percent).

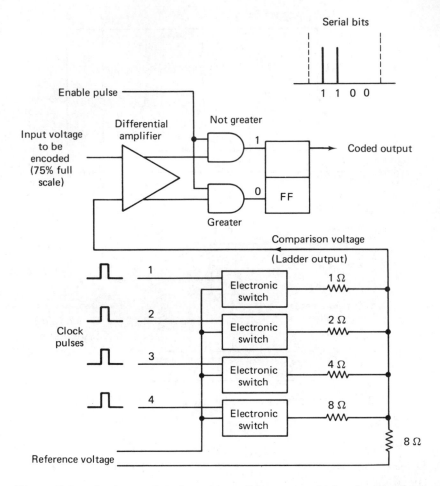

Figure 5-4.   Analog-to-digital converter (binary encoder) using the four-bit system.

When pulse 1 arrives, switch 1 is turned on and the remaining switches are off. The ladder output is a 50-percent voltage that is applied to the differential amplifier. The balance of this amplifier is set so that its output is sufficient to turn on one AND gate and turn off the other AND gate, if the ladder voltage is greater than the input voltage. Likewise, the differential amplifier will reverse the AND gates if the ladder voltage is not greater than the input voltage. Both AND gates are enabled by the pulse from the clock.

In our example (75 percent of full scale), the ladder output is less than the input voltage when pulse 1 is applied to the ladder. As a result, the *not greater* AND gate turns on, and the output FF is set to the 1 position. Thus, for the first of the four bits, the FF output is 1.

When pulse 2 arrives, switch 2 is turned on and switch 1 remains on. Both switches 3 and 4 remain off. The ladder output is now 75 percent of full-scale voltage. The ladder voltage equals the input voltage. However, the ladder output is still not greater than the input voltage. Consequently, when the AND gates are enabled, the AND gates remain in the same condition. Thus, the output FF remains at 1.

When pulse 3 arrives, switch 3 is turned on. Switches 1 and 2 remain on, while switch 4 is off. The ladder output is now 87.5 percent of full-scale voltage, and is thus greater than the input voltage. As a result, when the AND gates are enabled, they reverse. The not greater AND gate turns off, and the *greater* AND gate turns on. The output FF then sets to 0.

When pulse 4 arrives, switch 4 is turned on. All switches are now on. The ladder is now maximum (full-scale) and thus is greater than the input voltage. As a result, when the AND gates are enabled, they remain in the same condition. Likewise, the output FF remains at a 0.

The four binary bits from the output are 1, 1, 0, and 0, or 1100. This is a binary 12, which is 75 percent of 16.

In a practical converter, when the fourth pulse has passed, all the switches are reset to the off position. This places them in a condition to spell out the next four-bit binary word.

5–1.4 Digital-to-analog Conversion

A D/A converter performs the opposite function of the A/D converter just described. The D/A converter produces an output voltage that corresponds to the binary code.

As shown in Figure 5-5, a conversion ladder is also used in

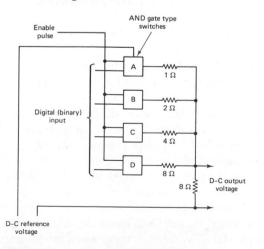

Figure 5-5. Digital-to-analog converter using the four-bit system.

the D/A converter. The output of the conversion ladder is a voltage that represents a percentage of the full-scale reference voltage. The output percentage depends on the switch positions. In turn, the switches are set to on or off by corresponding binary pulses. If the information is applied to the switch in four-line (parallel) form, each line can be connected to the corresponding switch. If the information is in serial form, the data must be converted to parallel by a shift register and/or storage register (see Chapter 4).

The switches in the D/A converter are essentially a form of AND gate. Each gate completes the circuit from the reference voltage to the corresponding ladder resistor when both the enable pulse and binary pulse coincide.

Assume that the digital number to be converted is 1000 (decimal 8). When the first pulse is applied switch *A* is enabled and the reference voltage is applied to the 1-Ω resistor. When switches *B*, *C*, and *D* receive their enable pulses, there are no binary pulses (or, the pulses are in the 0 condition). Thus, switches *B*, *C*, and *D* do not complete the circuits to the 2-, 4- and 8-Ω ladder resistors. These resistors combine with the 8-Ω output resistor to produce a 1-Ω resistance in series with the 1-Ω ladder resistance. This divides the reference voltage in half to produce 50 percent of full-scale output. Since 8 is one-half of 16, the 50-percent output voltage represents 8.

5-2 HIGH-SPEED D/A CONVERTERS

There are a number of D/A conversion methods. We cannot cover them all in any detail here. However, the following discussion covers the basic methods, and points out where the various techniques may fit into the system requirements of the logic designer.

5-2.1 Voltage Output D/A Converters

The output of a D/A converter can be an analog voltage or current. Voltage output types are used most commonly and are easiest to understand.

Figure 5-6 shows the block diagram of a three-bit voltage D/A using weighted resistors and a summing amplifier. The summing resistors of an operational amplifier (op-amp) are weighted in binary fashion, and are connected by an electronic switch to the reference or to ground, depending upon the state of each individual digital input. A digital 1 connects the resistor to the reference, and thus adds in the respective binary-weighted increment.

The circuit of Figure 5-6 has a significant disadvantage in that accuracy and stability depend upon absolute accuracy of the re-

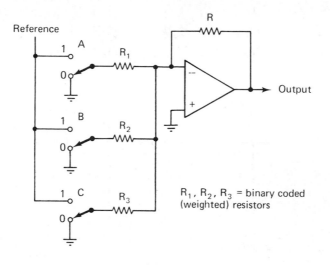

$$\text{Output} = (\tfrac{1}{2}A + \tfrac{1}{4}B + \tfrac{1}{8}C) \times \text{reference}$$

Figure 5-6.   Three-bit voltage D/A converter using weighted resistors and a summing amplifier.

sistors and their ability to track each other versus temperature. To overcome these problems, a *R-2R* resistive ladder as shown in Figure 5-7 is used in many current systems. Note that accuracy is not dependent upon the absolute value of all the *R* s, but rather *only their differences.* Temperature effects are only significant with respect to how well all the *R* s and 2*R* s *track* each other.

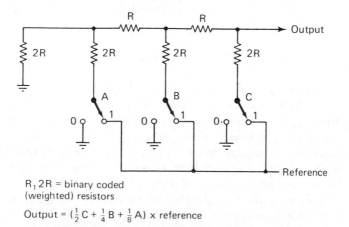

R, $2R$ = binary coded
(weighted) resistors

$$\text{Output} = (\tfrac{1}{2}C + \tfrac{1}{4}B + \tfrac{1}{8}A) \times \text{reference}$$

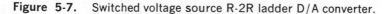

Figure 5-7.   Switched voltage source R-2R ladder D/A converter.

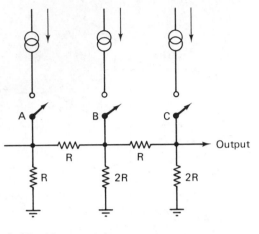

R, 2R = binary coded
(weighted) resistors

Output = $(C + \frac{1}{2} B + \frac{1}{4} A) \times 1R$

Figure 5-8.   Switched current source R-2R ladder.

Another type of *R-2R* ladder, voltage output D/A, is shown in Figure 5-8. In this circuit, called a *switched current source R-2R ladder*, equal value current sources are switched into the ladder, rather than switching the "legs" of the ladder between voltages. The effect of each current, at e_0, is binary weighted.

Currents may be switched much more rapidly than voltages. This gives the current source D/A an increase in speed by at least one order of magnitude. For this reason, and because the technique is generally easy to fabricate in monolith form, the circuit of Figure 5-8 is often found in ICs.

5–2.2   Current Output D/A Converters

Current output D/As can be implemented by generating binary-weighted currents, preferably from active sources, and summing these on a common line. Figure 5-9 shows such a D/A, called a *weighted current source* D/A. In a practical circuit the switches controlled by the digital word output are current steering circuits, not on-off switches. Current from a current source is either steered into the output line, or into another node of the circuit. The weighted current source method is the fastest method of current switching available. Switching speeds of less than 1 ns are possible with ECL.

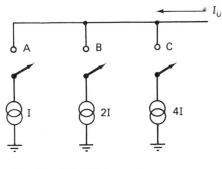

$I_0 = AI + B2I + C4I$

I, 2I, 4I = binary coded (weighted) currents

Figure 5-9. Current output D/A using weighted current sources.

The technique is also a method that can easily be implemented in monolithic form.

Figure 5-10 shows a current output R-2R ladder found in IC D/A converters (the Motorola MC1406 and MC1408). In this type of D/A, a constant current I_L is injected into an R-2R ladder. Each ladder leg is terminated with a current-steering switch. The state of these switches depends on the digital word input. In one state, the ladder current in each respective leg is steered into the output line. In the opposite state, the switch steers the leg current to ground. In

R, 2R = binary coded (weighted) resistors

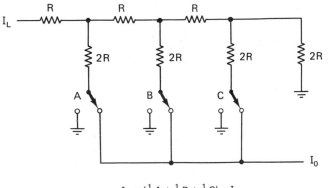

$$I_0 = (\tfrac{1}{2}A + \tfrac{1}{4}B + \tfrac{1}{8}C) \times I_L$$

Figure 5-10. Current output R-2R ladder D/A converter.

this way, an analog current is produced proportional to the digital word input. One advantage of this system is that current in all portions of the ladder is constant at all times, and not a function of the input digital word. Thus, the loss of speed due to the time constant of the ladder is eliminated.

5–3 HIGH-SPEED A/D CONVERTERS

A/D conversion can also be accomplished by a variety of techniques. However, A/D systems capable of high speed (less than 1 μs conversion time) are limited to a few basic conversion methods.

There are three basic types of high-speed A/D converters: parallel, serial, and combination. In parallel, all bits are converted simultaneously by many circuits. In serial, each bit is converted sequentially, one at a time. The combination A/D conversion includes features of both types. Generally, parallel is faster, but more complex, than serial. The combination types are a compromise between speed and complexity.

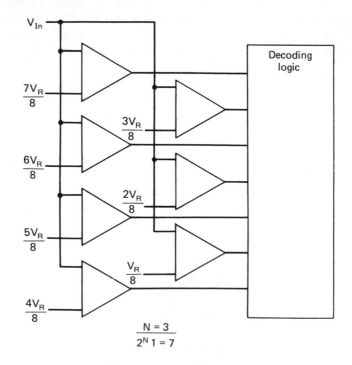

Figure 5-11. Parallel (flash) A/D block diagram.

5–3.1   Parallel (Flash) A/D

Figure 5-11 shows the basic parallel (or flash) A/D conversion circuit. In this circuit, all bits of the digital representation are determined simultaneously by a bank of voltage comparators. For N-bits of binary information, the system requires $2^N - 1$ comparators, and each comparator determines one LSB level. Until the development of MSI and LSI IC techniques, the parallel A/D method was prohibitive if N were very large.

Another disadvantage of the parallel system is that the output of the comparator bank is *not directly usable* information. The output must be converted to binary information in some sort of binary code. For large values of N, the massiveness of the conversion logic not only increases cost and complexity, but requires more successive stages, thus increasing conversion time.

5–3.2   Tracking A/D

Figure 5-12 shows the basic tracking A/D conversion circuit.

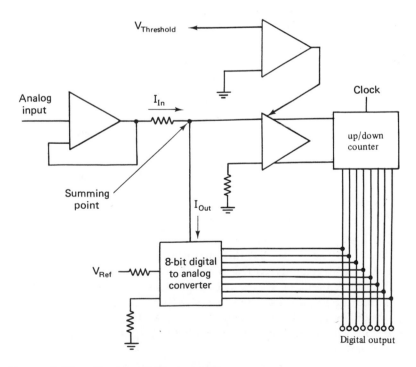

Figure 5-12.   Tracking A/D converter.

Tracking A/D derives its name from the fact that the digital output continuously "tracks" the analog input voltage. Tracking A/D is usually used in communications systems or some similar application where the input is a continuously-varying signal. As shown, tracking A/D uses a D/A in a feedback path. Accuracy of the system is no better than the D/A being used (typically six to ten bits).

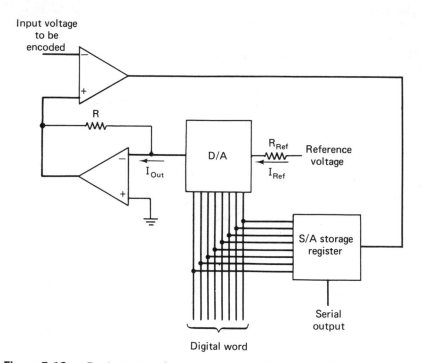

Figure 5-13.   Basic successive approximation A/D converter.

5–3.3 Successive Approximation A/D

Figure 5-13 shows the basic successive approximation A/D conversion circuit. Note that this circuit is essentially the same as the basic A/D converter described in Sec. 5–1.3 and illustrated in Figure 5-4. The D/A block of Figure 5-13 represents the electronic switches and ladder of Figure 5-4; the successive approximately (S/A) storage register of Figure 5-13 represents the logic AND gates and the FF of Figure 5-4. However, four bits are shown in Figure 5-4, whereas eight bits are shown in Figure 5-13.

The S/A type of A/D is relatively slow compared to other types of high-speed A/Ds, but the low cost, ease of construction, and system operational features more than make up for the lack of speed (in most applications). S/A is by far the most widely used A/D system. For that reason, we describe implementation of a complete S/A type of A/D, using IC components, in Sec. 5–4.

As shown in Figure 5-13, the system enables the eight bits of the D/A, one at a time, starting with the MSB. As each bit is enabled, the comparator gives an output signifying that the input signal is greater, or not greater, in amplitude than the output of the D/A. If the D/A output is greater than the input signal, the bit is "reset" or turned off. The system does this with the MSB first, then the next most significant bit, then the next, and so on. After all eight bits of the D/A have been tried, the conversion cycle is complete, and another conversion cycle is started.

The serial output of the system is taken from the output of the comparator. While the system is in the conversion cycle, the comparator output is either 0 or 1, corresponding to the digital state of the respective bit. In this way, the S/A type of A/D gives a serial output during conversion, and a parallel output between conversion cycles.

Speed and *accuracy* are the prime design factors of any A/D system. The speed and accuracy of the S/A type of A/D are directly dependent on the D/A specifications. Typical S/A systems will convert in about 200 to 500 ns per bit, and have bit accuracies of about 6 to 12 bits.

5–3.4   Parallel Ripple A/D

Figure 5-14 shows the basic parallel ripple A/D conversion circuit. The technique was developed to decrease the amount of hardware required to implement the standard parallel converter (Sec. 5–3.1) without increasing the conversion time dramatically. The system sacrifices some speed in return for a considerable reduction in cost and complexity.

Basically, parallel ripple A/D consists of two each, M-bit parallel converters, and an M-Bit D/A. The total system has an N-bit output, where $N = 2M$. In this system, both the parallel converters and the D/A subtraction system must be N-bit accurate. (Subtractors are discussed in Chapter 6.)

In operation, the parallel ripple A/D converts the first M-bit of the output by the standard parallel (flash) A/D. As in most A/D

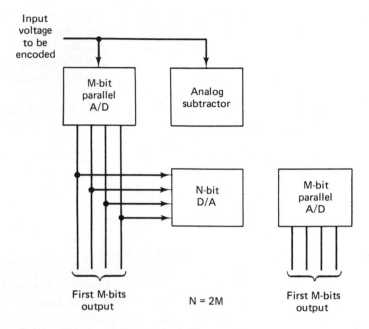

Figure 5-14.   Basic parallel ripple A/D converter.

systems, the output of the first M-bit encoder is a digital word representing the largest number of discrete quantities that does not exceed the input signal.

The output of the first parallel converter is used not only as the first M-bits of the output word, but is also used to access the D/A in the analog subtraction section. The output of the D/A gives a voltage output that is equal to the *highest* discrete level that does not exceed the input signal. This voltage is subtracted, by analog means, from the input signal. The remainder is then fed to another M-bit flash encoder which converts the remaining M-bits of the system. In a practical system, either the thresholds of the second set of $2^M - 1$ comparators must be scaled by a factor of 2^M, or the remainder signal must be amplified by 2^M.

5-4   SUCCESSIVE APPROXIMATION A/D CONVERTER

Figure 5-15 shows a schematic diagram of a successive approximation A/D, such as described in Sec. 5–3.3. In the Figure 5-15 circuit, the basic components include a monolithic IC D/A and the CMOS SAR (successive approximation register). The system requires a total of 4

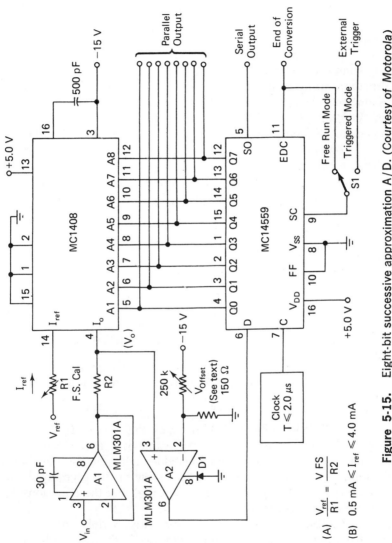

Figure 5-15. Eight-bit successive approximation A/D. (Courtesy of *Motorola*)

(A) $\dfrac{V_{ref}}{R1} = \dfrac{V\,FS}{R2}$

(B) $0.5\ mA \leqslant I_{ref} \leqslant 4.0\ mA$

ICs. As shown, the system operates on +5 and −15 V supplies, requires approximately 200 mW of power, and operates at 2 μs/bit conversion rates.

5–4.1   Theory of Operation

With the exception that a current output D/A is being used, the circuit shown in Figure 5-15 operates exactly as described in Secs. 5–1.3 and 5–3.3.

In operation, the input voltage V_{in} drives an MLM301A op-amp connected as a noninverting, unity-gain buffer. (For further information on op-amps of all types, refer to the author's *Manual for Operational Amplifier Users*. [Reston Publishing Company, Inc., Reston, Virginia, 1976.]) The op-amp translates impedances so that the impedance of the driving source has no effect on the A/D output.

The output of the D/A is a current sink proportional to the reference current I_{ref}, and the digital word on the address lines of the D/A (inputs A1 through A8). The digital word input to the D/A is represented by X.

(1) $I_o = I_{ref} \cdot X$

(2) $I_{ref} = V_{ref}/R1$

The voltage on the output of the D/A, V_o, is a function of V_{in} and the output current of the D/A.

(3) $V_o = V_{in} - R2I_o$

The comparator A2 compares V_o to V offset, which is −1/2 LSB.

If V_o is greater than V offset, the comparator output is a logic 1.

Full-scale voltage (11111111) of the system is 2.56 V. This gives each LSB a value of 10 mV. Any value of full scale can be chosen as long as the input buffer amplifier does not saturate. (In this case, the input voltage must stay about 1 V below the positive supply of the op-amp to keep the op-amp from saturating) and the equations A and B of Figure 5-15 are followed.

Calibration of the system is very easy. Simply put a voltage of full scale, minus ½ LSB, into the input, and adjust the full-scale calibrate potentiometer R1 to make the transition from 11111110 to 11111111 occur at this point. Next, put an input of +½ LSB into the system and adjust the offset adjust potentiometer (V_{offset}, 250 kΩ) to et the 00000000 to 00000001 transition to occur at this point. Since the o adjustments described are somewhat interactive, it may be neces-

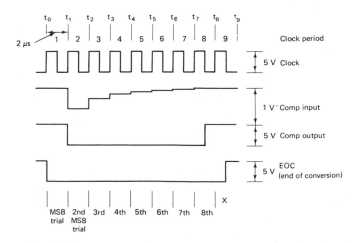

Figure 5-16.   S/A system waveforms showing comparator input and output. *(Courtesy of Motorola)*

sary to go through the procedure more than once.

The system operates at 2 μs/bit giving a total conversion time of $(N + 1) \times 2$ μs. In this case N is 8 so the system has a conversion time of 9×2 or 18 μs. The primary limit of speed in the system is the delay time of the comparator (MLM301A) and the SAR (MC14559). The comparator delay time is about 1 μs, whereas the SAR delay time is about 450 ns. Adding these delays gives about 1.5 μs. When the setting time of the D/A is added in (about 250 ns) the total is 1.75 μs. A safe operating figure is 2 μs. Typical operating waveforms for the system are shown in Figures 5-16 and 5-17.

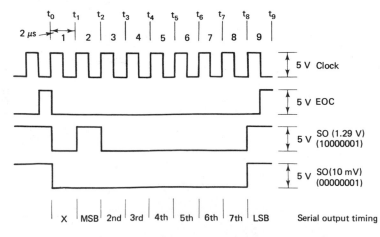

Figure 5-17.   S/A system waveforms showing serial outputs. *(Courtesy of Motorola)*

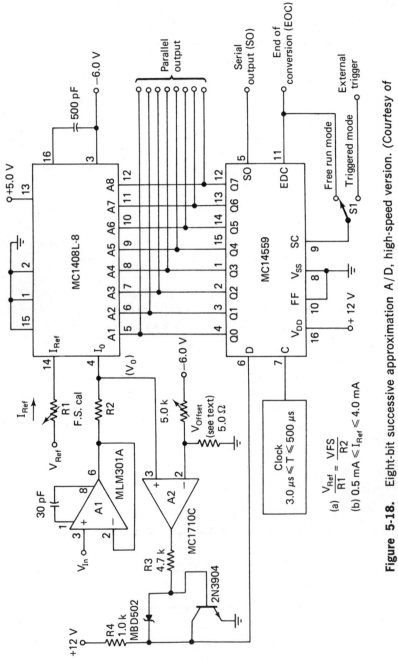

Figure 5-18. Eight-bit successive approximation A/D, high-speed version. (*Courtesy of Motorola*)

5–4.2   High-speed S/A System

Figure 5-18 shows a schematic of another system which is very similar to the one in Figure 5-15, except that the SAR is operating at +12 V, and an MC1710C comparator is used with a one-transistor level translator on its output. At 12 V, V_{DD} on the SAR, the delay time is typically 135 ns. The comparator and level translator have a total delay of about 50 ns. This produces a delay time of 135 ns for the SAR, 50 ns for the comparator, and 250 ns for the D/A, or about 435 ns total per bit. Rounding this off to 500 ns, and using the equation $(N + 1)$ bits times the delay-per-bit, we get 9×500 ns or about 4.5 μs for the system (compared to 18 μs for the Figure 5-15 system).

Cost of the high-speed system in Figure 5-18 is about the same as the lower-speed version, but the high-speed system requires several more components and the addition of one more power supply, as well as requiring about 400 mW of power. Accuracy calibration and operation of the high-speed system are exactly the same as described for the lower-speed version in Sec. 5–4.1.

For clock speeds up to 500 kHz, use the lower-speed system (Figure 5-15). When clock speeds must be increased from 500 kHz to 2 MHz, use the high-speed system (Figure 5-18).

5–4.3   Alternate S/A Systems

Both the lower-speed and higher-speed systems described thus far are eight-bit systems. If desired, a four-, five-, six-, or seven-bit system could be implemented using the same configuration as shown in Figure 5-15 or 5-18. The only change required would be to alter the SAR as shown in Figure 5-19, and to eliminate the unused D/A bits. For example, a six-bit system requires only a six-bit D/A.

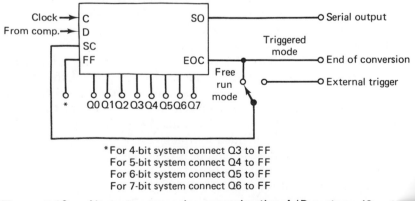

*For 4-bit system connect Q3 to FF
For 5-bit system connect Q4 to FF
For 6-bit system connect Q5 to FF
For 7-bit system connect Q6 to FF

Figure 5-19.   Alternate successive approximation A/D system. (Court
of Motorola)

If a system of more than eight bits is required, the MC14559 may be cascaded with another SAR (the MC14549) to make an SAR of anything from nine to sixteen bits. For twelve bits, the MC14559 must be altered to four bits, and the MC14549 used for the remaining eight bits. The datasheets of these ICs show methods for cascading.

5–4.4   System Accuracy

Any successive approximation A/D system has several sources of error: quanitization error, D/A accuracy, comparator gain, offset voltages of components, and D/A setting time. The following is a summary of these problems.

Quanitization error.   Every A/D has an inherent quanitization error. The error comes from the fact that the *smallest increment* the system can resolve is $\pm\frac{1}{2}$ of a unit. That is, an N-bit A/D has 2^N equal quanitization levels. There are 2^N possible digital words the A/D can give as an output, each representing one of the 2^N discrete words. Since there are no words in between these 2^N words, a voltage that is between two levels must be represented by one or the other, usually the closest one.

For example, the actual value of the input voltage could be exactly halfway between two levels, and the A/D would represent it with one or the other of the two words. In this case, the system would be in error $+\frac{1}{2}$ unit if the upper level were read out, and $-\frac{1}{2}$ unit if the lower level is read out. The maximum error here is thus $\frac{1}{2}$ of a unit. In most A/D systems (including the systems of Figures 5-15 and 5-18), the unit is equal to the LSB. Thus, the system has a built-in quanitization error of $\pm\frac{1}{2}$ LSB.

D/A accuracy.   The D/A converter gives an analog output dependent upon the reference, and the digital word on its input. The accuracy of the D/A depends on how closely the actual analog output of the D/A matches the ideal value described by the reference and the digital word input. In order for a D/A to be N-bit accurate, the analog output must not deviate from the ideal by more than $\pm\frac{1}{2}$ of the LSB. The value of the LSB is $\frac{1}{2^N}$ of reference.

Comparator gain.   The comparator is essentially a linear device and as such has a certain amount of voltage gain. If the voltage gain is anything less than infinity, the differential input voltage required to switch the comparator output from one state to the other, call it V_d, is greater than zero. The value of V_d is simply the logic swing of the comparator, divided by the open loop gain. If the differential input voltage to the comparator is less than V_d, the comparator's output cannot be guaranteed to be a logic 1 or 0. If the threshold of the comparator is halfway through this uncertainty region, then an error of up to $V_d/2$ must be allowed, due to the finite gain of the comparator.

Voltage offset of components. There are three sources of voltage offset error in the system of Figure 5-15. One is the offset voltage of the input buffer amplifier, the second is the comparator offset voltage, and the third is the offset produced by misadjustment of the V_{offset} adjustment potentiometer.

The first two offset voltages mentioned are inherent in the devices used and are fixed; usually they are on the order of about ±2 mV for commercial-grade components. These offsets are fixed, and can be easily compensated for by the V_{offset} potentiometer. Once adjusted, the only concern is changing value due to temperature or age.

D/A settling time. In practice, the settling time of the D/A is usually not considered an error source (although settling time is a definite factor in system speed). However, if the D/A is not given time to settle, an error can result. In D/A specifications, a figure of time is given for the D/A to settle to some specific amount of accuracy. This means that once the digital word on the input of the D/A has been changed, a certain minimum amount of time is required before the analog output voltage of the D/A can be guaranteed to fall within the given accuracy.

When designing any S/A system, the *clock period must be long enough* to give the SAR and comparator time to function, in addition to giving the D/A time to settle to the desired accuracy. Note also that all of these events are sequential. That is, the SAR must give the proper address to the D/A, then the D/A must be allowed to settle, and then time must be allowed for the comparator to react. All this must be allowed to happen within one clock period.

5-4.5   Estimating System Accuracy

Given the sources of error described in Sec. 5-4.4, let us examine the circuit of Figure 5-15, and try to estimate the total system accuracy.

First of all, there is the quantitization uncertainty of $\pm\frac{1}{2}$ LSB. In addition to this we must add the error due to the D/A converter. Usually, a D/A has an error specification of $\pm\frac{1}{2}$ LSB, although it could be better or worse, depending on the D/A.

In the example of Figure 5-15, the MC1408L can be purchased with accuracy specifications of six, seven, or eight bits. Eight-bit accuracy implies error of no more than $\pm\frac{1}{2}$ of one part out of 256, or ±1 part in 512. Thus, for an eight-bit system as shown, the D/A contributes a maximum of $\pm\frac{1}{2}$ LSB. Since the quantitization error an the D/A error are independent, worst case error is simply the sum, or ± 1 LSB.

There is also error contributed by the comparator and input buffer. The offset voltages of these devices can be zeroed by the V_{offset} potentiometer, therefore only the changes due to temperature and aging need be added. Typical offset voltage drifts of these components are about 5 $\mu V/°C$. Except for very wide temperature changes, these drifts may be neglected.

The error due to the finite voltage gain of the comparator is also negligible for standard components. The MLM301A has a typical voltage gain of 200,000. For a 5 V logic swing, the uncertainty region is about 25 μV. With an LSB magnitude of 10 mV (full scale of 2.56 V), and the ± 1 LSB error due to the A/D quantization and the D/A error, the error due to the comparator is virtually zero.

5–4.6   Advantages of the S/A System

The successive approximation system of A/D conversion offers several advantages. Let us see how some of these tie into design of current MSI systems.

Constant conversion.   When the S/A system is used in multiplexing operation (that is, when the A/D must convert multiple inputs on a time-sharing basis) the constant conversion time characteristic is very desirable. Some other A/D system conversion times are dependent upon the value of the input signal. This is undesirable in multiplexing because the worst case (the longest) conversion time must be allowed for each input. Since S/A gives constant conversion independent of input voltage, optimum use may be made of the system's speed.

Serial and parallel outputs.   In systems where the A/D output is to be sent to another location (such as in communications, computer terminals, and so on) the serial output of the S/A system has a natural advantage. Unless the designer desires to run multiple data lines, one for each bit of the A/D, the output of a parallel-only A/D must be changed from parallel to serial before the information can be sent to a remote location. This is not required for the S/A system.

Speed and accuracy versus cost.   The S/A system gives a very high-speed-accuracy product. For example, using the eight-bit S/A system of Figure 5-15, A/D conversion can easily be accomplished in less than 5 μS, at a total cost of less than $20. When these same parameters are considered for other types of A/Ds, such as tracking, parallel, parallel ripple, and so on, the S/A speed-accuracy product for a given system cost is considerably less.

6

Arithmetic Units

This chapter is devoted to the various arithmetic units found in current logic design. In this chapter we shall concentrate on explanations of basic principles and techniques used for addition, subtraction, multiplication, and division in logic circuitry. A logic circuit that can perform these functions can solve almost any problem since most mathematical operations are based on these four functions. The chapter also includes specific design examples of state-of-the-art arithmetic arrays.

6–1 ADDER CIRCUITS

Adder circuits are generally used in logic design to add binary numbers. There are two basic types of adders: the half-adder and the full-adder.

The *half-adder* is a device for adding two binary bits. The circuit, symbol, and truth table for the half-adder is shown in Figure 6-1. As shown, the half-adder has two inputs, for the two bits to be added; and two outputs, one for the sum bit S and the other for the

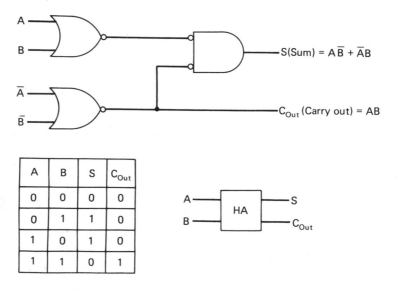

Figure 6-1.   Basic half-adder.

carry-out bit C_{out}. The letters HA are often used to denote a half-adder. The half-adder sum is called the EXCLUSIVE OR function, or sometimes the "ring-sum." The $\oplus$ symbol represents the EXCLUSIVE OR or ring-sum operation. Note that the half-adder of Figure 6-1 requires complemented inputs. In a typical logic adder system, the HA receives its inputs from an FF (or register) which has complemented outputs.

A *full-adder* is a device for adding three binary bits. The circuit, symbol, and truth table for the full-adder are shown in Figure 6-2. As shown, the full-adder has three inputs (A, B, C_I) where A and B represent the two bits to be added, and C_I represents the "carry-in" for that stage, which is the "carry-out" from the next lower-order addition. The two outputs from the full-adder are the sum-bit S, the sum of the three input bits, and the "carry-out" bit C_{out}. The letters FA are often used to denote the full-adder.

Note that there are two full-adder circuits shown in Figure 6-2. Circuit A requires both true and complemented inputs, while circuit B requires only true inputs. Both circuits are composed of half-adders, gates, and inverters (or gates used as inverters).

The FA function can also be implemented by using NOR gates only. Such a circuit is shown in Figure 6-3. This configuration

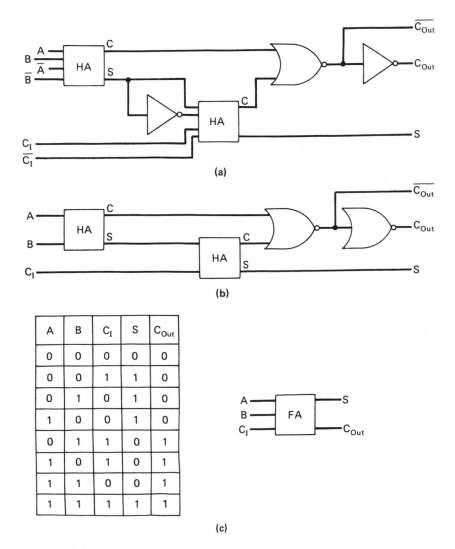

(a)

(b)

A	B	C_I	S	C_{Out}
0	0	0	0	0
0	0	1	1	0
0	1	0	1	0
1	0	0	1	0
0	1	1	0	1
1	0	1	0	1
1	1	0	0	1
1	1	1	1	1

(c)

Figure 6-2.   Basic full-adder.

is based on taking the negated input variables, and the negated S and C_{out}, and using them to determine the sum and carry-out directly. The approach is the equivalent of going from the simplest sum of products (Chapter 1) to the simplest product of sums.

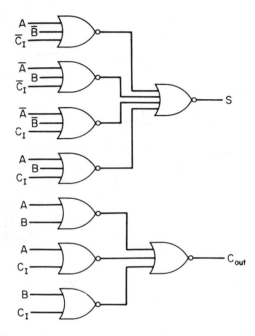

Figure 6-3.   NOR gate full-adder. *(Courtesy of Motorola)*

Another NOR gate implementation of the FA can be accomplished by the direct substitution of the NOR gate equivalent of the HA shown in Figure 6-2 for the sum, and driving the carry from the carry and $\overline{\text{carry}}$ outputs of the first HA. The resulting FA shown in Figure 6-4 can be implemented with an inverter and two-input NOR gates exclusively.

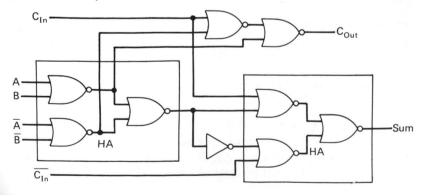

Figure 6-4.   NOR gate full-adder with inverter. *(Courtesy of Motorola)*

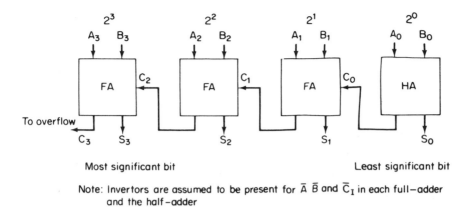

Figure 6-5.   Four-bit parallel ripple-carry adder. *(Courtesy of Motorola)*

In implementing an adder system, the designer may functionally substitute any of the three models for an FA. The FA of Figure 6-2 using HA elements is a direct result of the application of FA equations. This FA uses five logic levels for the sum, and six logic levels for the carry, with the carry output being available. The NOR gate equivalent FA shown in Figure 6-4 uses the same number of logic levels for the sum, but the carry requires only three logic levels. The FA of Figure 6-3 has the advantage of the higher speed due to the low number of logic levels (only two logic levels for sum and carry).

Binary numbers can be added in *parallel form* or in *serial form*. The following paragraphs describe typical adder systems.

6–1.1   Parallel Adder

The parallel adder is generally simpler to implement than the serial adder. Parallel addition is an *asynchronous* operation, independent of a clock signal. A four-bit parallel adder is shown in Figure 6-5. Here, two four-bit binary numbers (A and B) are applied to the inputs simultaneously. A preliminary sum is formed simultaneously at all positions. However, when a carry is generated, it appears at the input of the next higher-order state. Thus, in a worst case condition, a carry-bit could "ripple" the full length of the adder. Hence, the parallel adder is often referred to as a "ripple-carry adder." As a rule of thumb the maximum time required for a ripple to settle out of an N-bit parallel adder is N times the carry propagation time per bit.

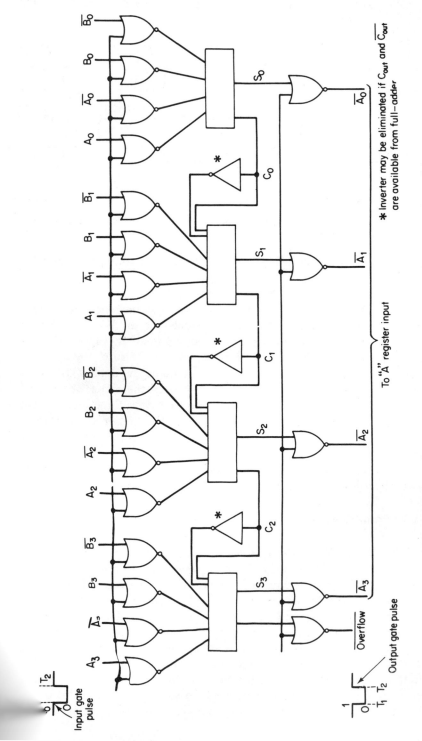

Figure 6-6. Gated four-bit ripple-carry adder. *(Courtesy of Motorola)*

* Inverter may be eliminated if C_{out} and $\overline{C_{out}}$ are available from full-adder

To "A" register input

A more useful type of parallel adder occurs when the input binary numbers (from a register) are gated into the adder and, at a predetermined time later, the sum number is gated out (to another register).

An example of a gated parallel adder circuit is shown in Figure 6-6. Here, the A register contains the augend number (the number to which another number is to be added) and the B register contains the addend number (the number which is to be added to the augend). At some time T_0, both numbers are "dumped" into the parallel adder and, at some later time T_1, the sum bits are gated into the augend register. Thus, in the time $T_2 - T_0$, the number in the B register has been added to the number in the A register, and the sum has been placed into the A register.

Note that the input gate pulse shown in Figure 6-6 goes to the 0 state at time T_0, and must remain at 0 until time T_2. This pulse width restriction on the input gate pulse is necessary so that the output sum-bits are correct and in a stable state when the output gate pulse goes to a low state at time T_1.

Note that the complements of the inputs and outputs are used instead of the true state of the bits to allow for the inversions involved in gating. Since registers are composed of FFs, both the true and complementary states are generally available.

6–1.2   Serial Adder

Binary serial addition is a time-sequential (synchronous) operation, performed one bit at a time. The two least significant bits (A_0 and B_0) are added first, forming a least significant sum bit (S_0) and a carry bit (C_0). The least significant carry bit (C_0) is then added with the two next higher-order bits (A_1 and B_1) and the sum and carry bits (S_1 and C_1) are formed, and so on. The two numbers to be added are each stored, in binary form, in separate registers. The two numbers are fed serially into an adding network, and the sum bits are fed serially into another register.

The basic serial adder concept is shown in Figure 6-7. Here, the A register contains the augend number and the B register contains the addend. The S register receives the sum number bits serially as they are generated from the full adder. The carry storage element provides temporary storage for C_{out} until C_{out} is used in the next addition, and the next C_{out} is placed in the storage element. The carry storage element is generally a FF, and must be clocked by the same timing pulse that is applied to the registers.

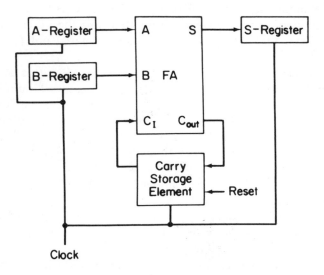

Figure 6-7.   Basic binary serial adder. *(Courtesy of Motorola)*

After the final two bits from A and B are added, the carry storage element contains the final C_{out}, which can represent an overflow-alarm bit, if desired. (Register circuits used in arithmetic arrays often include an overflow-alarm network to indicate that the register is full, and can accept no further inputs.) If the A and B registers are large enough that the most significant bit positions in both registers are zeros before the two numbers are added, no overflow condition will occur.

The A, B, and S registers operate in a shift-right mode. The augend and added numbers are loaded into the A and B registers in either serial or parallel mode.

The S register in Figure 6-7 can be eliminated by feeding the sum bits back into the A register. This concept is shown in Figure 6-8. Under these conditions, the A register is known as an "accumulator register."

In the circuit of Figure 6-8, the number in the B register is added to the number in the A register, and the sum appears back in the A register. The circuit of Figure 6-9 illustrates a four-bit register version of the binary serial adder, with serial input loading. The full adder shown in Figure 6-9 (in block form) can be any of the three FA circuits previously discussed (Figure 6-2, 6-3, or 6-4).

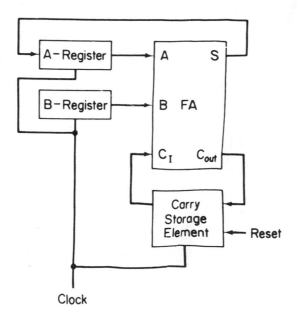

Figure 6-8. Basic binary serial adder with S register eliminated. (Courtesy of Motorola)

In Figure 6-9, the A and B registers consist of four FFs (each), together with the associated gating. For serial data entry, input gating is required on A_3 and B_3 FFs only. The load and store commands could conceivably be applied throughout load and store operations, with actual register timing being provided by the shift command. The same applies to the add command. Thus, the circuit of Figure 6-9 is functionally complete with serial load and store capability. Note that the simultaneous occurrence of "load A" and "add" will result in a meaningless outcome. It is assumed that such simultaneous application of inputs pulses is precluded by system design.

Parallel data entry into serial adder registers. The circuit of Figure 6-10 shows a simple four-bit serial adder with parallel loading capability. This provides a reduced processing time, at the price of additional logic elements. The parallel entry feature requires the capability of setting each FF in the A and B registers to the state of the input data bits. The FFs must have asynchronous inputs (preset and preclear). The gating circuits of Figure 6-10 are simplified since only the sum is shifted serially (as a result of parallel loading and storing).

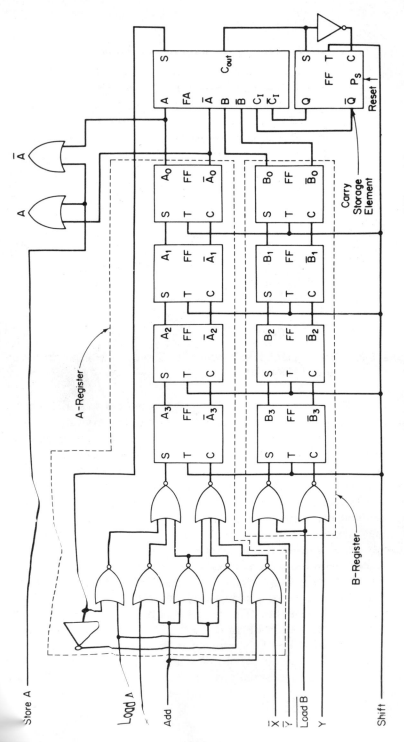

Figure 6-9. Four-bit serial adder with serial loading. (*Courtesy of Motorola*)

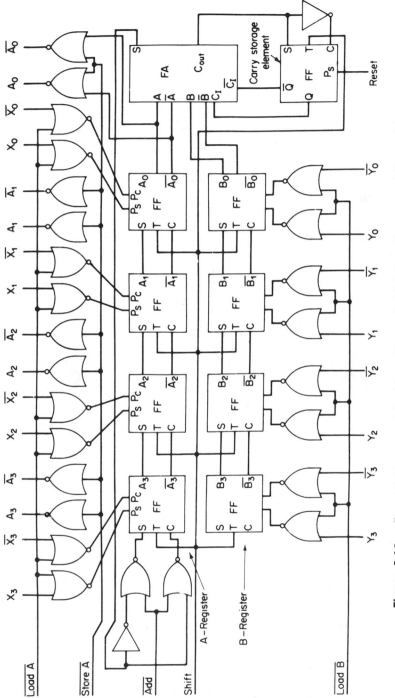

Figure 6-10. Four-bit binary serial adder with parallel loading. (Courtesy of Motorola)

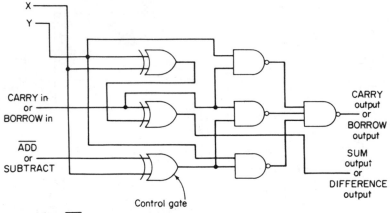

With control ($\overline{\text{ADD}}$/SUBTRACT) input at 0, SUM and CARRY appear as outputs.
With control ($\overline{\text{ADD}}$/SUBTRACT) input at 1, DIFFERENCE and BORROW appear as outputs

Figure 6-11. Universal adder/subtractor. *(Courtesy of Motorola)*

6–2 SUBTRACTOR CIRCUITS

Operation of a binary number subtractor circuit is essentially the same as that for the adder circuit. However, the "carry in" is replaced by a "borrow in," the "carry out" is replaced by a "borrow out," and the "sum" is replaced by a "difference."

The circuit of Figure 6-11 is a universal adder/subtractor in that it will perform either addition or subtraction, depending upon the state of the "control" input. As shown, if the control input is low (0), the sum and carry appear as outputs, and the circuit operates as a full adder. If the control input is high (1), the difference and borrow outputs are used, and the circuit functions as a full subtractor. The circuit of Figure 6-11 uses both EXCLUSIVE OR and NAND gates.

A full-subtractor circuit using only NAND gates is shown in Figure 6-12. Here, input B is subtracted from input A, and the borrow input is also subtracted from input A. When the circuit of Figure 6-12 is used in a system, the borrow output is connected to the borrow input of the next higher position.

6–3 SIGN AND EQUALITY COMPARATORS

Comparators have many uses in arithmetic arrays. Typically, a *sign comparator* is used to determine the sign of numbers (plus or minus) or to compare the polarity of dc voltages. *Equality detectors* compare bits in registers to detect errors. There are many comparator circuits. The following paragraphs describe two typical circuits.

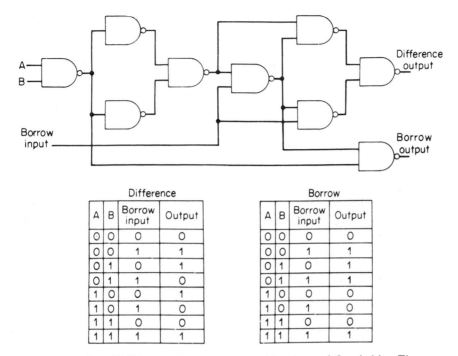

Difference

A	B	Borrow input	Output
0	0	0	0
0	0	1	1
0	1	0	1
0	1	1	0
1	0	0	1
1	0	1	0
1	1	0	0
1	1	1	1

Borrow

A	B	Borrow input	Output
0	0	0	0
0	0	1	1
0	1	0	1
0	1	1	1
1	0	0	0
1	0	1	0
1	1	0	0
1	1	1	1

Figure 6-12. NAND gate full-subtractor. *(Courtesy of Cambridge Thermionic Corporation)*

6–3.1   Sign Comparator

As shown in Figure 6-13, a typical sign comparator consists of two FFs, four AND gates, and two OR gates. The two FFs, each indicating the polarity of an applied voltage, can produce four possible combinations, two for the like polarities, and two for unlike polarities.

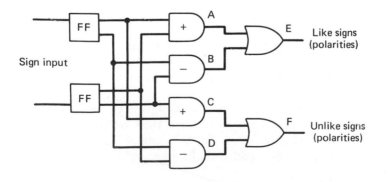

Figure 6-13. Typical sign (or polarity) comparator.

If the FFs have the same output, one of the AND gates (A for positive or B for negative) will have an output, and OR gate E provides an output signal. That is, there is an output indicating like polarities whether both FFs have a positive or negative output. One of the two AND gates provide an output signal, and since an OR gate requires only one input to produce an output, OR gate E will provide a like-polarity output.

If the two polarities are opposites (one positive and one negative), there are only two possible combinations. For either combination, one of two AND gates (C or D) will provide an output to OR gate F, indicating an unlike-polarity output.

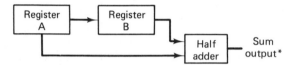

*Digits are alike if there is no sum output
Digits are unlike if there is a sum output

Figure 6-14. Half-adder used as comparison circuit (equality detector).

6-3.2 Equality Comparator

The half-adder can be used as a comparison circuit for detecting errors. If two unlike digits are fed into the half-adder, the output is a sum. If the digits are alike (0 and 0, or 1 and 1), there is no sum output (even though there may be a carry output). Thus, the HA is a detector of equality or inequality.

Figure 6-14 shows how a HA is used to compare the contents of two registers in an arithmetic array. The binary count in register A is transferred to register B, and the half-adder must make sure that both registers contain the same count. The HA compares the contents of the A and B registers, digit by digit. As long as the half-adder has no sum output, the digit in A is the same as its respective digit in B. If there is a sum output, the digit in A is unlike the digit in B.

6-4 ARITHMETIC ARRAYS

Arithmetic arrays (circuits) can be designed to perform all known arithmetic operations, dependent upon their intended use. Basic binary addition and subtraction can be accomplished using the circuits described in Secs. 6-1 and 6-2. In some arithmetic arrays, special circuits are developed to perform multiplication and division. In most arithmetic circuits, the basic adder, shift register, and storage register are combined to perform the four basic functions. One reason for this arrangement is that adders and registers are readily available in IC form. An arithmetic array can be made up entirely of these ICs, with proper interconnection of adders and registers. The following paragraphs describe the combination circuits that make up typical arithmetic arrays. Sec. 6–5 describes a typical arithmetic array using standard IC building blocks. Sec. 6–6 describes a high-speed multiplication circuit using an arithmetic logic unit (ALU) found in IC form.

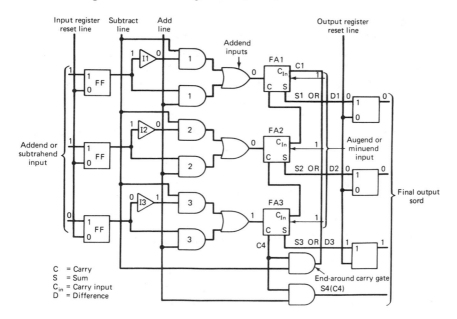

Figure 6-15.   Subtraction by complementing.

6–4.1 Subtraction By Complementing

It is possible to accomplish binary subtraction by complementing. In a practical circuit, this is done by complementing (or inverting) the subtrahend before it enters either of the inputs (augend or addend) of a full adder. An example of this method is shown in Figure 6-15, which is a combined parallel adder and subtractor circuit.

When the circuit is to be used for subtraction, the subtract line is made true by a control panel switch or by a programmed pulse. With the subtract line true, the subtract AND gates pass the inverted subtrahend through the OR gate to the corresponding input of the

FAs. At this time, the end-around carry gate is opened (by the subtract line being true), inserting C_4 (carry four) back into C_1 (carry one). The process of complementing and "adding back" the end-around carry produces the correct difference on $D_1, D_2,$ and D_3.

For example, assume that 011 is to be subtracted from 111. Using normal binary subtraction,

111 minuend

-011 subtrahend

100 difference

Now, using the complement method circuit of Figure 6-15, the 111 minuend is entered into the minuend input of FA 1, FA 2, and FA 3. The 011 subtrahend is entered at the inverters so that 0 is applied to $I3$, 1 is applied to $I2$, and the least significant 1 is applied to $I1$. The inverters produce the complement 100, which appears at the subtract AND gates (1 at gate 3, 0 at gate 2, and least significant 0 at gate 1). In turn, the normal addend input of all three adders receive this complement. This results in the addition of 111 and 100, or

111 minuend

$+100$ complement

1   011 sum before end-around carry

However, the end-around carry (C 4) of 1 is applied to the carry input (C 1) of FA 1. Since the sum-before carry of FA 1 is 1, the end-around carry of 1 changes the sum (now the difference) output of FA 1 to 0. Likewise, the carry output of FA 1 is applied to the carry input of FA 2, changing the sum-before carry of 1 to a difference output of 0. In turn, the carry output of FA 2 is applied to the carry input of FA 3, changing the sum before carry of 0 to a difference output of 1, with no additional carry output from FA 3. This can be shown as follows:

111 minuend

$+100$ complement

1   001 sum-before carry

$+$ 1 end-around carry

100 difference

Note that the registers shown in Figure 6-15 are used to hold the numbers or count at the input and output of the adder/subtractor circuit. Typically, such registers are made up of FFs that can be reset to zero (by the operator or a timing pulse) before and after arithmetic operations.

When the circuit of Figure 6-15 is to be used for addition, the addition line is made true, permitting the add AND gates to pass the noninverted addend through the OR gate to the corresponding addend input of the full adders. The circuit then acts as a conventional adder.

6-4.2   Multiplication By Shifting

It is possible to accomplish binary multiplication by adding and shifting, because multiplication is the process of adding the multiplicand to itself as many times as the multiplier dictates. For example,

```
      1111 multiplicand
  ×  1101 multiplier
      1111 A
      0000 B
      1111 C        partial products
      1111 D
  11000011 product
```

In binary multiplication (where only ones or zeros are used), the partial products are always equal to zero, or to the multiplicand. The final product is obtained by the addition of as many partial products that are equal to the multiplicand, as there are ones in the multiplier. In our example, there are three ones in the multiplier. Thus three partial products (A, C, and D) are equal to the multiplicand and must be added. Of course, the addition must take place in the proper order, since each partial product is shifted by one digit for each digit in the multiplier (where there is a one or zero). In our example, there are four digits in the multiplier. Thus, there must be four shifts.

Figure 6-16 shows a typical shift multiplicand circuit. Note that the circuit consists of three registers (multiplicand, multiplier, and accumulator) all made up of FFs, gates, and inverters; three parallel FAs; and a shift FF. The shift FF provides two bit times for each step of operation. Thus, one bit time is required for adding or not adding, and a second bit time is necessary for the shift. The shift FF alternately sets and resets, thus providing a true signal (S) to the various gates every other bit time. The output of the shift FF controls

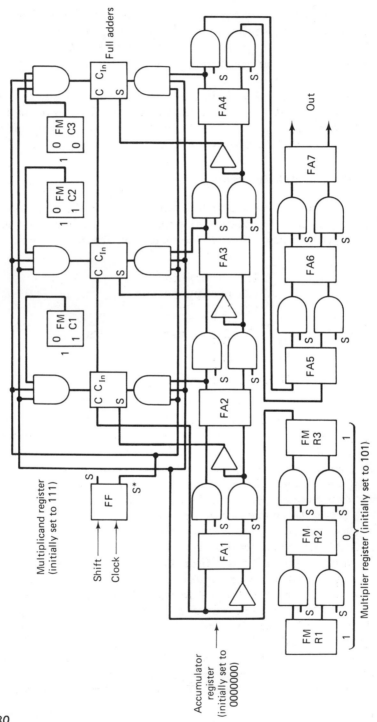

Figure 6-16. Multiplication by shifting.

the action of the "add" or "not add" during one bit time (S^*) and the shift during the next bit time (S).

Assume that the circuit of Figure 6-16 is to multiply binary 101 by 111 ($5 \times 7 = 35$). All the multiplier and multiplicand register FFs are reset to zero; then the appropriate digits (1 or 0) are set into each FF as applicable. (The reset and preset lines are omitted from Figure 6-16 for clarity.)

The multiplication process is started by a clock pulse, which resets the shift FF, making the S output false and the S^* output true. Thus, the first action that can occur (in conjunction with the first clock pulse) is an add or a not add, depending on the state of multiplier FF FMR3.

With the shift FF reset, the S^* output is true. The output of FMR3 is also true, since the multiplier register is initially set to 101. Under these conditions, all the parallel adder AND gates are enabled. The contents of the multiplicand register (FMC1–FMC3) and the accumulator register (FA 2–FA 4) are added. Then the sum and carry outputs of the full-adders are put back into the accumulator register FFs (FA 1–FA 4). Since the accumulator register is initially reset to all zeros, the sum is equal to the multiplicand (111), and this value is the sum answer in the accumulator FFs FA 2–FA 4.

At the next clock pulse, the S output is true, and the S^* output is false. Thus, no more adding can take place, but a shift occurs in both the multiplier and accumulator registers (since the AND gates for these registers are enabled by the S output).

At the next clock pulse, S^* is again true, and S is false. The full-adders again try to add the contents of the multiplicand register (FMC1–FMC3) to the accumulator register (FA 2–FA 4). However, the output of multiplier FF FMR3 is now false, keeping the full-adder gates closed. FF FMR3 is false because it has been shifted one digit by the clock pulse. Originally, FMR1 was at 1, FMR2 at 0, and FMR3 at 1. After the shift, however, both FMR1 and FMR3 are at 0 (false), with FMR2 at 1.

At the next clock pulse (S^* false, S true), another shift occurs. The accumulator register FFs are now at FA 1 = 0, FA 2 = 0, FA 3 = 0, FA 4 = 1, FA 5 = 1, FA 6 = 1, and FA 7 = 0.

At the next clock pulse (S^* true, S false), an add does take place since the MSD of the multiplier (101) has shifted into FMR3. The addition that takes place is:

$$
\begin{array}{ll}
111 & \text{multiplicand register} \\
\underline{0001110} & \text{accumulator register} \\
100011 & \text{final true product}
\end{array}
$$

Although the correct product now exists in the accumulator register, the next shift will take place automatically, placing the product's LSD into FA7.

In a typical arithmetic array (such as one found in simple computers), a *stop multiplication* pulse from the control section will halt action of the multiplication circuit at this time. In some arithmetic arrays, the contents of the accumulator register can be read out by front-panel lights connected to show the state of individual accumulator FFs (the light is on if the FF is in the 1 state). Typically, the next operation in the computer program is to transfer the contents of the accumulator register (now 0100011) to another register, or perhaps into the computer memory, for further processing.

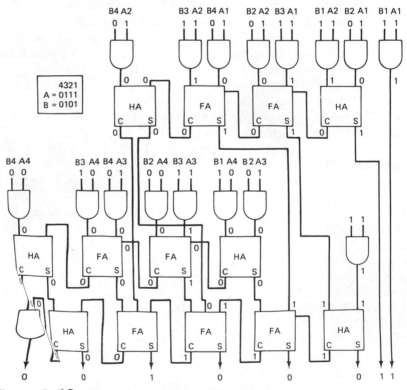

Figure 6-12. Simultaneous multiplication.

6–4.3 Simultaneous Multiplication

In some arithmetic arrays, particularly those used in special-purpose computer operations must be performed at high speeds. Instead of adding (or subtracting), complementing, and shifting two sets of binary digits, each digit is processed simultaneously with all other digits. A separate logic network (adder, subtractor, control gates, and so on) is required for each digit, similar to the circuit of Figure 6-15. Of course, the circuits that provide simultaneous arithmetic operations add many components to the array (thus increasing the size and cost).

Figure 6-17 shows a circuit for simultaneous multiplication of two four-digit binary numbers, producing an eight-digit output. The best way to understand operation of this circuit is to trace through the multiplication of two numbers. The binary numbers used in the explanation of Figure 6-16 (111 × 101) are again used in Figure 6-17. The input and output states of all gates and adders are shown. Note that the final product is 00100011.

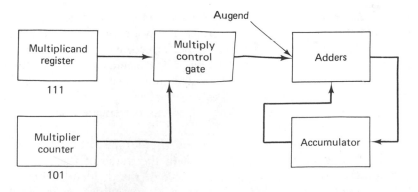

Figure 6-18.   Repeated-addition multiplication.

6–4.4   Repeated-addition Multiplication

Since multiplication is the process of adding the multiplicand to itself as many times as the multiplier dictates, it is possible to perform multiplication by repeated addition. For example, to multiply 111 by 101 (7 × 5), the multiplicand 111 is added to itself 101 (or 5) times.

Some arithmetic arrays use the repeated-addition system of multiplication. However, the trend today is toward the shifting method (for general-purpose use) or simultaneous multiplication (for high-speed, special purpose use).

Figure 6-18 is a block diagram of a repeated-addition multiplier circuit where the value of the multiplicand is put into the multiplicand register, and the value of the multiplier is put into the multiplier counter. The counter counts down (from the multiplier value) to 0. For example, if the multiplier is 101, the counter registers 101 before the first count, 100 after the first count, and 011 after the second count, and so on, down to 000. At count zero, the multiply control gate is closed, halting the repeated-addition process.

As long as the gate is open, the value of the multiplicand register is added to the existing value in the accumulator register. Note that the accumulator must have as many FFs as the sum of the multiplier and multiplicand. In our example (101×111) there is a total of six digits. Thus, the accumulator must have at least six digits. The multiplicand and accumulator outputs make up the augend and addend inputs to the adder circuit. The accumulator starts at 0 and holds the answer from the last addition. Thus, the accumulator value increases as the repeated addition continues.

6–4.5   Division Circuits

As with multiplication, a separate circuit is sometimes used in arithmetic arrays to perform division only. Such circuits are used where arithmetic operation must be performed at high speeds, without regard to increased size and cost. *High-speed circuits for simultaneous division* are similar to those for multiplication (Figure 6-17), except that subtractors and gates are used (instead of adders and gates) since division is essentially a process of repeated subtraction, or subtraction and shifting.

The basic repeated-subtraction divider circuit is shown in the block diagram of Figure 6-19, where the value of the divisor is put into the divisor register, and the value of the dividend is put into the accumulator. The divisor and accumulator output make up the subtrahend and minuend inputs to the subtractor circuit. The counter starts at zero and holds the partial quotients until the last subtraction takes place.

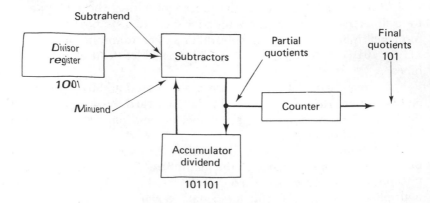

Figure 6-19. Repeated-subtraction divider.

As each subtraction occurs, the value in the accumulator becomes smaller, while the count in the counter becomes larger, finally producing the correct quotient. One problem with such a divider is that it does not stop when the accumulator reaches 0 (or some remainder). Instead, the subtractor trys to continue subtracting beyond the point of 0 remainder. When this occurs, the value in the accumulator register becomes a negative quantity. In a practical circuit, this problem is overcome with a sign comparator (Sec. 6–3.1). When the sign comparator notes a negative quantity in the accumulator, the divide process is stopped and the overage is added back to the accumulator total.

The basic subtraction-shifting divider is shown in the block diagram of Figure 6-20, where the dividend is put into the accumulator, and the divisor is put into the divisor register. The quotient register holds the quotients produced by repeated subtraction.

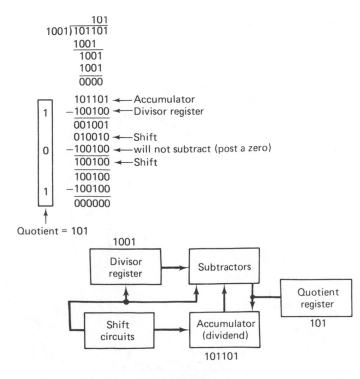

Figure 6-20.   Subtraction-shifting divider.

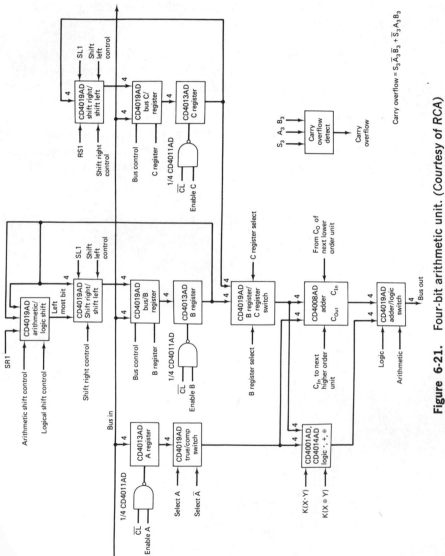

Figure 6-21. Four-bit arithmetic unit. *(Courtesy of RCA)*

For each step (or timing pulse) a shift occurs in the accumulator and quotient register. Thus, a 0 or a 1 is posted in the quotient register for each step. A 0 is posted when a subtraction can not be performed, and a 1 is posted when a subtraction can be performed. Note that the divisor is fed into the subtractor in a manner corresponding to the digits of highest order of the dividend. The ones and zeros posted in the quotient register become the final quotient, as shown by the sample division problem of Figure 6-20.

6–5   TYPICAL ARITHMETIC ARRAY DESIGN

This section describes design of an arithmetic array using RCA COS/MOS building blocks. The arithmetic array or unit is capable of adding, subtracting, multiplying, and dividing, using the basic principles described in Sec. 6–4. The unit is also able to perform the logic of OR, AND, and EXCLUSIVE OR of two four-bit words. Three four-bit registers are provided that permit either of two words to perform a desired operation with a third word. The system uses standard COS/MOS devices which include registers, AND-OR select gates, a full-adder, as well as NOR and NAND gates.

Figure 6-21 is a block diagram of the four-bit arithmetic unit. Figure 6-22 shows the required IC package count and the function performed by each unit. The following is a brief description of each IC device used. Full details, including electrical characteristics, are given in the related RCA Data Bulletin.

Function	Packages				
	CD4008 CD4008A	CD4019 CD4019A	CD4013 CD4013A	CD4011 CD4011A	CD4001 CD4001A
A register (includes A, Ā select capability and buffered outputs)		1	2		
B register (includes shift capability and buffered outputs for B and B̄)		3	2		
C register (includes shift capability and buffered outputs)		2	2		
Select B or C		1			
Add	1				
Perform logic		1			5
Select logic or addition		1			
Overflow detector					1
Clock inhibit (enable)				1	
TOTALS	1	9	6	1	6

Figure 6-22.   Package count for arithmetic unit. *(Courtesy of RCA)*

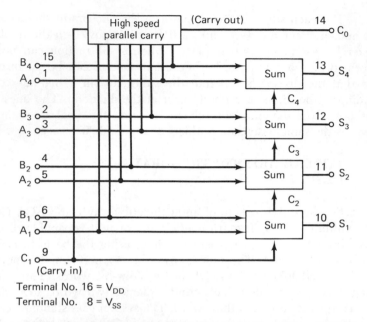

Figure 6-23. Four-bit full-adder, logic diagram. *(Courtesy of RCA)*

6–5.1 Four-bit Full-adder CD4008 or CD4008A

Figure 6-23 is the logic diagram of the four-bit FA, which consists of four FA stages with fast look-ahead carry provision from stage to stage. Circuits are included to provide a fast parallel-carry-out bit, which permits high-speed operation in arithmetic units that use several FAs.

The FA inputs include the four sets of bits to be added (A_1 to A_4 and B_1 to B_4), and the carry-in bit from a previous section. FA outputs include the four sum bits, S_1 to S_4, and the high-speed parallel-carry-out, which may be used as the input to a succeeding FA section.

6–5.2 Quad AND-OR Select Gate CD4019 or CD4019A

Figure 6-24 is the logic diagram of the select gate, which consists of four AND-OR select gate configurations. Each configuration consists of two two-input AND gates driving a single two-input OR gate. Selection is performed by control bits K_a and K_b. In addition to the selection of either channel A or channel B information, the control bits can be applied in combination to accomplish a third selection of data.

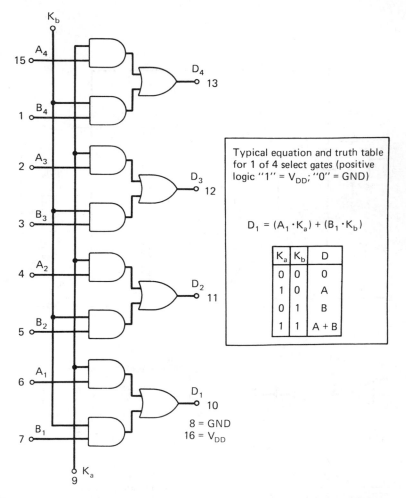

Typical equation and truth table for 1 of 4 select gates (positive logic "1" = V_{DD}; "0" = GND)

$$D_1 = (A_1 \cdot K_a) + (B_1 \cdot K_b)$$

K_a	K_b	D
0	0	0
1	0	A
0	1	B
1	1	A + B

Figure 6-24.   CD4019 or CD4019A, logic diagram. *(Courtesy of RCA)*

6–5.3   Dual *D*-type FF CD4013 or CD4013A

Figure 6-25 shows the logic diagram for one FF section. Each FF section consists of a *D*-type FF, with both (identical) sections on the same IC package. Each FF has independent DATA, RESET, SET, and CLOCK inputs, and complementary buffered outputs. The FFs can be used in shift register applications and in counter and toggle type FF applications, by connection of the $\overline{Q}$ output back to the DATA input. The logic level present at the *D* input is transferred to the $\overline{Q}$ output during the positive-going transistion of the clock pulse.

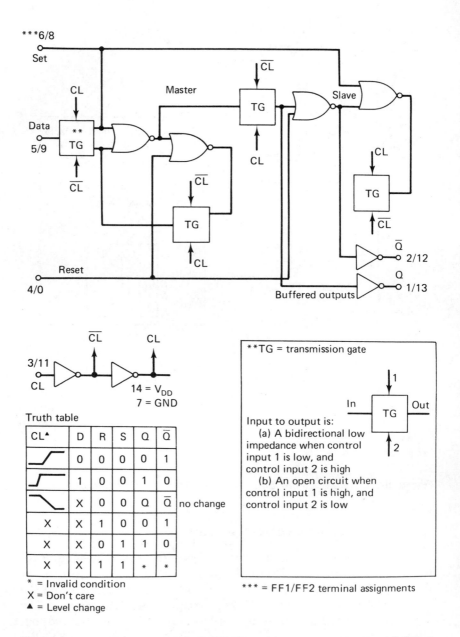

Figure 6-25. Logic diagram and truth table for one of two identical CD4013 or CD4013A FFs. *(Courtesy of RCA)*

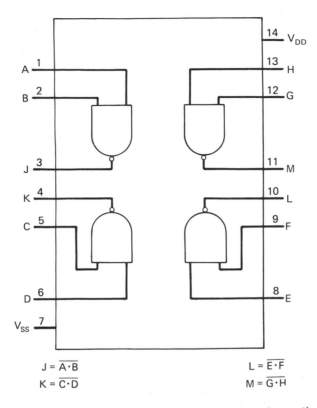

$$J = \overline{A \cdot B} \qquad\qquad L = \overline{E \cdot F}$$
$$K = \overline{C \cdot D} \qquad\qquad M = \overline{G \cdot H}$$

Figure 6-26. CD4011 or CD4011A logic diagram and equations. *(Courtesy of RCA)*

Resetting or setting is accomplished by application of a high logic level to the reset line or set line, respectively.

6–5.4   Quad Two-input NAND Gates CD4011 or CD4011A

Figure 6-26 is the logic diagram for the NAND gates. Four identical, independent two-input positive-logic NAND gates are included in the IC package.

6–5.5   Quad Two-input NOR Gates CD4001 or CD4001A

Figure 6-27 is the logic diagram for the NOR gates. Four identical, independent two-input positive-logic NOR gates are included in the IC package.

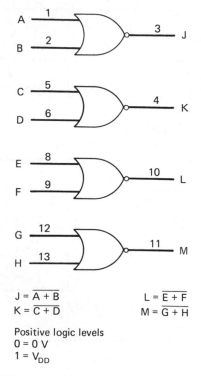

$$J = \overline{A + B}$$
$$K = \overline{C + D}$$

$$L = \overline{E + F}$$
$$M = \overline{G + H}$$

Positive logic levels
$0 = 0$ V
$1 = V_{DD}$

Figure 6-27.   CD4001 or CD4001A logic diagram and equations. *(Courtesy of RCA)*

6–5.6   Arithmetic Unit Operation

The A register uses a CD4019 or CD4019A quad AND-OR select gate to present either the true or the complemented data to the logic circuits and adder. Thus, the data from the A register can be either added or subtracted from the B or C registers.

The CD4019 or CD4019A at the input of the B register has two control lines which permit data to be shifted right or left, or which permit new data to be accepted from the bus lines. Figure 6-28 shows the interconnection diagram of two CD4019s or CD4019As that perform the shift-right, shift-left and arithmetic/logic shift functions (for shift multiplication and division).

The arithmetic shift mode is used only on the highest order bits and insures that the sign bit does not change during the shifting. On the lowest order bits the left shift input for the B register (BRSI) is tied to the B_0 output of the next higher set of bits.

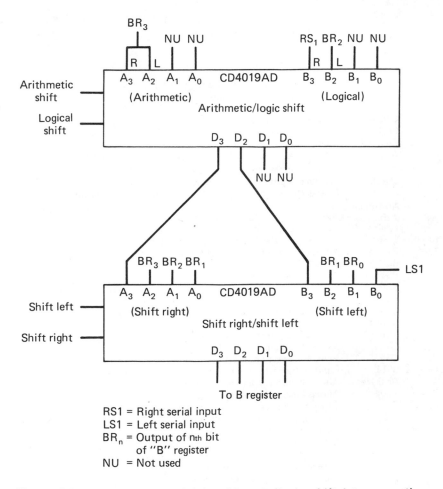

Figure 6-28. Shift-right/shift-left arithmetic/logic shift interconnection diagram. *(Courtesy of RCA)*

The *C* register is identical to the *B* register but does not provide for arithmetic mode shifting. Separate shift-right and shift-left controls are provided for the *B* and *C* registers. (BSL and BSR is provided for the *B* register, and CSL and CSR for the *C* register. See Figure 6-29 for functions and symbols.) Another CD4019 or CD4019A is used to select either the *B* or the *C* register information for the logic and arithmetic operations.

Symbols	Functions
V_{DD}	Input for positive power supply
GND	Ground
$\overline{CL}$	Inverted clock
In_0	Lowest order bus input
In_1	Second lowest order bus input
In_2	Third lowest order bus input
In_3	Highest order bus input
Enable A	Enables the A register
Enable B	Enables the B register
Enable C	Enables the C register
A	Selects the true of the A register data for half-output function
$\overline{A}$	Selects complement of A register data for half-output function
SEL B	Selects B register data for output function with A register data
SEL C	Selects C register data for output function with A register data
Arithmetic shift	Allows B register to shift in the arithmetic mode
Logic shift	Allow C register to shift in the logical mode
B Bus	Allows B register to accept data from bus inputs
B shift	Allows B register to shift right or left
BSR	Allows B register to shift right
BSL	Allows B register to shift left
BSRI	Input to lowest order bit of the B register when shifting right
BSLI	Input to highest order bit of B register when shifting left
C Bus	Allows C register to accept data from bus inputs
C shift	Allows C register to shift right or left
CSR	Allows C register to shift right
CSL	Allows C register to shift left
CSRI	Input to lowest order bit of C register when shifting right
CSLI	Input to highest order bit of C register when shifting left
Arithmetic	Allows logical sums to appear at S outputs
Lodgic	Allows logic functions to appear at S outputs
C_{In}	Carry in to lowest order bit of adder
C_{Out}	Carry out from highest order bit of adder
Overflow	Indicates if addition exceeds limit of adder
$K(\cdot)$	Generates logical AND of logic circuits
$K(\oplus)$	Generates logical EXCLUSIVE-OR of logic circuits
S_0	Lowest order sum output
S_1	Second lowest order sum output
S_2	Third lowest order sum output
S_3	Highest order sum output (sign bit)

Figure 6-29. Logic functions and symbols used in the arithmetic unit. *(Courtesy of RCA)*

6–5.7 Logic Selection

The logic block diagram for the AND/OR/EXCLUSIVE OR logic selector is shown in Figure 6-30. One and one-quarter CD4001s or CD4001As and one-quarter CD4019/CD4019A are used per bit, (or five CD4001 or CD4001A and the CD4019 or CD4019A per four-bit word). The control inputs are labeled $K(.)$ and $K(\oplus)$ (see Figure 6-29). An example of logic selection is as follows:

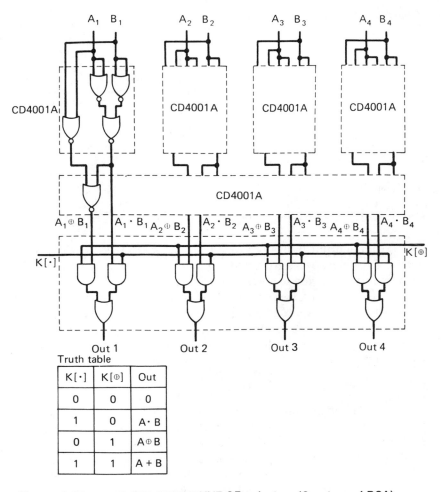

Figure 6-30.   AND/OR/EXCLUSIVE-OR selector. *(Courtesy of RCA)*

Assume that B_1 is an output from the B register and A_1 is a true output from the A register via the TRUE/COMP switch (Figure 6-21). When $K(.)$ is high and $K(\oplus)$ is low, the logic generated is AB. When $K(.)$ is low and $K(\oplus)$ is high, the logic generated is $A \oplus B = A\overline{B} + \overline{A}B$. When $K(.)$ and $K(\oplus)$ are both high, the logic is $AB + (A \oplus B) = AB + A\overline{B} + \overline{A}B = A + B$.

The adder is a single CD4008 or CD4008A. The carry input of this adder is tied to the carry output on the next-lower-order CD4008 or CD4008A. The carry input of the lowest-order bits will be a 0 for addition, and a 1 for the complement subtraction.

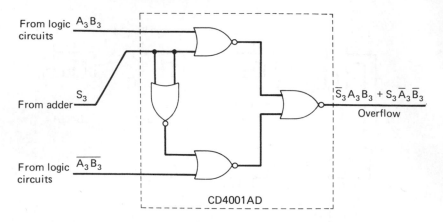

From logic $A_3 B_3$ circuits

From adder S_3

From logic $\overline{A_3 B_3}$ circuits

$\overline{S_3} A_3 B_3 + S_3 \overline{A_3} \overline{B_3}$
Overflow

CD4001AD

Figure 6-31. Overflow logic circuit. *(Courtesy of RCA)*

The output buffer is also a CD4019 or CD4019A. In this example, the device is used to select either the arithmetic or the logic outputs and to provide more output drive.

The output overflow circuit (Figures 6-12 and 6-31 will go high if the adder result exceeds the total bit capacity of the arithmetic unit. This overflow occurs only when two numbers that have the same sign bit are added and, the result is a sum which has the opposite sign.

6–5.8 Larger Arithmetic Units

A larger arithmetic unit of any desired word length can be made by cascading additional circuits. The interconnection of eight systems to form a thirty-two-bit arithmetic unit is shown in Figure 6-32. The three inhibit signals, common control signals, and inverted clock signal (which are common to all eight subsystems) are not shown. The functions and symbols of Figure 6-29 also apply to the circuits of Figure 6-32.

6–5.9 Manufacturer's Performance Data

In a thirty-two-bit arithmetic unit constructed in the laboratory by the manufacturer, the delay time (under worst-case logic conditions) for the inverted clock to be inverted, for data to be written into the register, and for that data to get to the adder and generate a carry-out from a four-bit system was found to be 782 ns. The delay time (worst case) for a carry-in to generate a carry-out was found to be 87 ns. The delay time (again worst case) for a carry-in to generate a sum at the adder, and for this sum to appear at the outputs, was found to be 623 ns. These numbers result in addition time of 1927 or $792 + (6 \times 87) + 623$ ns for two thirty-two-bit words and 1579, or $782 + (2 \times 87) + 623$ ns for two sixteen-bit words.

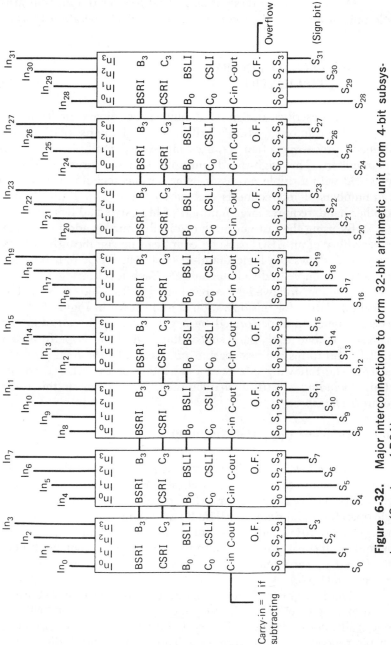

Figure 6-32. Major interconnections to form 32-bit arithmetic unit from 4-bit subsystems. *(Courtesy of RCA)*

297

Maximum power dissipation is 2350 μW for the four-bit arithmetic unit, using a 10-V power supply at $+25°$C operating temperature.

6–6   HIGH-SPEED BINARY MULTIPLICATION

This section describes design of a high-speed binary multiplication system using Motorola MECL arithmetic logic units (ALUs). Fast arithmetic processing often requires high-speed multiplication of binary numbers. This high-speed multiplication has usually been implemented with a ripple processor using various techniques (requiring large numbers of interconnects and parts). Parallel multiplication, on the other hand, requires large numbers of high-speed adders.

By using the fast add time (typically four-bit add in 7 ns) and the versatility of an ALU, package count can be significantly reduced. Figure 6-33 shows the logic diagram of a four-bit-by-four-bit multiplier using a simple add-shift technique. Each level of logic performs an add function, followed by a hardwired shift. The equivalent numerical tabulation for this operation is also shown in Figure 6-33. The first add operation is performed by a gate package because no carry can result at that level. Each succeeding logic level is performed by an ALU, and forms the next partial product, until the final product is reached. The following is a brief description of the ALU used. Full details, including electrical characteristics, are given in the related Motorola Datasheet.

6–6.1   Arithmetic Logic Unit MC10180

Figure 6-34 shows the negative logic diagram of the MC10181. Figure 6-35 shows the function select table for both positive and negative logic. The MC10181 is an ALU capable of performing sixteen logic operations and sixteen arithmetic operations on two four-bit operands. The MC10181 has inputs for a four-bit operand A, a four-bit operand B, and carry-in. The outputs provide for a four-bit function out $(F_0 - F_3)$, carry-out $(C_{n + 4})$, group carry propagate (P_G), and carry generate (G_G).

The logic or arithmetic operation of the ALU is determined by internal logic controlled by the mode (M) input. In the arithmetic mode, full internal lookahead carry is incorporated to determine the four-bit function-out and carry-out. The mode control disables the

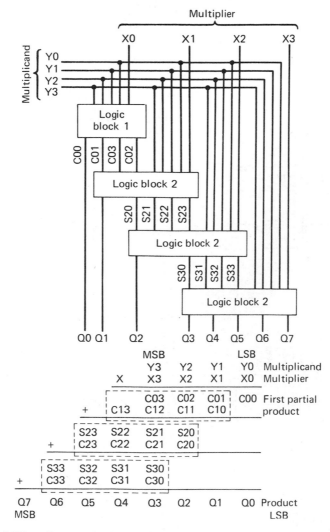

Figure 6-33.   Four-by-four-bit ripple technique for high-speed multiplication. *(Courtesy of Motorola)*

internal carry circuitry for logic mode operation, and carry is ignored. In the logic mode, the function-out is generated on a bit-per-bit basis.

Each mode of operation has sixteen function capability (thirty-two total). The four select lines $(S_0 - S_3)$ determine which function will be performed on operands A and B in either mode of operation.

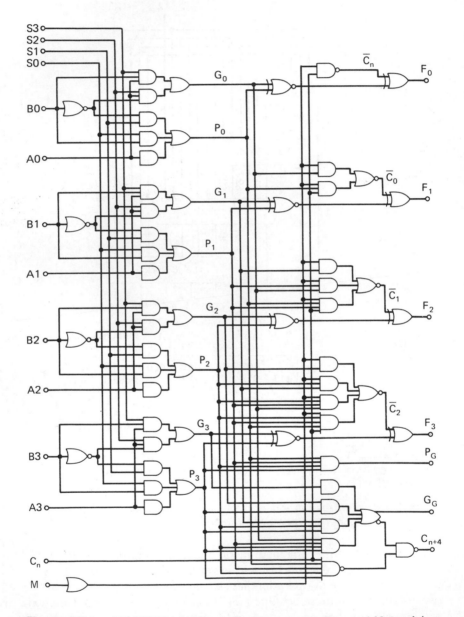

Figure 6-34.   MC10181 negative logic diagram. *(Courtesy of Motorola)*

Positive Logic

Function select $\overline{S3}$ $\overline{S2}$ $\overline{S1}$ $\overline{S0}$				Logic functions M is high C = D.C. F	Arithmetic operation M is low C_n is low F
L	L	L	L	$F = \overline{A}$	$F = A$
L	L	L	H	$F = \overline{A} + \overline{B}$	$F = A$ plus $(A \cdot \overline{B})$
L	L	H	L	$F = \overline{A} + B$	$F = A$ plus $(A \cdot B)$
L	L	H	H	$F = $ logical "1"	$F = A$ times 2
L	H	L	L	$F = \overline{A} \cdot \overline{B}$	$F = (A + B)$ plus 0
L	H	L	H	$F = \overline{B}$	$F = (A + B)$ plus $(A \cdot \overline{B})$
L	H	H	L	$F = A \odot B$	$F = A$ plus B
L	H	H	H	$F = A + \overline{B}$	$F = A$ plus $(A + B)$
H	L	L	L	$F = \overline{A} \cdot B$	$F = (A + \overline{B})$ plus 0
H	L	L	H	$F = A \oplus B$	$F = A$ minus B minus 1
H	L	H	L	$F = B$	$F = (A + \overline{B})$ plus $(A \cdot B)$
H	L	H	H	$F = A + B$	$F = A$ plus $(A + \overline{B})$
H	H	L	L	$F = $ logical "0"	$F = $ minus 1 (two's complement)
H	H	L	H	$F = A \cdot \overline{B}$	$F = (A \cdot \overline{B})$ minus 1
H	H	H	L	$F = A \cdot B$	$F = (A \cdot B)$ minus 1
H	H	H	H	$F = A$	$F = A$ minus 1

Negative Logic

Function select S3 S2 S1 S0				Logic functions M is high F	Arithmetic operation M is low C_n of LSB must be high F
L	L	L	L	$F = \overline{A}$	$F = A$ minus 1
L	L	L	H	$F = \overline{A} + B$	$F = A$ plus $(A + \overline{B})$
L	L	H	L	$F = \overline{A} \cdot B$	$F = A$ plus $(A + B)$
L	L	H	H	$F = $ logical "0"	$F = A$ times 2
L	H	L	L	$F = \overline{A} \cdot \overline{B}$	$F = (A \cdot B)$ minus 1
L	H	L	H	$F = \overline{B}$	$F = (A \cdot B)$ plus $(A + \overline{B})$
L	H	H	L	$F = A \oplus B$	$F = A$ plus B
L	H	H	H	$F = A \cdot \overline{B}$	$F = A$ plus $(A \cdot B)$
H	L	L	L	$F = \overline{A} + B$	$F = (A \cdot \overline{B})$ minus 1
H	L	L	H	$F = A \odot B$	$F = A$ minus B minus 1
H	L	H	L	$F = B$	$F = (A \cdot \overline{B})$ plus $(A + B)$
H	L	H	H	$F = A \cdot B$	$F = (A \cdot \overline{B})$ plus A
H	H	L	L	$F = $ logical "1"	$F = $ minus 1 (two's complement)
H	H	L	H	$F = A + \overline{B}$	$F = (A + \overline{B})$ plus 0
H	H	H	L	$F = A + B$	$F = (A + B)$ plus 0
H	H	H	H	$F = A$	$F = A$ plus 0

*F outputs of ALU are one's complement of function listed below

Figure 6-35. MC10181 function select table. *(Courtesy of Motorola)*

When the mode control is in the more positive (high) logic state, the MC10181 is in the logic mode of operation, and can perform any of sixteen possible logic operations on two variables. These functions are especially useful for applications (such as computer processors) where the processor must execute commands such as generation of all ones or all zeros, "mask" the contents of a register, OR or AND the contents of two registers, or similar logic operations.

The arithmetic mode is also composed of sixteen functions. These functions perform various forms of addition and subtraction on operands A and B (which, in turn, can be used for multiplication and division). The different functions are useful for a variety of operations. As examples, the A times 2 function may be used for an arithmetic shift in a processor, the A minus 1 function does a decrement (countdown) operation, the two's complement (or F minus 1) function may be generated, and the A plus 0 function may either ripple through word A, or increment word A, by adding a carry-in. (Figure 6-35 shows the necessary logic states to select a desired function.)

Among other functions, the MC10181 has the ability to sum (word A) + (word B), or to sum (word A) + (ZERO), depending on the state of the control inputs ($S_0 - S_3$). This feature is to "mask" the multiplicand (word B) and add (ZERO) to the previous partial product of the multiplier, and then add, to form the next partial product.

6–6.2   High-Speed Ripple Multipler Operation

Figure 6-36 shows the diagram of the high-speed ripple multiplier using three MC10181 ALUs and four logic gates (contained in an MC10101). Note that this diagram differs from the block diagram of Figure 6-33 in that the one's complement of the multiplicand and multiplier are used to provide the one's complement of the product. This is necessary because of the basic NOR function definition of the MC10181.

The typical multiply time can be calculated from the worst-case delay path through the multiplier. Multiply time is equal to one gate propagation delay, plus three times the "word-A-to-function-out" delay of the MC10181. The delay is typically

$$2 \text{ ns} + (3 \times 7 \text{ ns}) = 23 \text{ ns.}$$

6–6.3   Expanding the Multiplier

The multiplier may be expanded to accommodate larger numbers of bits. To multiply n bits by n bits, each adder must be n bits

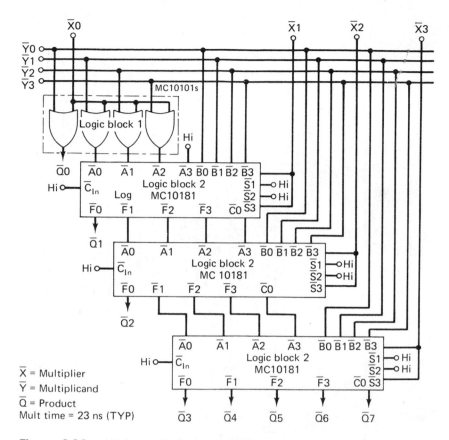

Figure 6-36. High-speed ripple multiplier using three MC10181 arithmetic logic units. *(Courtesy of Motorola)*

wide. Also, there will be n logic levels, with the first logic level implemented with gates, and $n - 1$ secondary logic levels implemented using ALUs.

A simplified block diagram eight-by-eight ripple multiplier is shown in Figure 6-37. Logic block 1 consists of two gate packages, and logic block 2 consists of two ALUs connected as an eight-bit adder. Multiply time for this circuit is typically 56 ns.

The techniques described here could be implemented with TTL, although multiplication times will be longer. As an example, using standard 54/74 TTL, multiplication time would be about 4 times longer for a 4×4 multiplier. Using high speed 74S TTL, multiplication time is still about twice as long as required for ECL.

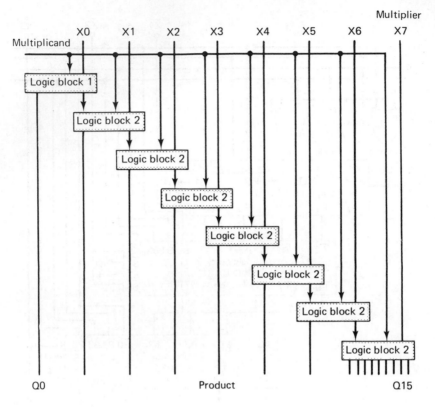

Figure 6-37. Simplified block diagram of 8-by-8 ripple multiplier. *(Courtesy of Motorola)*

Memory Units

This chapter is devoted to the various solid-state IC memory units, or memories, found in current logic design. At one time, the only memories available were those made up of discrete-diode matrices, or core memories, or combinations of both. Today memories are available in MOS, ECL, and TTL. In this chapter, we shall concentrate on such subjects as the basic principles of IC memories, how to program a typical programmable memory, and how to replace sequential logic with memories.

7–1 TYPICAL MOS MEMORIES

Typical MOS memories include read-only memories (ROMs), programmable logic arrays (PLAs), random-access memories (RAMs) and content-addressable memories (CAMs). The following descriptions apply primarily to Texas Instruments' MOS/LSI line of MOS memories.

7–1.1 Read-only Memories

The information stored in a ROM is permanently programmed into the circuit at the time of its manufacture. Once the information is entered, it cannot be changed. However, the information can be

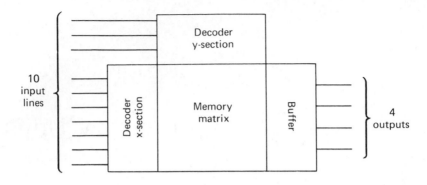

Figure 7-1. Single MOS/LSI read only memory (ROM) sections. *(Courtesy of Texas Instruments)*

read out as often as needed. The most obvious advantages of MOS ROMs are:

Cost: MOS typically one-tenth that of a diode matrix

Size: MOS can put 4096 bits in a twenty-four-pin package (chip size is 120 × 110 mil)

Speed: MOS techniques can provide access time as low as 50 ns.

A single MOS ROM device will be made up of three sections as shown in Figure 7-1. These sections are:

decoder in which the binary address is decoded and *X–Y* pairs of lines going to the memory matrix are enabled (one pair of *X–Y* lines if there is one bit per output word; two pairs of *X–Y* lines if there are two bits per output word, and so on);

memory matrix containing as many MOS transistor locations as there are bits in the memory;

buffer which supplies output levels for the external circuitry.

The physical arrangement of the memory matrix containing MOS transistor locations is shown in Figure 7-2. At the intersection of every *X*-line and *Y*-line, any MOS transistor can be either constructed, or omitted, by growing either a thin-gate oxide, or a thick-gate oxide. The absence of a MOS transistor is interpreted by the

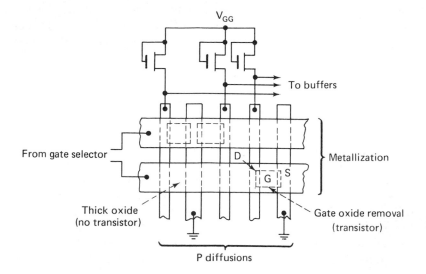

Figure 7-2.   Physical arrangement of MOS/LSI ROM memory matrix. *(Courtesy of Texas Instruments)*

buffer as a logic 0, and the presence of a thin-gate MOS transistor is interpreted as a 1. The programming of the memory (placement of the thin-gate oxide transistors) is performed during the manufacturing process.

Static versus dynamic.   Aside from organization of the ROM, which defines its bit capacity, the most important parameter in most applications is probably access time. *Access time* is defined as the time required for a valid output to appear after a valid input has been applied.

In a static ROM there are no clocks required. If a valid input address is applied to the memory, after the expiration of the required access time, a valid output will appear. The output will remain valid as long as the input address remains unchanged.

In a conventional dynamic ROM, the information is clocked in and clocked out. The output remains valid only for a certain period. Dynamic ROMs are advantageous in implementing synchronous logic. To take advantage of their logic flexibility, and to allow the output to be kept valid as long as desired, Texas Instruments has designed latches on the outputs of all dynamic ROMs. Also, Texas Instruments ROMs do not require clock drivers since these are incorporated on the chip.

Typical applications.   ROMs are well suited to such design applications as: *look-up tables,* where the output is a mathematical function of the input; *code conversion,* where the input is one code (such as EBCDIC) and the output is another code (USACII, for example); *character generator,* where an alphanumeric character is represented by a binary word; and *random logic,* where the ROM is used to perform Boolean algebra (programmed to provide outputs that are Boolean functions of the input variables).

7–1.2   Programmable Logic Arrays

A PLA is essentially a large ROM with a nonexhaustive decode section adapted to the implementation of random logic. The term random logic, as applied here, means a logic circuit that is not strongly structured, as opposed to circuits such as shift registers, ROMs, and so on. When a random logic circuit is implemented by the usual custom MOS process, a large part of the chip area is used for interconnection between basic cells, and is often wasted. The PLA approach minimizes this waste by providing the designer with a programmable circuit already fabricated on a chip (making the best use of space).

In order to program the PLA, only one photomask is modified, and this is accomplished easily and economically. The photomask programs the matrices of the PLA in accordance with logic equations written by the designer. From a user's standpoint, logic design with PLAs is easy. With this approach, the designer writes down the logic equations of each output in terms of external inputs and feedback inputs. Once this is done, the programming of the matrices (modification of the photomask) is handled by a computer. A software package bulletin describes in detail the mechanical aspects of the operation; therefore such details will not be repeated here.

In effect, from a user's standpoint, the PLA bridges the gap between a custom MOS logic array, and an off-the-shelf ROM.

7–1.3   Random-access Memory

MOS random-access memories (RAMs) are ideally suited for those applications requiring high speeds and low-power dissipations. MOS RAMs can replace core memories used for scratch pads or for computers, while offering faster access times and a simpler drive circuitry. RAMs presently on the market are either static or dynamic, as shown in Figure 7-3.

In the static RAM, an MOS FF is used to store the data. Clocks are not needed. The data will stay in storage as long as the power is maintained.

In the dynamic RAM, MOS capacitors are used as data storage elements. Data must be refreshed to ensure integrity.

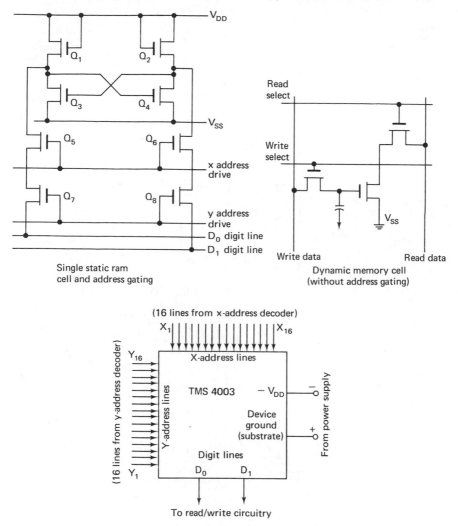

Figure 7-3.   MOS/LSI random-access memory (RAM) arrangement. *(Courtesy of Texas Instruments)*

RAMs can also be defined as:

Fully-decoded memories, where a binary address determines the location in which the Write or Read operation is performed;

Two-dimensional decode, where the address of the word is given by an X–Y select.

The static RAM cell of Figure 7-3 is an example of two-dimensional decode. Assume that the cell is used in a memory with 16 X-address and 16 Y-address lines. Any one of the 256 bits (cells)

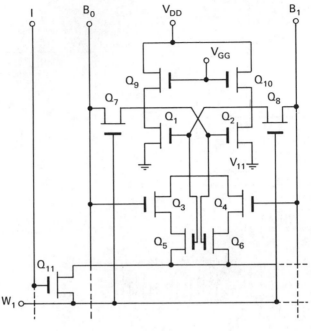

Single cam cell

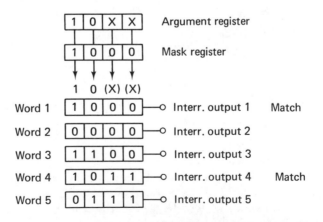

Function of content-addressable memory

Figure 7-4. MOS/LSI content-addressable memory (CAM) arrangement. (*Courtesy of Texas Instruments*)

can be selected by driving one X- and one Y-address line in coincidence. The cell is driven by the X-line through Q_5 and Q_6, and by the Y-line through Q_7 and Q_8.

The state of the FF ($Q_1 - Q_4$) is sensed by observing outputs D_0 and D_1. The same D_0 and D_1 digit lines are used to write information into an addressed bit.

7–1.4 Content-addressable Memory

A content-addressable memory (CAM) is a cell in which all the words contained can be matched simultaneously against an argument word, and outputs can be given wherever a *true match* is obtained. The CAM concept has been around for many years, but cost has made the technique impractical. With improved MOS technology, the CAM has become a working tool for logic design.

Figure 7-4 shows the functions of a CAM, as well as a single CAM cell. Each word has a Write input which can also be used to interrogate that word for a match. Two bit lines run through each column of cells allowing the word data to be written in. The cells may also be used for reading the content of a word, or for masking parts of the argument word where an irrelevant (don't care) state exists.

In the basic CAM cell, transistors Q_1, Q_2, Q_9, and Q_{10} compose a FF for data storage. Transistors Q_7 and Q_8 are selection transistors which, when turned on by the application of a negative voltage to the W (write) line, connect the FF to the B_0 and B_1 bit lines. When the W line is at 0, Q_7 and Q_8 are off, isolating the FF from the bit lines.

Transistors Q_3 through Q_6 and Q_{11} perform the Interrogation logic. For this mode, each W line is grounded through a small resistor and Q_{11} is turned on. Transistors Q_3 through Q_6 compare the state of the memory cell FF with the voltages externally applied to the bit lines. Thus, the word lines are controls for Write and Read operations, but are outputs for the Interrogate operation. The bit lines are inputs for Write and Interrogate, but outputs for Read.

7–2   PROGRAMMING A PROM

This section describes the basic procedure for programming a PROM (or programmable read only memory). The PROM selected for discussion is the Motorola MCM5003/MCM5004. These PROMs are 512 bit (64 × 8) TTL devices. PROMs are similar to ROMs (Sec. 7–1.1), except that a PROM can be easily programmed by the user.

7–2.1   Advantages of a PROM

Other types of ROMs typically require that the user send a truth table to the manufacturer, then wait several weeks for a custom mask and processing of parts. During manufacture, the information is fixed with each new pattern requiring a different custom mask. Using the MCM5003/5004 PROMs, the user can custom program memories without experiencing a mask charge for each pattern, along with associated lead time. Thus, the user can buy one device type and program it in any number of custom patterns, and do this when the devices are needed.

Some of the advantages of the MCM5003/5004 are:

Logic levels compatible with all MDTL and MTTL families (MDTL and MTTL are Motorola versions of DTL and TTL);

An access time of less than 75 ns;

Field programming;

A ninth bit designed into each word enabling the manufacturer to test the device properly without programming any of the normal 64 × 8 bit storage array.

7–2.2   Functional Description of the MCM5003/5004

Figure 7-5 is a block diagram of the 512-bit PROM. Six address lines (A_0 through A_5) are used to select one of sixty-four words. The memory consists of an array of sixty-four words by nine bits, although the customer normally uses only sixty-four words by eight bits. The eight-bit words appear on outputs B_0 through B_7. The ninth bit in the array (B_8) is used during manufacturing final test to determine if the address decoding logic is operating properly. Also, the ninth bit is used to assure that the links have fusing characteristics required by the normal 64 × 8 bit array. Since the ninth bit is physically farthest from the word line drivers, it is also used for worst case ac testing. The test output pin (14) in Figure 7-5 provides the ninth bit.

The PROM has two chip enables (CE1 and CE2) which are ANDED together internally. Both chip enables must be high to enable the selection of one-of-sixty-four words ($W_0 - W_{63}$) in the memory array. If one or both of the chip enables is low, then none of the word lines are selected and all the output bits would be in the high state. The MCM5003L has open collector outputs, whereas the MCM5004L has 2 kΩ pull-up resistors on the outputs.

Truth table

Word number	(1)E	Inputs						Outputs							
		A5	A4	A3	A2	A1	A0	B7	B6	B5	B4	B3	B2	B1	B0
(2) X	0	X	X	X	X	X	X	1	1	1	1	1	1	1	1
0	1	0	0	0	0	0	0	(3) *	*	*	*	*	*	*	*
1	1	0	0	0	0	0	1	*	*	*	*	*	*	*	*
2	1	0	0	0	0	1	0	*	*	*	*	*	*	*	*
—	—	—	—	—	—	—	—	—	—	—	—	—	—	—	—
—	—	—	—	—	—	—	—	—	—	—	—	—	—	—	—
—	—	—	—	—	—	—	—	—	—	—	—	—	—	—	—
63	1	1	1	1	1	1	1	*	*	*	*	*	*	*	*

Notes: (1) $E = (C_{E1}) \cdot (C_{E2})$
(2) X = "Don't care"
(3) * is a "0" or "1", depending on program.

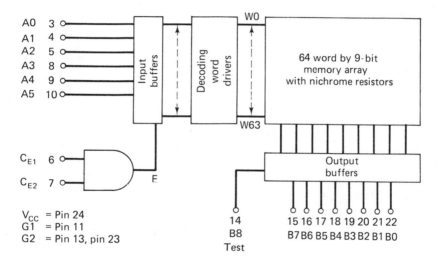

Figure 7-5.   MCM5003/5004 512-bit PROM. (*Courtesy of Motorola*)

The truth table is also shown in Figure 7-5. Note that the word number is the decimal equivalent of the six binary inputs A_0 through A_5; with A_0 being the LSB, and A_5 being the MSB. As shown in the truth table, if the enable lines are high, then the output word (B_0 through B_7) is selected by inputs A_0 through A_5.

The PROM has two isolated grounds, G_1 and G_2. In normal operation, G_1 and G_2 are tied together to ground, and the V_{cc} is connected to +5 V. To program the memory array, the grounds must be separated, G_1 connected to −6 V and G_2 connected to ground. The input buffers and substrate are internally referenced to G_1; therefore, in the programming mode, the input low voltage to the address and chip-enable lines is −6 to −5.2 V, and the input high voltage is −4 to +5 V. These larger voltage swings are made possible by the fact that the input Schottky diodes have a typical breakdown of 30 V.

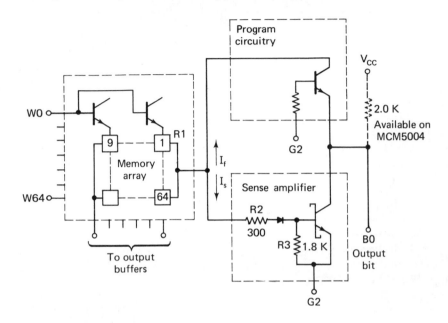

Figure 7-6. Memory array and output circuitry. (*Courtesy of Motorola*)

7–2.3 Memory Array and Programming Circuitry

Figure 7-6 illustrates the memory array and output circuitry of the PROMs. The memory array consists of a thin film nichrome resistor for each memory bit, labeled R1, having a nominal value of 150Ω. Vacuum deposition techniques are used to deposit the nichrome on a planar surface to ensure uniform thickness. Before the device is programmed, all the outputs are in a 0 state. Programming of the memory is done by "opening" the appropriate nichrome resistor elements where a 1 state is desired.

It takes 25 to 33 mA of fuse current (I_f) through R1 for a duration of typically 200 ms to "open" the nichrome link. However, it could take as long as one second to "open" the link in some devices.

The MCM5003/5004 is unique in that it provides additional program circuitry to supply the current gain necessary to insure high reliability when fusing the memory link. To program a one into a particular memory bit, the proper bit-output line is connected to −6 V. This causes the transistor in the program circuitry (Figure 7-6) turn on, allowing the fuse current to flow through the appropriate nichrome resistor R1. The sense amplifier Schottky transistor turns off when programming, although about 3 mA will flow through R3 and the base-collector junction of the Schottky transistor to the bit output.

Current may also flow out of the Schottky clamped output transistor due to breakdown. This is the reason why the total fuse current leaving the output terminal is specified as a maximum of 120 mA. For this reason, if current limiting is used, it should accommodate 120 mA to insure fusing.

In normal operation when a word line is selected, the sense current flows through R1 and R2, turning the Schottky output transistor on, and causing a zero at the bit output. If R1 has been fused, then current will not flow into R2 and the output transistor will be off, causing a one to appear at the bit output. The amount of sense current I_s, flowing through a nonfused link when selected in normal operation, can be calculated as follows:

$$I_s = \frac{V_{WO} - 3\,V_{BE}}{R1 + R2} = \frac{(3.5 - 2.1)\,\text{V}}{(150 + 300)\,\Omega} \approx 3.1\ \text{mA}$$

where V_{WO} appears at the base of the transistor in the memory array, and is nominally 3.5 V due to address decode circuitry.

Thus, the ratio of use current to sense current is about 8 to 1, assuming the minimum fuse current to be about 25 mA. This high ratio improves the reliability of the memory, and minimizes the probably of a memory link becoming fused during normal operation.

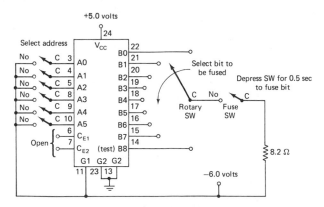

Figure 7-7.   Programming connections for 512-bit PROM. (*Courtesy of Motorola*)

7–2.4   Manual Programming of the 512-bit PROM

Programming the MCM5003/5004 is a simple operation, requiring a small amount of equipment. Figure 7-7 shows the programming connections necessary to program the device. Only two power

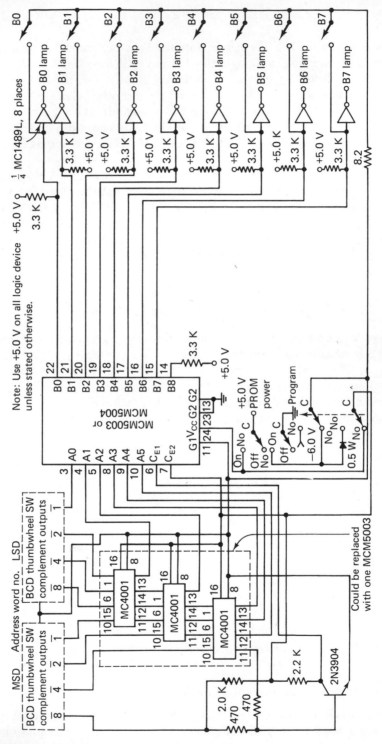

Figure 7-8. Manual 512-bit PROM programmer. *(Courtesy of Motorola)*

supplies, seven single-pole single-throw switches, and one eight-position rotary switch are required. Note that a switched voltage at the output bit is adequate and reliable for fusing. The small series resistance at the output bit is to limit current to 120 mA at -5 V, which appears at point C on the rotary switch. This level is adequate for fusing.

Only one memory link is fused at a time since the power dissipation of the device would be excessive if more than one bit were fused simultaneously. The programming procedures for the connections shown in Figure 7-7 are:

1. Select the address code by connecting the proper address bits to -6 V for a zero, and unconnected for a one. The zero can be from -6 to -5.2 V, whereas the one can be from -4 to $+5$ V, or an open condition (no voltage).

2. Set the rotary switch to the output bit in which a one is desired.

3. Press the fuse switch for approximately one second. This connects -6 V to the output and causes the fuse current to flow.

4. Repeat steps 2 and 3 until all the output bits for that word in which ones are desired have been fused.

5. Select the next address code and repeat steps 2, 3, and 4.

Although the connections in Figure 7-7 are straightforward and simple to implement for programming, some disadvantages are obvious. The programming of the total unit is cumbersome. Insuring that the device has been programmed properly could be difficult. For example, setting the six address switches to the proper binary code for address selection can be cumbersome. This could be overcome to some extent by using two thumbwheel switches with octal coding instead of 6 binary switches. Also, a display is needed to verify that the bit has indeed been fused.

It should be noted that the G1 terminal must be at ground (not at -6 V) when outputs are checked to see if the output bits have been fused properly.

7–2.5   Improved Manual Programming

Figure 7-8 shows the schematic of a simple manual programmer that overcomes some of the objections to a circuit such as Figure 7-7. The MC1489 quad line receivers illustrated in Figure 7-8 are used to show the contents of the output bits by driving 5 V, 20 mA lamps.

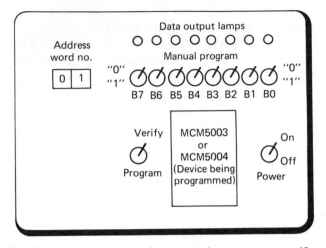

Figure 7-9. Front panel layout for manual programmer. *(Courtesy of Motorola)*

Figure 7-9 shows the panel layout of the manual programmer of Figure 7-8. The address word number is selected using two BCD thumbwheel switches. Three MC4001 TTL standard ROMs are used to convert the BCD code into the code that is required at the address inputs. The transistor is used to inhibit programming of the PROM if an illegal number is selected for the address inputs (such as a number 64 or greater).

Figure 7-10 shows a typical specification form which could be used for specifying the custom program. Place an X in each bit position that is to be a 1. This form can be used to set up the switches when programming the PROM.

The manual program instructions are given in Figure 7-11.

Address word no.		Data outputs							
		B7	B6	B5	B4	B3	B2	B1	B0
0	0								
0	1								
0	2								
0	3								
6	0								
6	1								
6	2								
6	3								

Figure 7-10. PROM specification form. *(Courtesy of Motorola)*

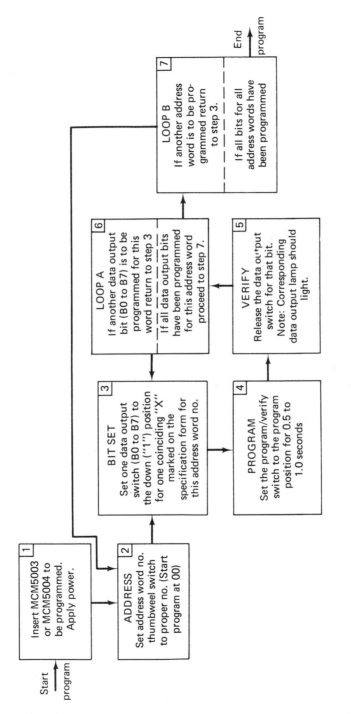

Figure 7-11. Manual programming instructions. *(Courtesy of Motorola)*

Start program

1
Insert MCM5003 or MCM5004 to be programmed. Apply power.

2
ADDRESS
Set address word no. thumbweel switch to proper no. (Start program at 00)

3
BIT SET
Set one data output switch (B0 to B7) to the down ("1") position for one coinciding "X" marked on the specification form for this address word no.

4
PROGRAM
Set the program/verify switch to the program position for 0.5 to 1.0 seconds

5
VERIFY
Release the data output switch for that bit. Note: Corresponding data output lamp should light.

6
LOOP A
If another data output bit (B0 to B7) is to be programmed for this word return to step 3

If all data output bits have been programmed for this address word proceed to step 7.

7
LOOP B
If another address word is to be programmed return to step 3.

If all bits for all address words have been programmed

End program

319

The PROGRAM/VERIFY switch must simultaneously be set along with the proper output switch for each fused bit. Remember, that if one bit is fused *in error*, the PROM *cannot* be used for that particular program. Once programmed, the PROM cannot be changed.

It is also possible to program the PROM automatically, using a circuit design available from the manufacturer (Motorola). However, the circuit is fairly complex, and is best suited for assembly line and/or mass production of logic circuits.

7–3   REPLACING SEQUENTIAL LOGIC WITH ROMs

This section describes the basic procedure for replacing sequential logic with a ROM (or PROM). As discussed in Chapter 1, sequential logic is any design that required feedback loops. The example given here shows how the PROM discussed in Sec. 7–2 (Motorola MCM-5003/5004) can be used to replace a divide-by-16 counter.

Before going into the specific design example, let us discuss how ROMs and PROMs can be converted into logic circuits.

7–3.1   Converting ROMs to Combinational Logic

Conversion of a ROM to a combinational logic design can be accomplished if any of the following are available: logic equations, truth table, or logic drawing.

If logic equations are given, the equations must be converted to a Karnaugh map (Chapter 1) in order to define each output for all combinations of inputs. Then the content of the Karnaugh maps is placed in a truth table to establish the pattern that the ROM must contain. If a logic drawing is provided, the equations for each output can be written, and then converted to a Karnaugh map.

7–3.2   Cost Factors of ROMs Versus Gates

To determine the number of gates that must be replaced before the ROM becomes economical, use the following formula:

$$\frac{\text{number of gates replaced}}{\text{per ROM}} \gtrsim \frac{\$ROM + \text{associated cost/ROM}}{\$GATE + \text{associated cost/gate}}$$

This formula states that the number of gates replaced for each ROM used should be greater than, or equal to, the cost of the

ROM, plus associated cost, divided by the cost of one gate with its associated overhead.

The associated cost per gate must include the cost of the PC card, power supply fans, insertion, inventory, connector, capacitor, checkout time, and number of interconnects affecting the reliability. Typically, a cost of $.50 per gate is realistic.

Assume that the cost of a gate (plus associated costs) is $.50, the cost of an MCM5003/5004 PROM is $15 (3 cents/bit) in quantities, and that the associated cost of the PROM is $2. Plugging these figures into the equation we find that the PROM must replace 34 (or more) gates before is it economical to use the PROM;

$$\frac{\$15 + \$2}{\$.50} = 34$$

A BCD to binary converter using the MCM5003/5004 replaces approximately 54 gates. This would be a savings of 20 gates, or approximately $10. Another advantage of the ROM is that the system size is reduced and reliability should improve as a result of fewer packages.

7–3.3   Converting ROMs to Sequential Logic

Unlike simple combinational circuits, many sequential logic systems require feedback loops, where the outputs are a function of the input signals and the present state of the outputs. There are basically two types of sequential circuit design, synchronous and asynchronous.

Synchronous sequential design requires data inputs that are synchronized with a clock. When using ROMs, there are two ways of designing synchronous sequential circuits, as shown in Figure 7-12. *Clocked storage feedback* or *direct feedback* can be selected.

The *clocked storage technique* shown in Figure 7-12a requires six additional *D*-type FFs, but results in a more efficient use of the ROM bits. As an example, the asynchronous circuit of Figure 7-12a can be used as a divide-by-64 (or less) counter with any sequence of states possible (more than one output can change at a time). The PROM is used to store the next state.

In contrast, the sequential circuit with *direct feedback* as shown in Figure 7-12b can be used as a divide-by-16 (or less) counter with three outputs. In turn, these outputs can be used as decoders, or

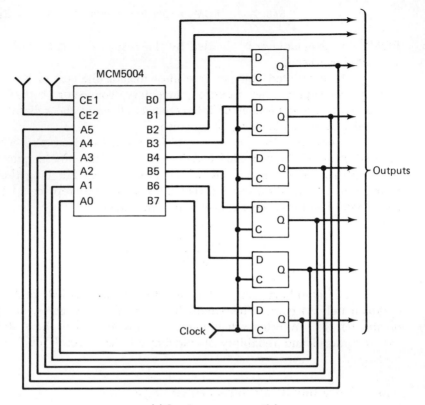

(a) Synchronous sequential
circuit using clocked
storage feedback

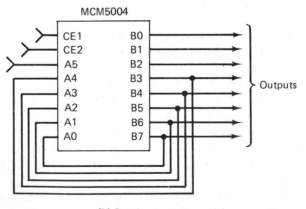

(b) Synchronous sequential
circuit with direct
feedback

Figure 7-12. Synchronous sequential circuit types. *(Courtesy of Motorola)*

to generate a programmable word sequence. Only one feedback output can change at a time so that race conditions will be avoided.

Synchronous data inputs could be added to the address inputs of the PROMs to produce other types of sequential circuits. Also, more than one ROM or PROM can be used where more storage is required.

7–3.4   A Divide-by-16 Counter Using a PROM

This example illustrates the technique for designing a synchronous sequential circuit using a PROM with direct feedback. The divide-by-16 counter is used as a programmable word generator. Since only one output can change for each state, a minimum change, reflected code must be used.

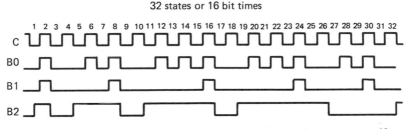

Figure 7-13.   Timing diagram for programmable word sequences. (Courtesy of Motorola)

Timing diagram.   The first step is to draw a timing diagram of the word statement as shown in Figure 7-13. Note that thirty-two states exist when both the low and the high states of the clock are included even though this is a divide-by-16 counter. The programmable word sequence, $B0$, $B1$, and $B2$ were chosen arbitrarily. The $B0$ output generates a number of pulses of 1, 2, 3, 3, and 2, during each sixteen-bit word. The $B1$ output generates a pulse synchronized with the last pulse occurring on the $B0$ output. Output $B2$ generates a pulse width that is 1, 2, 3, and 4 clock periods, in sequence, for each sixteen-bit word.

Primitive flow table.   The next step is to generate a primitive flow table as shown in Figure 7-14. The circles around the numbers represent stable states, and the numbers without circles represent unstable stages. If the circuit is in stable state 1, and the clock changes from 0 to 1, the circuit goes to unstable state 2, and locks up.

Transition map.   Next, a transition map is prepared as shown in Figure 7-15. A merger diagram is not required in this example, since merging does not occur. Each of the thirty-two states in the flow table (Figure 7-14) are represented by a five-variable Karnaugh map in Figure 7-15. The five variables $y1$, $y2$, $y3$, $y4$, and $y5$ represent the

C		Outputs			C		Outputs		
0	1	B2	B1	B0	0	1	B2	B1	B0
(1)	2	1	0	0	(17)	18	0	0	0
3	(2)	1	1	1	19	(18)	0	0	0
(3)	4	0	0	0	(19)	20	1	0	0
5	(4)	0	0	0	21	(20)	1	0	1
(5)	6	1	0	0	(21)	22	1	0	0
7	(6)	1	0	1	23	(22)	1	0	1
(7)	8	1	0	0	(23)	24	1	0	0
9	(8)	1	1	1	25	(24)	1	1	1
(9)	10	0	0	0	(25)	26	1	0	0
11	(10)	0	0	0	27	(26)	1	0	0
(11)	12	1	0	0	(27)	28	0	0	0
13	(12)	1	0	1	29	(28)	0	0	1
(13)	14	1	0	0	(29)	30	0	0	0
15	(14)	1	0	1	31	(30)	0	1	1
(15)	16	1	0	0	(31)	32	0	0	0
17	(16)	1	1	1	1	(32)	0	0	0

Figure 7-14.   Primitive flow table. *(Courtesy of Motorola)*

secondaries of the circuit; these are the counter FF output in this example. State 1 is placed in the square of the map where $y1$ through $y5$ are all at 1. This provides the initialization state when the chip enable is disabled. Since state 1 must go to state 2 when the clock switches from 0 to 1, the state 2 is placed in a square of the map where there is a *one variable change*. If the state 2 was placed in a square of the map where there was more than one variable change, a race condition could occur. Race conditions must be avoided since an indeterminate state could result.

All of the remaining states are placed on the transition map where there is a minimum change reflection code (only a one variable change). For example, state 3 changes only one variable from state 2, state 4 changes only one variable from state 3, and so on.

Flow table with secondary assignments and timing diagram. Next, the secondary assignments from the transition map (Figure 7-15) are transferred to the flow table with secondary assignments (Figure 7-16). The timing diagram for the secondary assignments is then

$y3$	$y4$	$y5$	$y1, y2$ 00	01	11	10
0	0	0	22	27	6	11
0	0	1	21	28	5	12
0	1	1	24	25	8	9
0	1	0	23	26	7	10
1	0	0	19	30	3	14
1	0	1	20	29	4	13
1	1	1	17	32	1	16
1	1	0	18	31	2	15

Figure 7-15.   Transition map. *(Courtesy of Motorola)*

Figure 7-16. Flow table with secondary assignments. (Courtesy of Motorola)

States 1–16

	Secondary variables					C		Outputs		
	y¹	y²	y³	y⁴	y⁵	0	1	B2	B1	B0
	1	1	1	1	1	(1)	2	1	0	0
	1	1	1	1	0	3	(2)	1	1	1
	1	1	1	0	0	(3)	4	0	0	0
	1	1	1	0	1	5	(4)	0	0	0
	1	1	0	0	1	(5)	6	1	0	0
	1	1	0	0	0	7	(6)	1	0	1
	1	1	0	1	0	(7)	8	1	0	0
	1	1	0	1	1	9	(8)	1	1	1
	1	0	0	1	1	(9)	10	0	0	0
	1	0	0	1	0	11	(10)	0	0	0
	1	0	0	0	0	(11)	12	1	0	0
	1	0	0	0	1	13	(12)	1	0	1
	1	0	1	0	1	(13)	14	1	0	0
	1	0	1	0	0	15	(14)	1	0	1
	1	0	1	1	0	(15)	16	1	0	0
	1	0	1	1	1	17	(16)	1	1	1

States 17–32

	y¹	y²	y³	y⁴	y⁵	C 0	C 1	B2	B1	B0
	0	0	1	1	1	(17)	18	0	0	0
	0	0	1	1	0	19	(18)	0	0	0
	0	0	1	0	0	(19)	20	1	0	0
	0	0	1	0	1	21	(20)	1	0	1
	0	0	0	0	1	(21)	22	1	0	0
	0	0	0	0	0	23	(22)	1	0	1
	0	0	0	1	0	(23)	24	1	0	0
	0	0	0	1	1	25	(24)	1	1	1
	0	1	0	1	1	(25)	26	1	0	0
	0	1	0	1	0	27	(26)	0	0	0
	0	1	0	0	0	(27)	28	0	0	0
	0	1	0	0	1	29	(28)	0	0	1
	0	1	1	0	1	(29)	30	0	0	0
	0	1	1	0	0	31	(30)	0	1	1
	0	1	1	1	0	(31)	32	0	0	0
	0	1	1	1	1	1	(32)	0	0	0

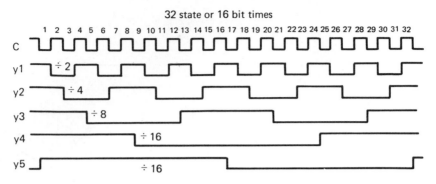

Figure 7-17. Timing diagram for counter outputs. *(Courtesy of Motorola)*

drawn, as shown in Figure 7-17. This shows the counter outputs which are the secondary variables.

Excitation map. Next, the excitation map is prepared as shown in Figure 7-18 to define the output excitation variables $y1$ through $y5$ in terms of the input secondary variables and the clock input. A six-variable Karnaugh is required with the word number located in the upper right-hand corner of each square. Each stable state in the map has been circled, and has the same code as the corresponding code of the secondary state. Each unstable state is represented with the code of the next secondary state that will make it a stable state.

For example, word 59 in Figure 7-18 has a circle around the code 11111 which represents stable state 1 in Figure 7-16. When the clock switches to the 1 state, word 63 is selected. Word 63 has the code 11110 which is an unstable state 2, and represents the next transfer to a stable state, which is word 55 representing stable state 2. The rest of the codes are filled in the squares of the excitation map, using Figure 7-16 as a reference.

Output map. Next, the output map is prepared, as shown in Figure 7-19, in order to define the $B2$, $B1$, and $B0$ outputs in terms of inputs. Each stable state in the map has been circled, and represents the same state as was circled in the excitation map (Figure 7-18). As an example, word 59 in Figure 7-19 has a circle around the code 100 which represents stable state 1 in Figure 7-16. When the clock switches to the 1 state, word 63 is selected which is unstable state 2. Since stable state 2, located at word 55, has an output of 111, a 111 code is placed in the square for word number 63. The last two outputs for this code in word 63 are actually optional conditions, and could be either a 0 or 1. For minimum delay time, the outputs are chosen to switch immediately during the unstable state condition.

C = 1

y1 y2	0 0	0 1	1 1	1 0
	00000 4	01001 5	11000 7	10001 6
	00000 12	01001 13	11000 15	10001 14
	00011 28	01010 29	11011 31	10010 30
	00011 20	01010 21	11011 23	10010 22
	00101 36	01100 37	11101 39	10100 38
	00101 44	01100 45	11101 47	10100 46
	00110 60	01111 61	11110 63	10111 62
	00110 52	01111 53	11110 55	10111 54

Y1 Y2 Y3 Y4 Y5

4
C = 0

32 y3	16 y4	8 y5	2 1 y1 y2 0 0	0 1	1 1	1 0
0	0	0	00010 0	01000 1	11010 3	10000 2
0	0	1	00001 8	01101 9	11001 11	10101 10
0	1	1	01011 24	01011 25	10011 27	10011 26
0	1	0	00010 16	01000 17	11010 19	10000 18
1	0	0	00100 32	01110 33	11100 35	10110 34
1	0	1	00001 40	01101 41	11001 43	10101 42
1	1	1	00111 56	11111 57	11111 59	00111 58
1	1	0	00100 48	01110 49	11100 51	10110 50

Figure 7-18. Excitation map for counter. (Courtesy of Motorola)

$C = 0$

y3 (32)	y4 (16)	y5 (8)	y1y2 00	01	11	10
0	0	0	1 0 0 (0)	0 0 0 (1)	1 0 0 (3)	1 0 0 (2)
0	0	1	1 0 0 (8)	0 0 0 (9)	1 0 0 (11)	1 0 0 (10)
0	1	1	1 0 0 (24)	1 0 0 (25)	0 0 0 (27)	0 0 0 (26)
0	1	0	1 0 0 (16)	0 0 0 (17)	1 0 0 (19)	1 0 0 (18)
1	0	0	1 0 0 (32)	0 0 0 (33)	0 0 0 (35)	1 0 0 (34)
1	0	1	1 0 0 (40)	0 0 0 (41)	1 0 0 (43)	1 0 0 (42)
1	1	1	0 0 0 (56)	1 0 0 (57)	1 0 0 (59)	0 0 0 (58)
1	1	0	1 0 0 (48)	0 0 0 (49)	0 0 0 (51)	1 0 0 (50)

$C = 1$

y1y2 00	01	11	10
1 0 0 (4)	0 0 1 (5)	1 0 1 (7)	1 0 1 (6)
1 0 1 (12)	0 0 1 (13)	1 0 1 (15)	1 0 1 (14)
1 1 1 (28)	1 0 0 (29)	1 1 1 (31)	0 0 0 (30)
1 1 1 (20)	1 0 0 (21)	1 1 1 (23)	0 0 0 (22)
1 0 1 (36)	0 1 1 (37)	0 0 0 (39)	1 0 1 (38)
1 0 1 (44)	0 1 1 (45)	0 0 0 (47)	1 0 1 (46)
0 0 0 (60)	0 0 0 (61)	1 1 1 (63)	1 1 1 (62)
0 0 0 (52)	0 0 0 (53)	1 1 1 (55)	1 1 1 (54)

Figure 7-19. Output map for counter defining the outputs B2, B1, and B0. (*Courtesy of Motorola*)

The rest of the codes are similarly placed in the output map (Figure 7-19) using the flow table of Figure 7-16 as a reference.

Program sheet truth table.   Finally, the codes in the excitation and output maps are transferred to the program sheet truth table (Figure 7-20) for the MCM5003/5004 PROM (Refer to Sec. 7–2 for data on PROM programming.) The excitation variable $y1$, $y2$, $y3$, $y4$, and $y5$ are chosen to represent the $B7$, $B6$, $B5$, $B4$, and $B3$ PROM outputs respectively. The $B2$, $B1$, and $B0$ outputs are represented by the same notation at the PROM outputs.

As an example, word 0 of Figures 7-18 and 7-19 is transferred to word 0 in Figure 7-20. Word 0 in Figure 7-18 is 00010, and in Figure 7-19 is 100, while Figure 7-20 is marked with the combination 100010100. Similarly, the rest of the words are transferred to the program sheet.

PROM connections.   Figure 7-21 shows the pin numbers and connections for the MCM5003 used in the counter design. Figures 7-22 and 7-23 show the actual waveforms at clock frequencies of 100 kHz and 1 MHz, respectively. Note that the waveforms are the same as the ones shown in Figure 7-13 (timing diagram).

Problem areas.   There are some possible problem areas that should be understood when using direct feedback. Although only one address input changes at a time, differences in the internal delays in the address decoding of the PROM can cause a hazard condition to exist in which a false word location is momentarily selected. In the MCM5003, the false word location will be selected for approximately 10 ns, or 1 gate delay during the hazard condition.

PROMs using ECL logic internally for the decoding, such as an MCM10149 (which is a Motorola 1024-bit ECL PROM), minimize the hazard condition. The ECL logic has approximately equal delays through the OR and NOR outputs of a gate, whereas TTL logic requires an extra gate (and delay) to perform the complement function.

There are ways around the hazard condition by sacrificing speed. The decoding "glitches" are evident in Figure 7-23 on the y1 output when in the low state. The decoding "glitches" are approximately 0.5 V, and are a possible problem only when the output is in the low state. The amplitude of the decoding "glitches" varies from device to device. If the amplitude is large enough to be detected when fed back at the input, the PROM could switch to an improper state.

The solution to the problem for the MCM5003 is the addition of capacitance C_L, for the outputs that are fed back as shown in

No.	A5 y3	A4 y4	A3 y5	A2 C	A1 y1	A0 y2	B7 Y1	B6 Y2	B5 Y3	B4 Y4	B3 Y5	B2	B1	B0
0	0	0	0	0	0	0	0	0	0	X	0	X	0	0
1	0	0	0	0	0	1		X						
2	0	0	0	0	1	0	X					X		
3	0	0	0	0	1	1	X	X		X		X		
4	0	0	0	1	0	0						X		X
5	0	0	0	1	0	1		X			X			X
6	0	0	0	1	1	0	X				X	X		X
7	0	0	0	1	1	1	X	X				X		X
8	0	0	1	0	0	0					X	X		
9	0	0	1	0	0	1		X	X		X			
10	0	0	1	0	1	0	X	X			X	X		
11	0	0	1	0	1	1	X	X			X	X		
12	0	0	1	1	0	0						X		X
13	0	0	1	1	0	1		X			X			X
14	0	0	1	1	1	0	X				X	X		X
15	0	0	1	1	1	1	X	X				X		X
16	0	1	0	0	0	0				X		X		
17	0	1	0	0	0	1		X						
18	0	1	0	0	1	0	X					X		
19	0	1	0	0	1	1	X	X		X		X		
20	0	1	0	1	0	0				X	X	X	X	X
21	0	1	0	1	0	1		X		X		X		
22	0	1	0	1	1	0	X			X				
23	0	1	0	1	1	1	X	X		X	X	X	X	X
24	0	1	1	0	0	0		X		X	X	X		
25	0	1	1	0	0	1		X		X	X	X		
26	0	1	1	0	1	0	X			X	X			
27	0	1	1	0	1	1	X			X	X			
28	0	1	1	1	0	0				X	X	X	X	X
29	0	1	1	1	0	1		X		X		X		
30	0	1	1	1	1	0	X			X				
31	0	1	1	1	1	1	X	X		X	X	X	X	X
32	1	0	0	0	0	0			X			X		
33	1	0	0	0	0	1		X	X	X				
34	1	0	0	0	1	0	X	X	X			X		
35	1	0	0	0	1	1	X	X	X					
36	1	0	0	1	0	0			X		X	X		X
37	1	0	0	1	0	1		X	X				X	X
38	1	0	0	1	1	0	X	X	X			X		X
39	1	0	0	1	1	1	X	X	X		X			
40	1	0	1	0	0	0					X	X		
41	1	0	1	0	0	1		X	X		X			
42	1	0	1	0	1	0	X	X	X		X	X		
43	1	0	1	0	1	1	X	X			X	X		
44	1	0	1	1	0	0			X		X	X		X
45	1	0	1	1	0	1		X	X				X	X
46	1	0	1	1	1	0	X	X	X			X		X
47	1	0	1	1	1	1	X	X	X		X			
48	1	1	0	0	0	0			X			X		
49	1	1	0	0	0	1		X	X	X				
50	1	1	0	0	1	0	X		X	X		X		
51	1	1	0	0	1	1	X	X	X					
52	1	1	0	1	0	0			X	X				
53	1	1	0	1	0	1		X	X	X	X			
54	1	1	0	1	1	0	X		X	X	X	X	X	X
55	1	1	0	1	1	1	X	X	X	X		X	X	X
56	1	1	1	0	0	0			X	X	X			
57	1	1	1	0	0	1	X	X	X	X	X	X		
58	1	1	1	0	1	0			X	X	X			
59	1	1	1	0	1	1	X	X	X	X	X	X		
60	1	1	1	1	0	0			X	X				
61	1	1	1	1	0	1		X	X	X	X			
62	1	1	1	1	1	0	X		X	X	X	X	X	X
63	1	1	1	1	1	1	X	X	X	X		X	X	X

"X" represents a bit to be programmed to a logic "1"

Figure 7-20. MCM5003 program sheet for counter. *(Courtesy of Motorola)*

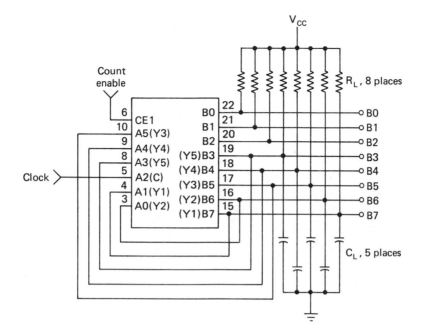

Figure 7-21. PROM connections for counter design with direct feedback. *(Courtesy of Motorola)*

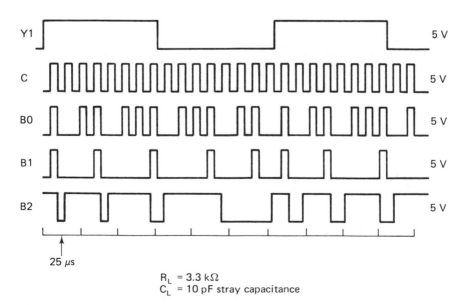

$R_L = 3.3 \text{ k}\Omega$
$C_L = 10 \text{ pF stray capacitance}$

Figure 7-22. Counter and output waveforms for clock frequency of 100 kHz. *(Courtesy of Motorola)*

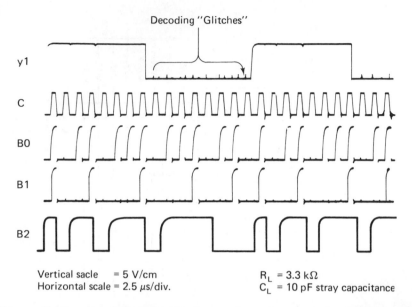

Figure 7-23. Counter and output waveform for clock frequency of 1 MHz. *(Courtesy of Motorola)*

Figure 7-21. If a 510-Ω load resistor R_L is used, then a 100 pF load capacitor C_L should be used. If $R_L = 3.3$ kΩ, a load capacitor is not required for a typical device, although 25 to 50 pF must be used for some devices.

    If the MCM5004 (with a 2 kΩ internal collector resistor, Figure 7-6) is used instead of the MCM5003, a 50 pF capacitor should be used. The *RC* time constant is enough to squelch the decoding "glitches" with a slight sacrifice in speed. The decoding "glitches" are smallest in amplitude at higher V_{CC} voltages and higher temperatures. The synchronous sequential circuit using clocked storage feedback (Figure 7-12a) does not have the problem of squelching the decoding "glitches," since the storage devices (FFs) see only the information that is present when the clock changes from a 0 to 1.

8

Interfacing

As discussed in Chapter 2, no matter what drive and load characteristics are involved for a logic IC, it may be necessary to provide the necessary drive current, to change the logic levels, and so on with an interfacing circuit or device. It is not practical to have a universal interfacing circuit for all logic families of all IC manufacturers. Likewise, it is not even possible to have a universal circuit for interface between a given logic family of one manufacturer, and all other logic families of the same manufacturers. Nor is it possible to have a single interface circuit that will accommodate the same logic family of different manufacturers. For one thing, the logic voltage levels (for 0 and 1), the supply voltages, and the temperature ranges vary with manufacturers, and with logic families. For that reason, most logic IC manufacturers publish interfacing data for their particular lines.

We will make no attempt to duplicate all this data here, but instead, concentrate on the interface requirements for the most popular lines, and the most used logic families. A careful study of this information should provide the designer with sufficient background to understand the basic problems involved with interfacing logic systems, and to interpret interfacing data that appears on logic IC datasheets.

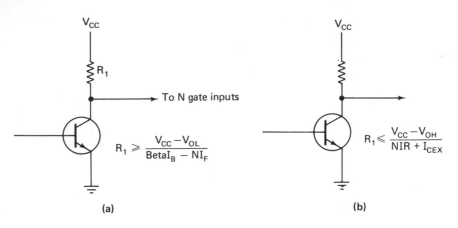

Figure 8-1.   Basic interfacing circuit for logic IC elements.

8–1   BASIC INTERFACING CIRCUIT

Although it is not practical to have a universal interfacing circuit for all ICs, the basic circuits of Figure 8-1 should provide enough information to design interfacing between some IC logic elements. The equations shown in Figure 8-1 are used to find the approximate or trial value of pull-up resistor R_1. The following is an example of how to use the equations.

Assume that the common-collector circuit is used, that V_{CC} is 5 V, that V_{OH} (high or logic-1 state voltage) is 3 V, that there are 7 gate inputs, and each has a forward current of 1 mA, and that I_{CEX} is 3 mA. Note that I_{CEX} is the input leakage current as tested with a voltage between base and emitter. I_{CEX} is a worst-case or active datasheet value, and is not to be confused with I_{CER}, which is input leakage current with no signal (tested with a fixed resistance between emitter and base).

Using the equation of Figure 8-1, the value of R_1 is:

$$R_1 = \left(\frac{5 - 3 \text{ V}}{(7 \times 1 \text{ mA}) + 3 \text{ mA}} \right) \quad \left(\frac{2 \text{ V}}{10 \text{ mA}} \right) = 200 \ \Omega$$

The next lowest standard value is 180Ω.

Two cautions must be observed when using the circuits of Figure 8-1. First, do not try to increase the output voltage of an IC (with an interface) to a level higher than the V_{CC} of either IC (input or output). For example, if both ICs have a V_{CC} of 5 V, keep the output of the interface transistor below 5 V, even though the IC data-sheet may show that inputs greater than 5 V are safe. This is because many ICs have a diode between resistors in the circuit and the power supply terminal (with the cathode of the diode connected to the power supply side).

Also, do not pull the interface circuit output below ground. In many ICs, each internal transistor has a diode connected between the collector and ground terminal. The diode polarity is such that the diode is normally reverse-biased (anode connected to ground). However, the diode will be forward-biased if the output is pulled below ground. Unfortunately, the datasheet schematics do not always show these diodes, even though they exist in the circuit. To be safe, keep the interface inputs and outputs above ground and below V_{CC}.

Sometimes, there are signal problems with interface circuits, particularly when high speeds are involved. Interface circuits slow down rise and fall times of the logic pulses. This slowdown can cause saturated logic elements (which affects just about every family but ECL) to break into oscillation on the leading and trailing edges of the pulses.

The basic rule to follow is that the *rise and fall times of an input to any logic IC must be shorter than the typical propagation delay time of the IC*. No oscillation should occur if this rule is followed. Note that the terms rise and fall time refer to the time required to go from a 1-state to a 0-state, and vice versa. These rise and fall times are somewhat longer than the conventional 10 and 90 percent values for pulse measurement.

8–2   MOS INTERFACE

Many designs presently under consideration use both MOS and two-junction (DTL, TTL, ECL, HTL) technologies in order to take full advantage of the low-cost and high-packaging density of MOS (such as MOS/LSI), as well as the flexibility of two-junction techniques for low-complexity functions. The interface of different logic families requires that the circuits operate at a *common-supply voltage,* and have *logic-level compatibility.* In addition, the devices must maintain *safe power dissipation levels,* and good *noise immunity* over the operating temperature range. In short, there must be *system compatibility* with all logic families used in the overall system.

8–2.1   MOS/LSI System Compatibility

The following notes apply to the MOS/LSI line of Texas Instruments.

Power supplies. Two manufacturing technologies are common in MOS/LSI, and common throughout the industry: high-threshold, and low-threshold MOS. The power supply requirements are:

	V_{SS}	V_{DD}	V_{GG}
High threshold	0	−12 V	−24 V
Low threshold	0	−5 V	−17 V

where V_{SS} is the substrate supply, V_{DD} is the drain supply, and V_{GG} is the gate supply.

The drain supply will draw most of the current. Some circuits are designed to use only one power supply (saturated logic). V_{DD} and V_{GG} are then common.

To use MOS in a system, it is often convenient to translate all of the power supply voltages to a certain voltage. A common arrangement is:

	V_{SS}	V_{DD}	V_{GG}
High threshold	+12 V	0 V	−12 V
Low threshold	+5 V	0 V	−12 V

Some high-threshold devices are specified at $V_{GG} = -28$ V and $V_{DD} = -14$ V.

Input compatibility. Referencing all voltages to V_{SS}, the input swing on most MOS circuits is as follows:

	High Level	Low Level
High threshold	0 to −3 V	−9 V to −24 V
Low threshold	0 to −1.5 V	−4.2 to −17 V

In relation to the translated power supplies, the input swing becomes:

	High Threshold	Low Threshold
V_{SS}	+12 V	+5 V
V_{DD}	0 V	0 V
V_{GG}	−12 V	−12 V
High level	+9 V to + 12 V	+3.5 V to +5 V
Low level	+3 V to −12 V	+0.8 V to −12 V

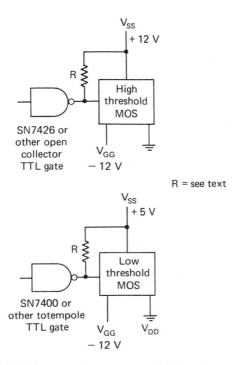

Figure 8-2. MOS/LSI system input compatibility. *(Courtesy of Texas Instruments)*

In all cases, the input of the MOS circuit will look like a very high impedance, and input compatibility is achieved by the circuits of Figure 8-2. The value of resistor R varies, depending on speed-power requirements. In many cases, this resistor R is diffused on the MOS chip. For low-threshold MOS, the resistor assures that the worst-case TTL output is pulled up to at least 3.5 V for proper MOS circuit operation.

Output compatibility. Three types of buffers are commonly used on MOS devices; *open-drain, internal pull-up,* and *push-pull.* These buffer arrangements are shown in Figure 8-3.

With *open-drain* and *internal pull-up,* the buffer is simply a current switch. In the OFF state, the impedance of the buffer is extremely large, whereas in the ON state, impedance is typically under 1 kΩ. A discrete resistor or MOS transistor may be used as a load with an open-drain buffer. This resistor or the transistor may be internal to the MOS circuit. When the load transistor is internal to the MOS, the buffer is called an *internal pull-up buffer.*

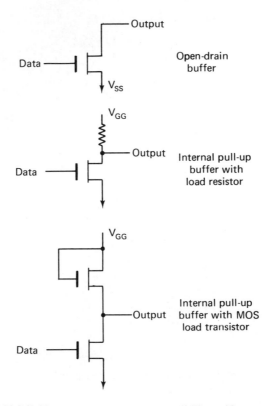

Figure 8-3. MOS/LSI system output compatibility. (*Courtesy of Texas Instruments*)

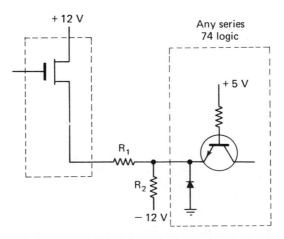

Figure 8-4. MOS/LSI high-threshold (with an open-drain buffer) to TTL interface. (*Courtesy of Texas Instruments*)

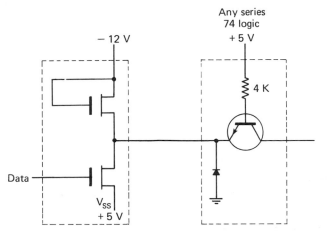

Figure 8-5. MOS/LSI low-threshold to TTL interface. *(Courtesy of Texas Instruments)*

If the MOS is high-threshold with an open-drain buffer, the output can be made compatible with TTL, as shown in Figure 8-4. Resistor R_2 provides the necessary current sink for the TTL input. Resistor R_1 limits the positive swing to +5 V.

If the MOS is low-threshold, V_{ss} is translated up to +5 V instead of to +12 V, eliminating the need for R_1. Further, if R_2 is on the chip (in resistor form or a MOS load resistor), no external components are necessary, permitting direct coupling of the MOS output to the TTL input, as shown in Figure 8-5.

There are two common types of push-pull buffer, as seen in Figure 8-6. The unsaturated push-pull buffer is the most commonly

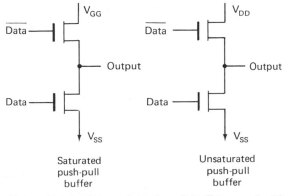

Figure 8-6. Two common types of push-pull buffers used with MOS/LSI. *(Courtesy of Texas Instruments)*

used for low-threshold circuits, and permits direct TTL compatibility without external components (as well as direct compatibility with other low-threshold MOS circuits).

Clocks.   MOS clock requirements depend on the circuit, as is the case with other logic families. For example, no clocks are required for static RAMs, ROMs, and so on. Some MOS devices require only one clock, with all other clocks generated internally. Most shift registers require two clocks. High-speed, low-power dissipation shift registers may require four clocks.

For one-clock operation, an internal circuit generates the clocks from a single outside clock. This external-clock signal has the same swing as the data input signal, and the compatibility is identical. Generally, single-clock low-threshold MOS circuits will accept a TTL clock without adding components.

When two or four clocks are required, the clock signals must swing between V_{ss} and V_{GG}. To go from a single TTL-level clock to a multiple MOS-level clock, two circuits are required. First, a *clock generator* is necessary to generate the basic clock pulses. Second, a *clock driver* is necessary to bring the clock levels to the required values. In most cases, only one basic clock circuit is needed for an entire MOS/LSI system.

8–2.2   COS/MOS—TTL/DTL Interface

The following notes apply to RCA COS/MOS. When interfacing one logic IC family with another, attention must be given to logic swing, output drive, dc input current, noise immunity, and speed of each family. Figure 8-7 shows a comparison of these for COS/MOS, medium power TTL, and medium power DTL. The supply voltage column of Figure 8-7 shows that both saturated bipolar (two-junction) and COS/MOS devices may be operated at a supply voltage of 5 V. Both logic forms are directly compatible at this supply voltage (with certain restrictions).

Figure 8-7 also shows that the voltage characteristics required at the output and input terminals of saturated logic devices, as well as the COS/MOS input and output characteristics at $V_{DD} = 5$ V. The COS/MOS devices are designed to switch at a voltage level about one-half of the power supply voltage. However, TTL/DTL devices are designed to switch at about $+1.5$ V, which is not one-half of the supply voltage.

COS/MOS driven by bipolar.   When a bipolar device is used to drive a COS/MOS device, the output drive capability of the driving device, as well as the switching levels and input currents of

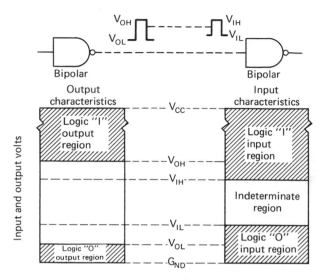

Interface voltage characteristics required at the output and input terminals of saturated logic devices.

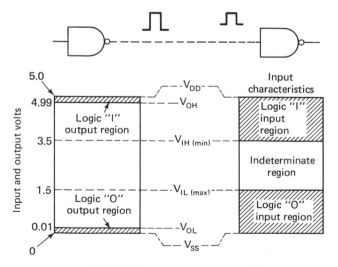

COS/MOS input and output characteristics at a power-supply voltage of 5 volts

Figure 8-7.   COS/MOS to DTL/TTL interface characteristics. (*Courtesy of RCA*)

Family	Supply voltage (volts)	Logic swing/output drive capability	DC input current	Noise immunity	Propagation delays
COS/MOS	3.0 to 15	V_{SS} to V_{DD} (driving COS/MOS) Output drive is type dependent (see text)	10 pA (typical) 1 and 0 state	1.49 at V_{DD} = 5 V The switching point occurs from 30% to 70% of V_{DD} which is 1.5 V to 3.5 V at V_{DD} = 5 V	35 ns (typical) for inverter C_L = 15 pF
DTL and TTL	5	<u>0 state:</u> 0.4 V max. at I_{sink} = 16 mA <u>1 state:</u> 2.4 V min. at I_{load} = − 400 μA	<u>0 state:</u> − 1.6 mA max. <u>1 state:</u> 40 μA max.	at V_{CC} = 5 V 0.4 V guaranteed The switching point occurs from 0.8 V to 2 V	20 ns (typical) for inverter C_L = 15 pF

Comparison of COS/MOS, TTL, DTL interfacing parameters

Logic voltage	Description	Voltage (volts)
V_{OL}	Maximum output level in low-level output state	0.4
V_{OH}	Minimum output level in high-level output state	2.4
V_{IL}	Maximum input level in low-level input state	0.8
V_{IH}	Minimum input level in high-level input state	2.0
V_{CC}	Positive supply voltage	5.0 ± 0.5
	Operating temperature range: − 55° to + 125°C—full temperature range product 0° to + 85°C—limited temperature range product	

Common logic voltages, supply voltage, and operating temperature range required to interface with DTL/TTL circuits

Figure 8-7.   (Continued)

the driven device, are important considerations. Figure 8-7 shows that only 10 pA of dc input current are required by a COS/MOS device in either the 0 or 1 state. The input thresholds (for the drive COS/MOS device) are 1.5 and 3.5 V. Thus, the output of the TTL/DTL driver must be no more than 1.5 V (0 logic voltage) and no less than 3.5 V (1 logic voltage) in order to obtain some noise immunity.

Current sinking.   Figure 8-8 shows the low state operation of a loaded bipolar driver stage. When the output drive circuit of the bipolar stage is in the low state, the collector is essentially at ground

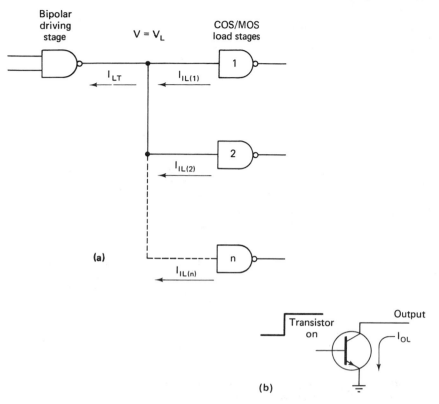

Figure 8-8.   (a) Low-state operation of a loaded bipolar driver stage; (b) typical bipolar output-drive circuit in the low state.

potential. The ON transistor must go into saturation in order to assure a reliable logic 0 level (0 to 0.4 V). To attain this voltage level, there should be a high impedance path from the output to the power supply. Current sinking capability is not a problem in this configuration because the COS/MOS devices have extremely high input impedances (typically 10^{11} ohms). Neither is the voltage level a problem; the COS/MOS devices have high noise immunity (greater than 1 V).

    Current sourcing.   Current flows from the V_{CC} terminal of a saturated logic (two-junction) output device into the input stages of the load. That is, the output device acts as a current source for the load. Figure 8-9 shows high-state operation of a loaded bipolar driver stage. Whenever a typical bipolar driver circuit is in the high state, a pull-up configuration (resistor or transistor) ties V_{CC} to the output pin.

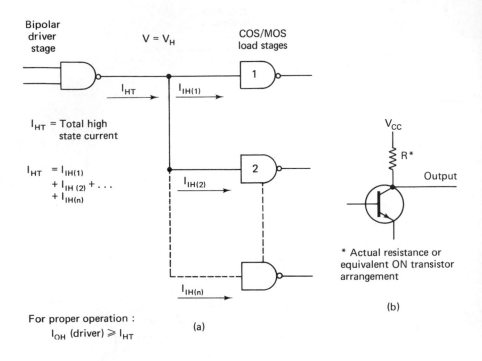

I_{HT} = Total high state current

$$I_{HT} = I_{IH(1)} + I_{IH(2)} + \ldots + I_{IH(n)}$$

For proper operation :
I_{OH} (driver) $\geqslant I_{HT}$

(a)

* Actual resistance or equivalent ON transistor arrangement

(b)

I_{OH} = Maximum permissible output driver current in high state ("1") (driver leakage)
I_{OL} = Maximum output driver sinking current in low state ("0")
I_{IL} = Low state input current drawn from the load stage (to the driver)
I_{IH} = High state input current flowing into the load stage from driver

Figure 8-9. (a) High-state operation of a loaded bipolar driver stage; (b) typical bipolar output-driver circuit in the high state.

The total load configuration should not draw sufficient current to reduce the output voltage level below the V_{IH} required by the COS/MOS devices. (V_{IH} is the maximum acceptable input level for the device in the high-level input state, as shown in Figure 8-10.)

There are three bipolar output configurations to consider: *resistor pull-up, open collector,* and *active pull-up.*

Resistor pull-up. Devices with resistor pull-ups, as shown in Figure 8-9, present no problem in interface with COS/MOS.

Open collector. Devices with open collectors require an *external pull-up* resistor, as shown in Figure 8-11. The selection of the external pull-up resistor requires consideration of fan-out, maximum allowable collector current in the low state (I_{OL} max), collector-emitter leakage current in the high state (I_{CEX}), power consumption, power-supply voltage and propagation delay times.

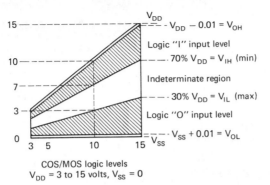

COS/MOS logic levels
V_{DD} = 3 to 15 volts, V_{SS} = 0

Logic voltage symbol	Description
V_{OH}	Minimum guaranteed noise free output level of device in high-level output state
V_{IH}	Minimum acceptable input level for device in high-level input state
V_{IL}	Maximum acceptable input level for device in low-level input state
V_{OL}	Maximum guaranteed noise free output level of device in low-level output state
V_{NL}	Maximum (positive) noise level tolerated at low level state
V_{NH}	Maximum (negative) noise level tolerated at high level state
$V_{OL} + V_{NL} \geq V_{IL}$	
$V_{OH} + V_{NH} \leq V_{IH}$	Operating temperature range
V_{DD}	Positive supply voltage — 55°C to + 125°C (full temperature prod).
V_{SS}	Negative supply voltage — 40°C to + 85°C (limited temperature prod).

Figure 8-10. Common logic voltages; supply voltage and temperature range for COS/MOS devices. *(Courtesy of RCA)*

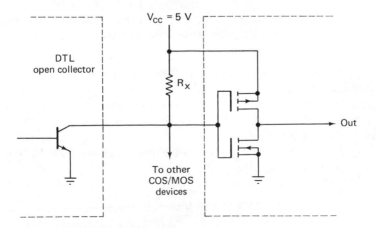

Figure 8-11. Example of DTL/TTL circuit with open collectors that require a resistor between the output and V_{CC}. *(Courtesy of RCA)*

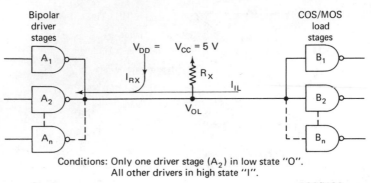

Conditions: Only one driver stage (A_2) in low state "O".
All other drivers in high state "I".

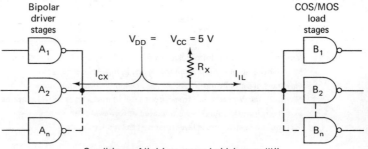

Conditions: All driver stages in high state "I"

$$R_X \text{ (min)} = \frac{V_{DD} - V_{OL \text{ (max)}}}{I_{OL}}$$

$$R_X \text{ (max)} = \frac{V_{CC \text{ (min)}} - V_{IH \text{ (min)}}^*}{(N) \, I_{CEX \text{ (max)}}}$$

*V_{IH} is the value for the COS/MOS device

Figure 8-12.   Bipolar output (with pull-up resistor) driving COS/MOS in low-state and high-state operation. *(Courtesy of RCA)*

The equations of Figure 8-12 provide guidelines for selection of pull-up resistor maximum and minimum values. These equations neglect the values of I_{IH} and I_{IL} for the MOS device because such values (typically 10 pA) are insignificant when compared with the value of the bipolar currents.

Assume that the equations of Figure 8-12 are used to find the value of R_X where conditions are:

$$V_{CC} = 5 \text{ V} \pm 0.5 \text{ V}$$

$$V_{DD} = 5 \text{ V}$$

$$V_{OL} = 0.4 \text{ V max}$$

$$V_{IH} \text{ (for the MOS)} = 3.5 \text{ V}$$

$$I_{OL} = 20 \text{ mA}$$

$$I_{CEX} = 100 \text{ }\mu\text{A max}$$

For one driver (bipolar) and one load (MOS), the values of R_X are:

$$R_{X(\text{min})} = \frac{(5.0 - 0.4)}{0.020} = 230 \text{ }\Omega$$

$$R_{X(\text{max})} = \frac{(4.5 - 3.5)}{0.0001} = 10 \text{ k}\Omega$$

If more than one driver is used with the open collector drive arrangement, the values of I_{OL} and I_{CEX} must be increased accordingly (I_{OL} and I_{CEX} multiplied by the number of drivers). However, the values of R_X will not be affected by an increase in the number of MOS loads. Of course, if an infinite number of MOS loads are added so that I_{IL} and I_{IH} become significant compared to I_{OL} (for the minimum value of R_X) or to I_{CEX} (for the maximum value of R_X), then the MOS input currents must be included in the calculations as follows:

$$R_{X(\text{min})} = \frac{V_{DD} - V_{OL(\text{max})}}{I_{OL} - I_{IL(\text{MOS})}}$$

$$R_{X(\text{max})} = \frac{V_{CC(\text{min})} - V_{IH(\text{min}) \text{ (MOS)}}}{I_{CEX} + I_{IH}}$$

For short propagation delay times with an external pull-up resistor, it is best to keep R_X small. That is, use the $R_{X(\text{min})}$ value, or the next higher standard value. However, power consumption increases rapidly at values below about 1 kΩ so some compromise is usually necessary. Of course, final selection of the pull-up resistor depends on what is most important for the intended application: high speed or low power.

Figure 8-13 shows typical speed–power relationships as a function of R_X for two popular bipolar open collector drivers and illustrates the tradeoff between speed and power. The power figure shown is the power dissipated in both the bipolar driver and the pull-up resistor.

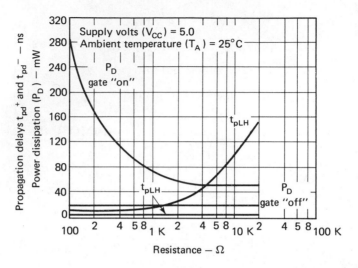

Buffer

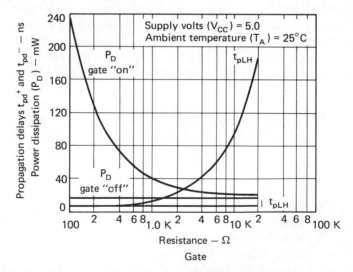

Gate

Figure 8-13.   Typical speed-power tradeoff of open collector TTL buffer or gate, and pull-up resistor. *(Courtesy of RCA)*

Active pull-ups.   When an active pull-up is used, such as the transistor-plus-diode arrangement shown in Figure 8-14, there can be a problem in the 1 state. This is because the minimum output level (2.4 V) cannot assure an acceptable 1 state input for the COS/MOS

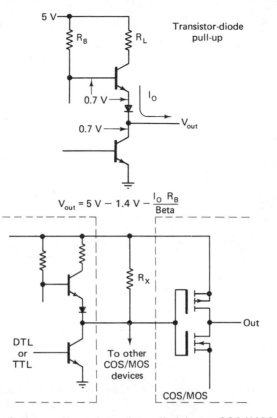

Figure 8-14.   Active pull-up (transistor-diode) to COS/MOS interface.
(Courtesy of RCA)

device. For example, assume that the 2.4 V minimum TTL/DTL output level is specified for a load current of 400 μA (which is typical). This would require approximately 40 MOS devices (with a typical input current of 10 pA).

   If only one, or a few, MOS devices are used as a load, the minimum TTL/DTL high-output level will rise to about 3.4 to 3.6 V. There is no noise immunity in such a configuration. It is recommended that a pull-up resistor be added to V_{CC} from the output terminal of the bipolar device. The selection of this resistor should be based on the calculations of Figure 8-12.

   When driving a COS/MOS device from an output arrangement as in Figure 8-14, the driver should not fan-out to TTL/DTL circuits; but only to COS/MOS devices. This is because it is not accepted practice to tie devices with active pull-ups together (Sec. 1–19.1).

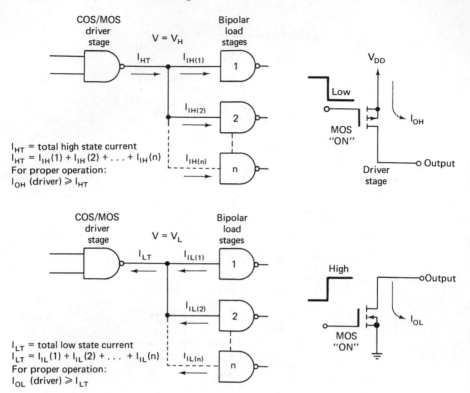

Figure 8-15. Logic diagram for a COS/MOS device driving a bipolar device, low-state and high-state. *(Courtesy of RCA)*

COS/MOS driving bipolar. Figure 8-15 shows COS/MOS devices driving bipolar devices. The current sinking capacity of the COS/MOS device must be considered when the device is a medium-power DTL and TTL circuit. Figure 8-7 illustrates that the TTL/DTL devices requires no more than 1.6 mA in the 0 input state, and a maximum of 40 μA in the 1 input state.

The COS/MOS device must be capable of sinking and sourcing the currents while maintaining voltage output levels required by the TTL/DTL gate. Any given TTL/DTL gate will switch state at a voltage that ranges from 0.8 to 2 V. Thus, the output drive capability of the COS/MOS driver must be at least 40 μA for a given 1-state output voltage of 2 V, and at least 1.6 mA for a given 0-state output voltage of 0.8 V. In order to provide a noise margin of 400 mV for the driven bipolar device, the COS/MOS device must sink 1.6 mA at a 0 logic state voltage of 0.4 V, and 40 μA at a logic 1 level of 2.4 V.

Current sourcing. In the high-state operation (Figure 8-15), V_{DD} is normally connected to the driver output through one or more ON P-channel devices which must be able to source the *total leakage current* of the bipolar load stages. The published information for the particular COS/MOS and bipolar devices must be consulted in order to determine the leakage currents (for the logic 1 state), and drive fan-out to be used in the equations shown in Figure 8-15. Saturated logic devices will not reach their required switching levels unless this equation has been satisfied.

Current sinking. When the output of a COS/MOS driver is in the low-state, an N-channel device is ON and the output is approximately at ground. The COS/MOS device sinks the current flowing from the bipolar input-load stage. The published data for the COS/MOS device must be consulted to determine the maximum output low-level sinking current, and the published data for the bipolar device must be consulted to determine its input low-level current.

Not all COS/MOS devices can sink the required current for all bipolar logic families. This problem can be overcome by connecting several COS/MOS devices in parallel. Likewise, some COS/MOS MSI devices (such as counters and shift registers) have limited drive capability. Their outputs may require buffering if these COS/MOS devices are to drive TTL/DTL. The COS/MOS line contains several buffers. In some cases, COS/MOS drive and current sinking capability can be increased if the devices are operated at higher supply voltage. These capabilities are described in COS/MOS literature.

8–2.3   Level Shifters (Level Translators)

When interfacing DTL and TTL devices with COS/MOS devices which are operated at a higher voltage supply, the same resistor interface shown in Figure 8-11 can be used. The resistor is tied to the higher level (V_{DD}). The maximum supply voltage for the DTL and TTL gates is generally specified at about 8 V. Thus, not all DTL/TTL gates may be used for interface applications that require higher supply voltages (V_{DD}).

Guaranteed operation at these higher supply voltages can be accomplished by selection of DTL/TTL units with breakdown voltages $V_{(BR)CER}$ exceeding the COS/MOS operating voltage, or by using a level shifter circuit (also known as a level translator) shown

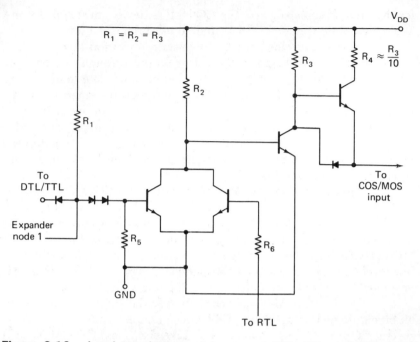

Figure 8-16.   Level translator used to convert DTL, TTL, and RTL input logic levels and voltages compatible with COS/MOS circuitry. *(Courtesy of RCA)*

in Figure 8-16. This circuit converts DTL, TTL, and even RTL input logic levels to voltages compatible with COS/MOS circuitry. In interface applications, the supply voltage for the translator should be equal to the supply voltage required by the COS/MOS circuitry.

The speed consideration is most important when a separate interface circuit is used. It is desirable (unless high noise immunity is a prime consideration) for the speed of the interfacing circuit to be maximum (or at least no slower) than either type of logic joined by the interface. No interfacing device other than a pull-up resistor is required, however, between the COS/MOS and TTL logic at a supply voltage of 5 V. Speeds involved when COS/MOS drives TTL (which can be found in the published data for COS/MOS devices) are comparable to the COS/MOS propagation delays. Speeds involved when COS/MOS is driven by TTL, even with a large external resistor, are no slower than delay times for COS/MOS logic circuits. As a result, speed is not a problem in COS/MOS–TTL interfacing, provided *clock rates are within the COS/MOS range.*

8–2.4   COS/MOS-HTL Interface

HTL circuits operate at voltage levels between 14 and 16 V. COS/MOS circuits can operate at these voltages as well, but generally are limited to voltages no higher than 15 V. HTL circuits have more limited temperature range, and dissipate much more power than COS/MOS circuits. Thus, care should be exercised when using the combinations in extreme temperature environments.

Typically, HTL resistance values vary by about 20 percent from one end of their temperature range to the other. In addition, the transistors used in HTL are sensitive to temperature, and are subject to thermal runaway. The V_{OL} level, propagation delay, and noise immunity of HTL circuits vary widely across the temperature range. COS/MOS circuits show almost negligible variation for these same parameters over a temperature range that is approximately 75 percent wider than that of HTL. In COS/MOS–HTL interface, the main concern in regard to temperature is the HTL parameters, not the COS/MOS parameter.

The same general rules described for COS/MOS-TTL/DTL (Sec. 8–2.2) apply to COS/MOS-HTL. Figure 8-17 shows the voltage characteristics required at the output and input of an HTL device for a $V_{CC} = 15$ V, as well as the same characteristics for a COS/MOS device.

The HTL types either have a built-in pull-up resistor (typically about 15 kΩ) or an active pull-up. An external pull-up resistor is unnecessary when COS/MOS devices are being driven by HTL. The dc noise immunity in the high state (logic 1) is 3.5 V for an active pull-up and 5 V for a resistor pull-up.

The published data should be consulted to be sure that the rise and fall times and the pulse widths of the HTL output are compatible with the required pulse width and input rise and fall times of the COS/MOS circuits. The rules for selection of external pull-up resistors are the same as described in Sec. 8–2.2, except that the values are different. For example, a typical HTL will have values such as:

$$V_{IH} = 8.5 \text{ V (min)}$$

$$V_{IL} = 6.5 \text{ V (max)}$$

$$I_{IH\text{max}} = 2 \ \mu\text{A}$$

$$I_{IL\text{max}} = 1.2 \text{ mA}$$

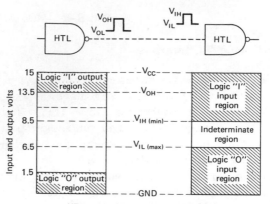

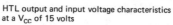

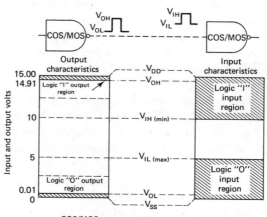

HTL output and input voltage characteristics
at a V_{CC} of 15 volts

COS/MOS output and input voltage characteristics
at a V_{CC} of 15 volts

HTL common logic voltages, supply voltage
and operating temperature ranges

Logic voltage symbol	Description	Voltage
V_{OL}	Maximum output level in low-level output state	1.5 V
V_{OH}	Minimum output level in high-level output state	13.5 V
V_{IL}	Maximum input level in low-level input state	6.5 V
V_{IH}	Minimum input level in high-level input state	8.5 V
V_{NL}	Worst case positive noise level tolerated at low level state	5.0 V
V_{NH}	Worst case negative noise level tolerated at high level state	5.0 V
V_{CC}	Positive supply voltage	15.0 ± 1 V
Operating temperature range − 30°C to + 75°C		

Figure 8-17. COS/MOS and HTL interface characteristics. (*Courtesy of RCA*)

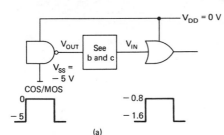

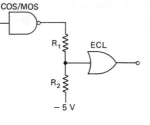

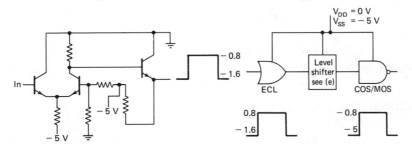

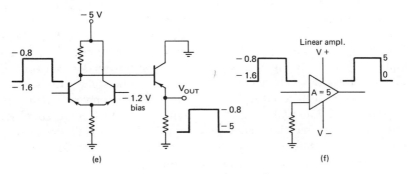

Figure 8-18.   COS/MOS and ECL/ECCSL interface characteristics. *(Courtesy of RCA)*

8-2.5   COS/MOS-ECL Interface

Figure 8-18 shows the interface of COS/MOS devices with ECL devices. The V_{CC} to V_{EE} voltage range is fixed from a ground level to −5 or −5.2 V. Logic 1 to logic 0 values are separated by only 0.3 to 0.5 V, depending on the particular type of ECL family used. Figure 8-19 shows some typical ECL values. However, since each manufacturer shows different logic levels for a number of ECL families, care should be taken to use only the applicable value taken directly from the published data.

A logic 1 is the most positive frame of reference, and a logic 0 is the most negative. For example, for positive logic, an RCA type

Logic voltage symbol	Description	Voltage range*** From	To
V_{OL}	Maximum output level low-level state	− 1.6	− 1.45*
V_{OH}	Minimum output level high-level state	− 0.8	− 0.795*
V_{IL}	Minimum input level low-level state	− 1.4	− 1.7*
V_{IH}	Maximum input level high-level state	− 0.75	− 1.1
V_{NL}	Worst case positive noise level tolerated at low-level state	0.20*	0.35*
V_{NH}	Worst case negative noise level tolerated at high-level state	− 0.235	− 0.305*
V_{CC}	Positive supply voltage	0	0
V_{EE}	Negative supply voltage	− 5.5	− 5.0
	Temperature range + 10 to + 60°C		

* At T = + 25°C
** These values are representative of the range for several ECL families

Figure 8-19.   ECL common logic voltages, supply voltages, and operating temperature range. *(Courtesy of RCA)*

CS2150 OR/NOR gate is at a logic 1 when its voltage is −0.8 V, and at a logic 0 when its voltage is −1.6 V (more negative value).

The interfacing of COS/MOS devices driving ECL devices requires a method to reduce the output voltage swing from 0.3 to 0.9 V. This can be accomplished with a precise resistor-divider network arrangement (Figure 8-18b), an emitter-follower (Figure 8-18c), or numerous combinations of resistor, diode, and transistor configurations. For example, if the COS/MOS output for a logic 1 is 5 V, and the input for an ECL logic 1 is 1 V, the resistor-divider network of Figure 8-18b can be chosen to provide a five-to-one voltage division.

Care must be taken to meet the necessary current sinking requirements of the ECL device, particularly in the 0 logic state. An external circuit may be required for COS/MOS to sink the current from the ECL load. Likewise, the published data must be consulted to be sure that the rise and fall times, and pulse widths, of the ECL are compatible with the COS/MOS.

The interfacing of COS/MOS devices that are being driven by ECL generally include an amplifier. Amplification is necessary because the COS/MOS requires a greater voltage swing. That is, the ECL −0.8 to −1.6 V swing must be amplified to the COS/MOS 0 to 5 V swing. The use of a separate transistor, such as that shown in Figure 8-18e, is recommended. Proper biasing of the transistor is essential. It is suggested that the V_{SS} level for the COS/MOS circuit be the same as the V_{EE} level of the ECL circuit. This will minimize the number of power supplies as well as provide better interface conditions.

8–2.6   COS/MOS to MOS Interface

There are a number of MOS devices which function at the same V_{DD} and V_{SS} ranges of COS/MOS. These devices can be interfaced directly, provided V_{DD} and V_{SS} are the same. Direct interface

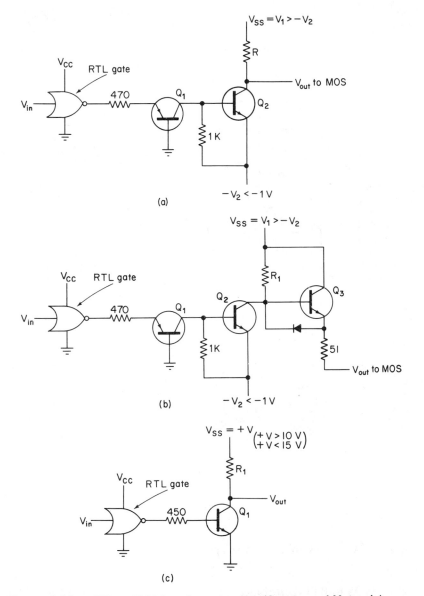

Figure 8-20.   RTL to MOS interface circuits. *(Courtesy of Motorola)*

applies (generally) to N-channel MOS devices. If P-channel devices are involved, there may be a problem of logic polarity. If the P-MOS device uses negative logic, the COS/MOS positive logic must be converted. The techniques involved for simple transposition from positive logic to negative logic are discussed in Sec. 1–9.2 and 1–9.3.

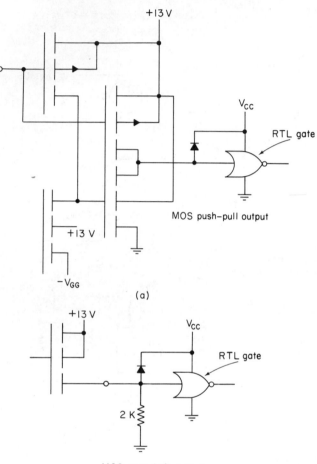

(a)

MOS open drain output

Note: Use germanium diode when $V_{CC} = 3.6$ V

(b)

Figure 8-21. MOS to RTL interface (with MOS substrate at $+13$ V). *(Courtesy of Motorola)*

8–2.7   MOS to RTL Interface

Since RTL is most similar to discrete transistor logic, discrete interface circuit are generally required when translating between RTL and MOS. Figure 8-20 gives three discrete circuits that can be used in converting RTL to MOS. The circuits of Figure 8-20a and 8-20b are particularly useful when the MOS substrate is at the RTL power supply voltage, or at ground.

The circuit of Figure 8-20b has an active pull-up, and should be used when driving large capacitive loads. Resistor R_1 should be selected on the basis of power dissipation and the desired rise and fall time at the output. If the MOS substrate is at a voltage in the range of +10 V to +15 V, the circuit of Figure 8-20c provides an efficient interface. Again, the value of R_1 is selected on the basis of available drive from the RTL gate, desired rise and fall times, and power dissipation.

The circuit of Figure 8-21 illustrates conversion from MOS to RTL when the MOS substrate voltage is at +13 V. The diode clamps the gate input to the RTL power supply voltage, which prevents the MOS output driver from pulling the gate input above the maximum specified rating at +4 V. When V_{CC} is 3.6 V, the clamp diode should be germanium.

In the circuit of Figure 8-22, the MOS substrate is at the RTL power supply voltage. In this mode of operation, the clamp diode is used to prevent the voltage on the RTL gate input from falling below the −4 V rating.

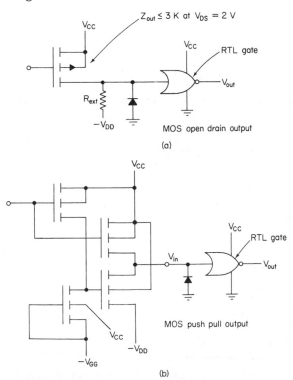

Figure 8-22.   MOS to RTL interface (with MOS substrate at V_{CC}). (*Courtesy of Motorola*)

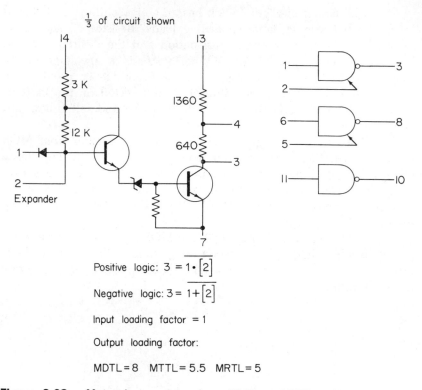

$\frac{1}{3}$ of circuit shown

Positive logic: $3 = \overline{1 \cdot \boxed{2}}$

Negative logic: $3 = \overline{1 + \boxed{2}}$

Input loading factor $= 1$

Output loading factor:

MDTL = 8   MTTL = 5.5   MRTL = 5

Figure 8-23.   Motorola translator from MHTL to MRTL, MDTL, or MTTL.

8–3   ECL AND HTL LEVEL TRANSLATORS

In general, level translators (or level shifters) are used when ECL and HTL must be interfaced with any other logic family. This is because of the special nature of ECL and HTL (ECL is nonsaturated logic, HTL operates with large logic swings).

8–3.1   HTL Interface

Most HTL logic lines include at least two translators; one for interfacing from HTL to RTL, DTL and TTL; and another for interfacing from these families to HTL. This is the case with the Motorola HTL line (designated as MHTL).

The translator for interface from MHTL to the other three logic families is shown in Figure 8-23. For conversion to DTL and TTL, a 5-V supply is connected to a 2-kΩ pullup resistor through pin 13. (Note that only ⅓ of the schematic is shown. However, all three logic elements are contained in one IC package.) For interface with

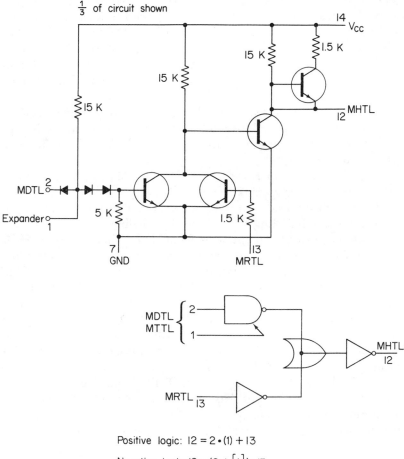

Positive logic: $12 = 2 \cdot (1) + 13$

Negative logic: $12 = (2 + [1]) \cdot 13$

Input loading factor:

$MDTL = 1$ $MTTL = 0.8$ $MRTL = 1$

Output loading factor $= 10$

Figure 8-24.   Motorola translator from MRTL, MDTL, or MTTL to MHTL.

RTL, pins 4, 9, and 12 are connected to the RTL supply voltage (nominally 3.6 V). Expander points (pins 2 and 5) without diodes are present at the inputs of two units, but not on the third unit. This is because of the need for additional leads for the RTL power supply, and pin limitation of the fourteen-lead package.

The translator for interface from RTL, DTL, and TTL to MHTL is shown in Figure 8-24. This translator is also a triple unit.

Signals from DTL/TTL sources are applied to one set of input terminals, while signals from RTL sources are applied to another terminal. The different inputs provide threshold levels and characteristics compatible with MDTL/MTTL and MRTL (Motorola) families. Each DTL/TTL section is also an input expander terminal (pins 1, 6, and 9) without diodes. These terminals may be used to expand input logic capability or to use high-voltage diodes to readily interface high-voltage relay or switch circuits to HTL levels. (A further discussion of such problems is given in Sec. 8–8.) Both types of inputs may be applied simultaneously, with one output going high if the logic function or either input goes high.

If the RTL input is used by itself, the DTL/TTL input must be grounded for proper operation. This is not necessary if the DTL/ TTL input is being used by itself, but is advisable under this condition to ground the RTL input to reduce any possible noise pickup.

Driving discrete components from HTL. In some cases, HTL must be used to drive discrete transistors. This creates a problem because of the high output voltage from HTL. The output in the high state is not as much of a problem as the output of an HTL in the low state. Assuming positive logic, the output of an HTL in the low state (or 0) is about 1 V. (This is generally listed as V_{OL}.) If 1 V is applied to an NPN in the low state, the NPN will probably never turn off. However, an HTL with an active pull-up output can be used for driving NPN transistors if the circuit of Figure 8-25 is used.

The higher V_{OL} voltage of the HTL is partially due to the extra diode D_1 on the output. However, if the gate is not required to sink current, then the voltage on the base of the NPN transistor is equal to the leakage current of the collector-base junction times the

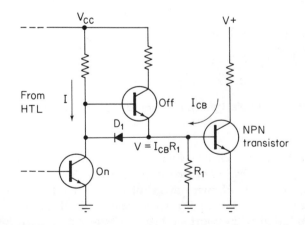

Figure 8-25. Driving discrete transistors from HTL logic IC outputs.

value of R_1. Thus, resistance R_1 should be chosen to provide a base voltage of about 0.2 V (or less), using I_{BC} leakage current as the factor. With such a circuit, the HTL should not be required to sink any current when the output is in the low state.

8–3.2   ECL Interface

As in the case of HTL, ECL level translators are available for interfacing with the other logic lines. However, if the only problem is one of interfacing from an input to ECL, it is possible that the circuit of Figure 8-26 will solve the problem. The input (either discrete component or IC) must be on the order of 1 V. As shown by the equations, the values of R_1 and R_2 are selected to give a logic 0 (-1.5 V) at the output (to the ECL input). The value of C can be determined by the fact that the RC time constant should be several times greater than the duration of the input pulse.

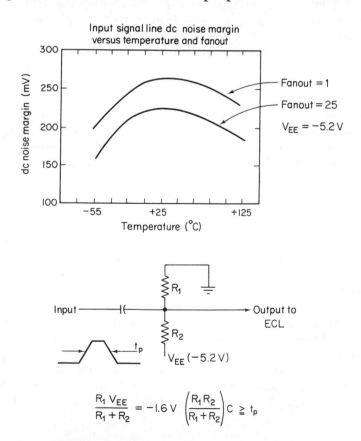

$$\frac{R_1\,V_{EE}}{R_1+R_2} = -1.6\,V \qquad \left(\frac{R_1 R_2}{R_1+R_2}\right) C \geq t_p$$

Figure 8-26.   Interfacing from non-ECL input to ECL.

8–4   RTL INTERFACE

The input signals to a typical RTL element should meet the following:

$$\text{high level: } E_{in} \geqq +1 \text{ V}$$

$$0.5 \text{ mA} \leqq I_{in} \leqq 0.5 \text{ mA per load}$$

$$\text{low level: } E_{in} \leqq 0.3 \text{ V}$$

When RTL is interfaced with MOS, follow the recommendations of Sec. 8–2.6. When an RTL output drives a DTL or TTL type input, there is a current sink requirement on the RTL output. To insure operation under all conditions, a maximum sink current of 5 mA is permissible. When interfacing RTL at the input, the high level output current of the driver is important. For a DTL driver with an output of 6 kΩ pull-up resistor to a + 5 V supply, one RTL load is allowed. Five gate loads can be driven if the DTL output resistor is reduced to 1.5 kΩ, for a typical RTL unit.

These specifications and resistor values are both typical and conservative requirements, and are given to insure operation over the normal temperature range.

The specifications of most RTL and DTL/TTL are not compatible, so worst-case interfacing cannot be guaranteed between the two families without performing special test and/or adding discrete components. However, by using conservative design principles and making one relatively simple test, interfacing can be accomplished.

The recommended circuitry for driving RTL gates from DTL outputs is shown in Figure 8-27a. In order to insure that this technique is valid for operation under worst-case conditions, it is necessary to test V_{OL} of the DTL circuit at an I_{OL} of 6.5 mA. Use only those DTL components with a V_{OL} less than 0.4 V for this application.

The circuit of Figure 8-27a should not be used to drive both RTL and DTL from a common output. If both must be driven from a common point, use the circuit of Figure 8-27b. The load current seen by the output transistor of the DTL gate in Figure 8-27b is greater than that of Figure 8-27a. The leakage current of the RTL gate may increase slightly also. Neither of these factors should cause significant problems.

Driving DTL from RTL outputs is not a situation where conservative design can result in no-test circuit limits. The primary reason for this is that the high level output of RTL is tested only to the level

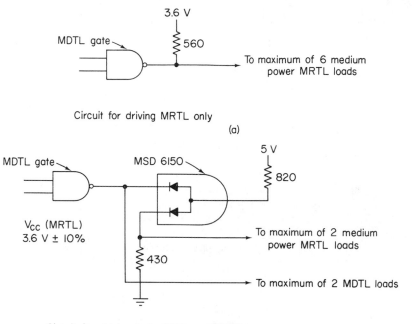

Circuit for driving MRTL only

(a)

Circuit for driving both MDTL and MRTL

(b)

Figure 8-27.   DTL to MRTL interface. *(Courtesy of Motorola)*

of V_{on} (about 0.67 to 1 V, depending on type and temperature). Although a typical RTL gate exhibits a high level output well above the DTL threshold, a test is always recommended.

The test required is relatively simple. With the datasheet value of V_{off} applied to all inputs of the driving RTL gate, the unloaded output voltage should be a minimum of 2.5 V at the maximum operating temperature. If so, RTL can be used to drive DTL directly without special interface.

In cases where driving both RTL and DTL from an RTL output is required (including the case of an unbuffered FF output driving a DTL component), use the circuit of Figure 8-28a. This is also the recommended alternative when the RTL fails the 2.5-V output test.

There are several techniques for insuring that an RTL gate can sink enough current to drive a number of DTL inputs. The most common method is to double the input loading factors of the driving

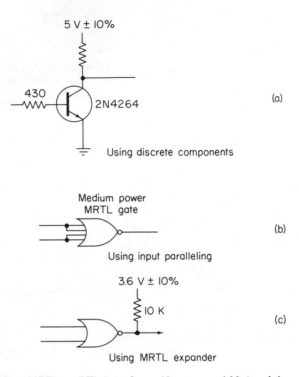

Figure 8-28. MRTL to DTL interface. *(Courtesy of Motorola)*

(RTL) gate (which is equivalent to allowing up to a typical 5 mA of additional sink current, or enough for 3 DTL gates). This is not a guaranteed condition, of course, but the incidence of devices which will not perform in this manner is very small. Even so, the conservative designer will drive at least two inputs of the RTL gate as shown in Figure 8-28b, thus insuring low-level drive capability under worst-case conditions.

Another worst-case design technique is shown in Figure 8-28c. This circuit requires an RTL expander, rather than an RTL gate, as an external resistor.

8–5 SUMMARY OF INTERFACING PROBLEMS

As discussed in preceding sections of this chapter, each logic IC line or family has its own set of design or usage problems. Most of these problems can be resolved using the datasheets and brochures supplied with the device. One area where the datasheets leave a gap is the interfacing between IC logic forms. For that reason, we shall summarize the interfacing problems here.

If MOS is involved, follow the recommendations described in Sec. 8–2.

If either ECL or HTL are involved, the best bet is to use the various level translators supplied with most ECL and HTL lines, or use the specific translator circuits shown in the datasheets and brochures. In the absence of such information, use the circuits described in Sec. 8–3.

If RTL must be interfaced, follow the recommendations of Sec. 8–4.

DTL and TTL are compatible with each other, thus eliminating the need for special translators or interface circuits. However, the load and drive characteristics may be altered somewhat. These changes are generally noted on the datasheets. In the absence of any DTL/TTL interface data, use the information of Sec. 8–1 through 8–4.

8–6 LINE DRIVER AND RECEIVER CONSIDERATIONS

Line drivers and receivers are used where it is necessary to transmit high-speed data from one system to another. The distances involved may vary from a few feet to several thousand feet. This section summarizes the various problems associated with IC line drivers and receivers.

8–6.1 System Description

A basic driver-receiver transmission system is shown in Figure 8-29a. Here, an input data stream (V_{in}) feeds a driver, which in turn drives a line. Information at the other end of the cable is detected by the receiver that provides an output data stream (V_{out}), usually of the same logic level as V_{in}.

The line involved could be a single line, a coaxial cable, a twisted pair line, or a multiline cable (ribbon cable type, multitwisted pair type, and so on). The line could be operated in a *single-ended* or *differential* mode.

Another common driver-receiver system, shown in Figure 8-29b, is commonly called a "party-line" or "bus" system. Here, the line is shared by drivers and receivers. It should be pointed out that although any driver can be used to drive the line, only one driver is used at one time.

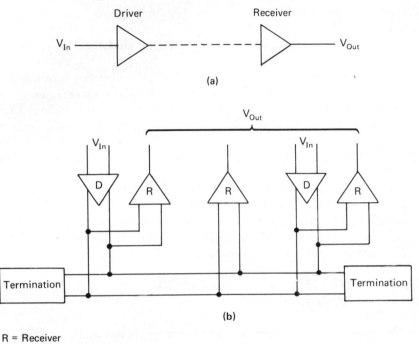

R = Receiver
D = Driver

Figure 8-29. Typical line driver-receiver systems.

8–6.2 Line Driver

The line driver commonly translates the input logic levels (TTL, MOS, CMOS, RTL, and so on) to a signal more suitable for driving the line. An important exception to this is found in the ECL logic family, where ECL gates may be used to drive the line directly.

Figure 8-30 shows two popular methods of driving lines between TTL systems. Figure 8-30a shows the basic pull-only circuit, whereas Figure 8-30b illustrates the basic push-pull circuit. Examples of both techniques are given in the following paragraphs.

8–6.3 Line Receiver

The line receiver provides the reverse function of a line driver. With a line receiver, the voltage previously applied to a line is now detected and restored to an output logic level. This function may be performed by various digital and linear ICs. Simple examples of this include comparators and gates. The latter includes a group of

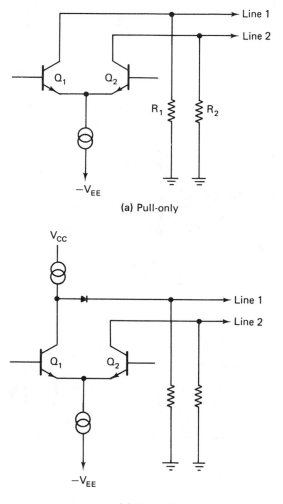

(a) Pull-only

(b) Push-pull

Figure 8-30.   Basic pull-only and push-pull line driver circuits.

ICs specifically designed as line receivers that combine linear and digital techniques. These ICs use linear circuitry to detect the output signal from the cable. This signal level may be quite small (less than 100 mV) with considerable noise. The digital circuits provide the necessary drive to interface with any of the common digital logic families.

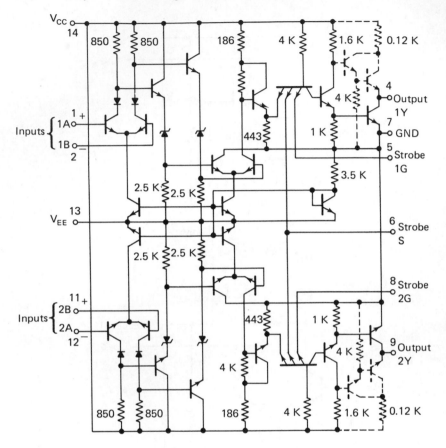

Figure 8-31. MC75107 line receiver. *(Courtesy of Motorola)*

An example of a line receiver is the Motorola MC75107 shown in Figure 8-31. This receiver uses a differential input stage to provide a high-input impedance and common-mode noise rejection. This stage is followed by a level-shifting stage and a second differential amplifier. The output stage provides TTL logic levels.

Since each receiver shunts the line, it is important to know the amount of loading contributed by the receiver. That is, how much resistance and capacitance is shunted across the line due to the input impedance of the receiver. Any significant loading of the line results in undesired reflections (as is the case with any transmission line).

Typical input impedance of the MC75107 is shown in Figure 8-32. Data for two frequencies, 5 and 10 MHz, versus input level is also shown. From this data, it can be seen that the input impedance increases with signal level. Even at minimum signal levels (30 mV) and maximum operating frequency (10 MHz), the equivalent parallel

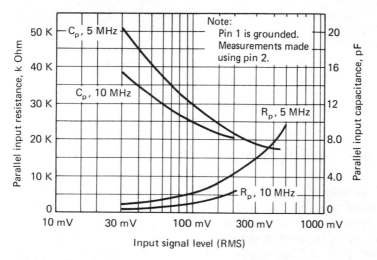

Figure 8-32. Parallel input impedance of MC75107 versus signal level. *(Courtesy of Motorola)*

resistance is typically 1000Ω. This is more than double the typical characteristic impedance of many lines.

8–6.4   Noise Considerations

Noise is an important consideration in the line transmission of digital data. Where noise appears and how it influences the system depends largely on the overall transmission line system used. This section discusses noise in several single-ended and differential transmission line systems.

Noise can appear in the ground system, or directly on the line, in driver/receiver systems. How much this noise degrades the signal-to-noise ratio appearing at the input terminals of the line receiver depends on the transmission system selected.

A simple single-ended system using one driver and one receiver is shown in Figure 8-33. Here, a single wire is used to transmit

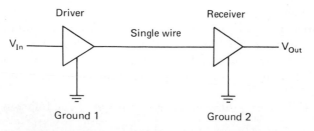

Figure 8-33. Basic single-ended data transmission system.

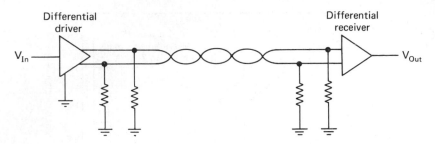

Figure 8-34.   Basic differential data transmission system.

the data signal. This system is particularly prone to both ground and induced noise. (Induced noise refers to that noise induced from nearby equipment, such as relays, and so on). Any induced noise on the line adds to or subtracts from the signal. Likewise, any ground noise between ground 1 and ground 2 adds or subtracts from the signal. Regardless of the noise source, the overall signal-to-noise ratio appearing at the input of the receiver is degraded.

One solution to the degradation of signal-to-noise ratio is to increase the signal level. For example, HTL can be used. Of course, this approach requires higher supply voltages as well as increased power.

Another possible solution is to use a shielded center conductor, coaxial cable. This approach successfully shields the line from induced noise. However, the problem of ground noise still remains with increased chance for data transmission error.

The second basic transmission system is the differential configuration shown in Figure 8-34. Since terminating the line at both ends is necessary, two sets of terminating resistors are shown. Lines are balanced in a differential system. Thus, externally-induced noise will appear equally on both inputs to the line receiver. The receiver uses a differential input stage to respond primarily to a differential signal, rejecting common-mode noise. This rejection of common-mode noise (as well as a reduction of ground noise) offers a significant advantage over the single-ended approach.

Although the rejection of common-mode noise is of paramount importance, the differential system also offers some immunity to differential-mode noise. The degree of differential noise immunity depends on several factors, such as line impedance, the matching of termination resistors, driver current capability, line length, and the clock frequency.

Calculating differential noise margin.   The following example

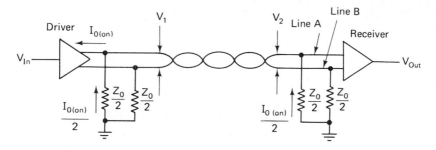

Figure 8-35. Differential data transmission system showing currents.

demonstrates how differential-mode noise margin may be calculated for a given system. For simplicity, an ideal system is considered where only the characteristic impedance of the line (Z_o) and the driver current capability $I_{o(on)}$ are included. With this system the transmission line is terminated in Z_o at both ends of the line. The assumption is also made that the driver is a pull-only type. Thus, the total drive current $I_{o(on)}$ for a particular logic state is effectively split between one terminating resistor at the driving end and another at the receiving end.

The voltage driving line A (V_1 of Figure 8-35) is:

$$V_1 = \frac{I_o \text{ (on)}}{2} \times \frac{Z_o}{2}$$

or

$$V_1 = \frac{I_o Z_o}{4}$$

Since the line is assumed to be short enough to neglect line loss, V_1 is also the input to the receiver ($V_1 = V_2$). When line B goes low during the other half of the input logic swing, V_1 assumes essentially the same value with an opposite polarity.

If the following assumptions are made, we can plot noise margin (see Figure 8-36) for a typical line receiver, such as the MC75107: $Z_o = 170\Omega$, $I_{o(on)} = 6.9$ mA (minimum value at 25°C), receiver threshold = 50 mV. Using these values:

$$V_1 = \frac{I_o Z_o}{4} = \frac{(6.9)\,(170)}{4} = 293 \text{ mV}$$

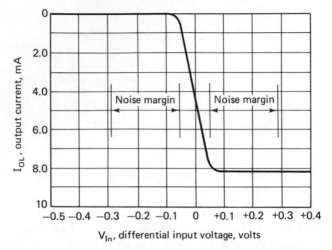

Figure 8-36.   Noise margin of MC75107. *(Courtesy of Motorola)*

The noise margin from Figure 8-36 is:

$$\text{Noise margin} = 293 \text{ mV} - 50 \text{ mV}$$
$$= 243 \text{ mV}$$

These values of noise margin assume that the clock frequency and/or line length is small enough to neglect signal attenuation by the line.

8–6.5 Line Attenuation and Impedance

Figure 8-37 shows the attenuation of various cables used as data transmission lines. A *twisted-pair* line is included with the coaxial cables. The twisted-pair line has more attenuation than the coaxial cables, but shows the same square-law attenuation characteristics. That is, the resistive loss varies in accordance with the square root of frequency.

8–6.6 Driver/Receiver Applications

Applications involving line drivers and receivers may vary from a simple transmission system between two points to a more involved party-line system. The distance in either system may vary from a few feet to thousands of feet. Cables used to transmit the data may vary from a simple two-wire twisted-pair to a multiline cable. Thus, there are numerous ways to implement a transmission system depending on the overall system requirements and the discretion of the designer.

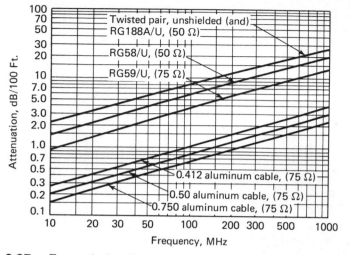

Figure 8-37. Transmission line attenuation versus frequency. (*Courtesy of Motorola*)

The following sections describe several line driver and receiver examples. The examples shown are divided into two groups: differential and single-ended systems.

8–6.7   Differential

This section describes several circuits using TTL or MECL compatible ICs. Since the differential mode is the most popular approach, a brief design analysis is included for one circuit.

Figure 8-38 uses the Motorola MC75110L and MC75107L (line driver and receiver respectively). The transmission line is the same twisted-pair shown in Figure 8-37.

The attenuation of the twisted-pair line is given as 2.6 dB/100 feet at 10 MHz. The value of $I_{o(on)}$ is typically 12 mA for the MC75110L. The voltage driving *one line* of the twisted-pair is:

$$V_1 = \frac{I_{o(on)}\, Z_o}{4}$$

$$= \frac{(12\ \text{mA})\,(100)}{4}$$

$$= 300\ \text{mV}$$

Where Z is the differential-mode impedance. (Note that the differential-mode impedance is considered to be twice the normal characteristic impedance.)

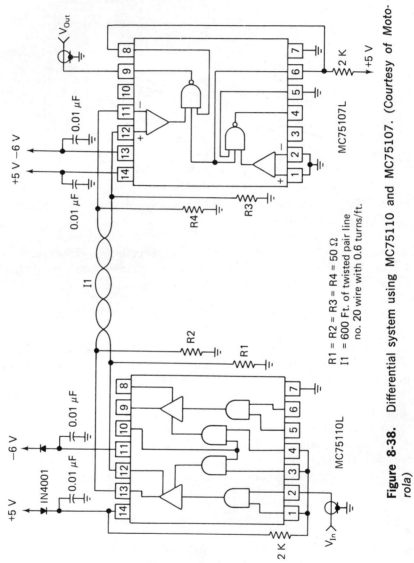

Figure 8-38. Differential system using MC75110 and MC75107. *(Courtesy of Motorola)*

Since two lines are involved in a twisted pair, the *differential* input voltage to the cable is 600 mV (or twice that of one line).

The threshold voltage of the MC75107 is 25 mV. Thus, a good design value (to offset practical system variables such as termination resistor match, and so on) is 6 dB higher or 50 mV. The allowable cable attenuation (A) is then:

$$A = \frac{600 \text{ mV}}{50 \text{ mV}} = 12 \text{ or } 21.6 \text{ dB}$$

The clock rate is chosen to be 18.5 MHz. For this clock rate, attenuation is approximately 3.53 dB/100 feet, as shown in Figure 8-37. Under these conditions, the maximum line length should be:

$$\text{length} = \frac{21.6 \text{ dB}}{3.53 \text{ dB}/100 \text{ feet}} = 612 \text{ feet.}$$

Single pulse problems. A single pulse corresponds to the case where a logic 1 with a long series of zeros is being transmitted. For the pull-pull driver used in this example (MC75110L), this means that one line will be at ground, and the other line at -300 mV. This negative offset voltage provides an additional restriction on the receiver (MC75107L). Depending on how the line is connected to the receiver, the receiver will either be "on" or "off" until the input signal exceeds approximately 300 mV.

For example, where the 300 mV offset biases the receiver "off," the differential signal to the receiver must be at least 325 mV. This value consists of 300 mV to overcome the offset voltage and 25 mV to overcome the threshold of the receiver. Thus, for an input differential signal of 600 mV the maximum allowable pulse attenuation is:

$$\frac{325 \text{ mV}}{600 \text{ mV}} = 0.54 \text{ or } 54 \text{ percent}$$

The single-pulse problem can be corrected by reducing the clock frequency, thus allowing the pulse to remain long enough for the line voltage to reach the desired level. The required pulse duration can be calculated using the graph of Figure 8-39, and a simple equation.

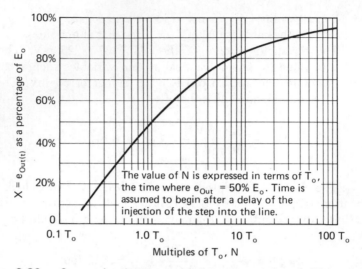

Figure 8-39. Curve showing $e_{out(t)}$ (as a percentage of E_o) versus N. *(Courtesy of Motorola)*

Figure 8-39 shows the output voltage $e_{out(t)}$ as a percentage of the full voltage, in terms of T_o. One T_o is the time required to reach 50 percent of full voltage, and is found by the equation:

$$T_o \text{ (in ns)} = \frac{(4.56)\,(a)^2}{f}$$

where a is the total attenuation of the line in dB and f is frequency in MHz.

As shown in Figure 8-37, the attenuation per hundred feet of the twisted-pair line is 2.6-dB, at a frequency of 10 MHz. Since there are 612 feet, the total line attenuation in dB is: 6.12×2.56, or about 15.6 dB. Applying this attenuation and frequency to the T_o equation, we get:

$$T_o = \frac{(4.56)\,(15.6)^2}{10} \approx 109 \text{ ns}$$

Referring to Figure 8-39, we need about 55 percent (anything above 54 percent) of full output to get 325 mV. 55 percent of full output requires a T_o of about 1.25. 109 ns $\times$ 1.25 = about 136 ns.

Assuming that the single 136-ns pulse represents 50 percent of the duty cycle, the full duty cycle is 272 ns, and the required clock frequency is about 3.68 MHz ($\frac{1}{272}$ ns).

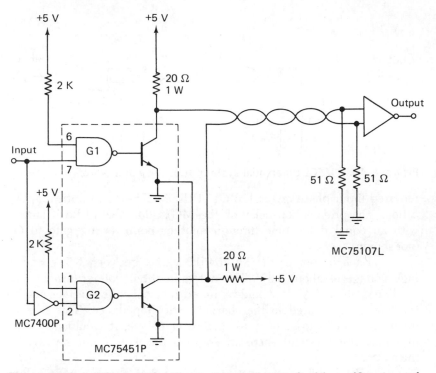

Figure 8-40. Differential system using single supply driver. *(Courtesy of Motorola)*

The differential system shown in Figure 8-38 uses only the two ICs, one line driver and one receiver. The strobe feature of the ICs allows these particular units to be used in party-line or bus applications. For example, if several drivers (MC75110L) are used to drive the line at different times, each can be enabled when desired. In the meantime, the other drivers not in use are disabled. Thus, loading by the unused drivers is held to a minimum.

During party-line operation, it is recommended that each positive and negative feedline to each line driver be fed through a diode (1N4001). Consequently, during power shutdown, a particular driver will not have its outputs pulled down below ground (substrate) when another driver pulls the line low.

Differential system using single supply driver. Figure 8-40 shows a differential system using the MC7541P and the MC75107L. The MC7541P with a few external components provides a differential signal from a single +5 V supply. The MC75107L requires two supplies. The single external gate shown in Figure 8-40 provides the

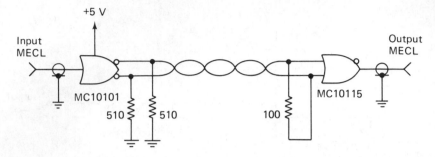

Figure 8-41. MECL differential system. *(Courtesy of Motorola)*

required input phase reversal to G_2 of the MC7541P. In critical applications, the propagation delay of the MC75400P should be balanced with an equal delay time through another noninverting gate to G_1 (not shown).

Each output of the MC75451P varies between 0.5 V and a high voltage of about 3.6 V. The net differential voltage driven into the line is about 6 V. It should be noted that only the receiver end of the line is terminated in the characteristic impedance. This is adequate for a point-to-point transmission; of course, it would not work in a party-line system where an enabled receiver is not necessarily the end of the line.

MECL differential system. Figure 8-41 shows a MECL line driver and receiver using gates. This system can be operated from a single supply (such as +5 V shown). The circuit shown uses MECL logic levels. If TTL levels are desired, the MC10124 and MC10125 can be substituted for the driver and receiver, respectively. The MC10125, however, requires an additional negative supply (−5.2 V).

8–6.8 Single-ended

Figure 8-42 shows a point-to-point, single-ended system using a positive supply only. This circuit supplies a 4.2 V input pulse to the transmission line. Two important considerations here are *rise time* and *dc offset*. The frequency for a given length of cable must be low enough to insure adequate "edges" (rise time or rising edge of the pulse) and dc offset for the MC7400P. Excessive offset, as previously discussed, results in the receiver always being "on" or "off."

Figure 8-43 shows a single-ended system using the MC1488 and MC1489 line driver and receiver, respectively. These ICs are intended primarily as interfacing devices that meet EIA specification

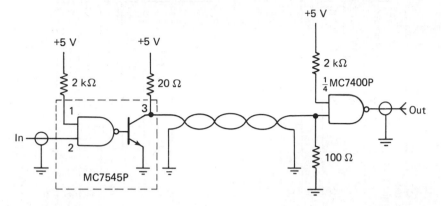

Figure 8-42. Single-ended, single-supply system. *(Courtesy of Motorola)*

RS–232C. Both the driver and receiver are flexible in terms of use with DTL, TTL, MOS, and MECL logic levels. The system of Figure 8-43 is intended for an input and output data stream using TTL. The typical operating frequency is 2 MHz.

8–6.9   Product Selection

The selection of line driver and receivers will depend on many factors including cost, power supply voltages, logic level, clock rate, length of time, and so on. Available Motorola line drivers and receivers are summarized in Figure 8-44 through 8-46.

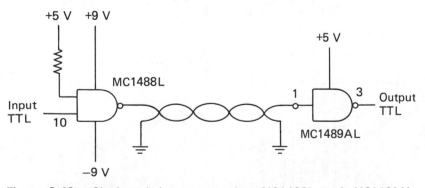

Figure 8-43. Single-ended system using MC1488L and MC1489AL. *(Courtesy of Motorola)*

Functional description	Logic compatibility	Output current ON mA typ.	Output current OFF µA max.	Prop. delay input to output ns. typ.	Prop. delay inhibit to output ns typ.	Device number
Dual driver with inhibit inputs	MTTL	6.0 mA	100	9.0	16	MC75109
Dual driver with inhibit inputs	MTTL	12 mA	100	9.0	16	MC75110
Driver, push-pull	MDTL/MTTL	±20 mA	—	20	—	MC75113
Quad driver with strobe	MTTL	11 mA	100	9.0	18	MC3453
Quad driver, single-ended	MDTL/MTTL	Conforms to EIA specification RS-232C				MC1488

TTL line drivers

Functional description	Logic compatibility	Output current ON mA	Prop. delay input to output ns typ.	Device number
Dual NAND driver with non-connected output transistors	MTTL	300	21	MC75450
Dual NAND driver	MTTL	300	71	MC75451
Dual AND driver	MTTL	300	18	MC75452
Dual NOR driver	MTTL	300	17	MC75453
Dual OR driver	MTTL	300	25	MC75454

TTL peripheral drivers

Figure 8-44. TTL line and peripheral drivers. (*Courtesy of Motorola*)

Functional description	Logic compatibility	Input threshold mV max	Input common mode range V min	Prop. delay input to output ns typ.	Prop. delay input to output ns typ.	Device number
Dual line receiver with active pull-up	MTTL	±25 mV	±3 V	18	10	MC75107
Dual line receiver with open collector output	MTTL	±25 mV	±3 V	19	13	MC75108
Quad line receiver with active pull-up (tri-state strobe)	MTTL	±25 mV	±3 V	25	10	MC3450
Quad line receiver with open collector output	MTTL	±25 mV	±3 V	25	13	MC3452
Quad receiver, (RS-232C) single-ended	MTTL	–	–	120	–	MC1489, A
Dual line receiver single-ended	MTTL	±100 mV	–	22	10	MC75140

Figure 8-45. TTL receivers. (Courtesy of Motorola)

Single ended	Propagation delay	Gate power	Edge speed	Typical frequency capability
10,000 series (all outputs)*	2.0 ns	25 mW	3.5 ns	150 MHz
10123 bus driver	2.0 ns	25 mW	3.5 ns	150 MHz
MECL III (all outputs)*	1.0 ns	6.0 mW	1.0 ns	350 MHz
Differential (complementary outputs)				
10 K series	2.0 ns	25 mW	3.5 ns	150 MHz
MECL III	1.0 ns	60 mW	1.0 ns	350 MHz
MC10124 quad TTL-MECL transistor	5.0 ns	85 mW	2.5 ns	75 MHz

*Note:
All MECL 10,000 and MECL III outputs specified to drive line impedances as low as 50 Ω (terminated to −2 volts).

MECL compatible line drivers

Figure 8-46. MECL compatible line drivers and receivers. (Courtesy of Motorola)

	Propagation delay	Gate power	Edge speed	Typical frequency capability	Common mode
Single ended					
10,000 series (all inputs)*	2.0 ns	25 mW	3.5 ns	150 MHz	—
MECL III (all inputs)*	1.0 ns	60 mW	1.0 ns	350 MHz	—
Differential					
MC10114	2.0 ns	25 mW	3.5 ns	100 MHz	±1 V
MC10115, quad	2.0 ns	25 mW	3.5 ns	125 MHz	+0.7 V, −1.5 V
MC10116, triple	2.0 ns	25 mW	3.5 ns	125 MHz	+0.7 V, −1.5 V
MC10216, triple	1.2 ns	25 mW	2.5 ns	150 MHz	+0.7 V, −1.5 V
MC1692	1.0 ns	60 mW	1.0 ns	325 MHz	+0.7 V, −1.0 V
MC10125, quad MECL-TTL translator	5.0 ns	90 mW	2.5 ns	75 MHz	±2.5 V

*Note:
Input impedance is typically 50 kΩ shunted with 3.5 pF

MECL compatible line receivers

Figure 8-46. (Continued)

8–7   INTERFACE CONSIDERATIONS FOR NUMERIC DISPLAY SYSTEMS

This section describes several methods of multiplexing multidigit seven-segment displays. The logic devices illustrated are primarily CMOS and TTL. The displays discussed are liquid crystal displays (LCD), light-emitting diodes (LED), gas discharge, fluorescent, and incandescent. The following paragraphs describe both the interface considerations, and how to interface between specific logic devices and the displays.

8–7.1   Display Systems

There are two basic types of display systems: direct drive and multiplex. Generally, it is more economical to multiplex displays of greater than four digits. Thus, multiplexing is emphasized in the following discussions. Several examples of time sharing are included.

Direct drive displays.   The simplest type of display system, shown in Figure 8-47a, consists of the four lines of BCD information feeding the decoder/driver whose outputs drive the display. This direct drive system does not have information storage capability, and thus reads out in real time. Another display system, particularly one that has less than four digits, is shown in Figure 8-47b. This system

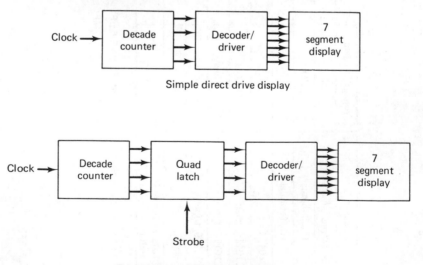

Simple direct drive display

Typical direct drive display system with storage

Figure 8-47.   Basic direct-drive display and direct-drive display system with storage. *(Courtesy of Motorola)*

contains a decode counter, quad latch, decoder/driver and display, one channel for every digit. This system has storage capability (the latches) which allows the counter to recount during the storage time.

The decoders discussed throughout this section are the BCD-to-seven-segment decoders used for driving the popular seven-segment displays. The older one-of-ten decoders have generally been replaced by the seven-segment devices.

Multiplex displays. The most commonly used system for multidigit displays is the multiplexed (or time-shared or strobed) system as shown in Figure 8-48. By time-sharing the one decoder/

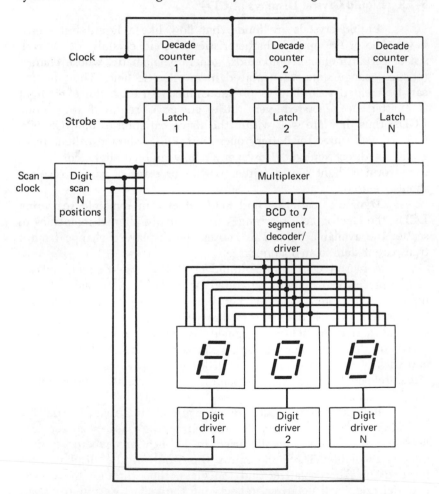

Figure 8-48.   Basic multiplexed display system. *(Courtesy of Motorola)*

driver, the parts count, interconnections, and power can be saved. The
N stage data register (one stage for each digit) feeds a scanned multiplexer whose sequenced BCD output drives like segments of the
display. The digit select elements are sequentially driven by the scan
circuit which also synchronously drives the multiplexer. Thus, each
display is scanned or strobed in synchronism with the BCD data presented to the decoder at a sufficiently high rate, usually greater than
50 scans/second, to appear as a continuously energized multidigit
display.

8–7.2   Liquid Crystal Displays (LCD)

Liquid crystals are fluids that flow like a liquid but which
have some of the optical characteristics of solid crystals. LCDs consist of certain organic compounds, generally, nematic, whose characteristics change state when placed in an electric field. Thus, images
can be created according to predetermined patterns. Since no light
is emitted or generated, very little power is required to operate
LCDs; thus they are well within the drive capabilities of MOS ICs,
and are well suited for battery operation. LCDs show excellent read-
ability in direct sunlight. However, in low ambient light conditions,
some form of light source (either within or external to the display)
should be used.

Dynamic scattering and field effect.   In dynamic scattering
LCDs, the electric field rearranges the normally aligned molecules to
scatter the available light. This causes the display to change from a
transparent state to an opaque state.

A more recent form of liquid crystal, the field effect, consists
of two pieces of plate glass separated by a glass spacer/seal with the
liquid crystal material injected between the two plates. A metalliza-
tion pattern is etched onto the glass. The glass and liquid crystal
material under the selected segment is activated when a field is
placed on that particular segment. Polarizers are attached to the front
and back of the display. The light striking the lower polarizer will be
either reflected or absorbed, depending on the relative direction of
the polarizers.

Reflective and transmissive operation.   If a reflective display
is desired, reflective material is adhered to the back polarizer. Reflective displays are generally used in applications where ambient
light is available. The light enters the front of the display, goes
through the front polarizer and the glass, then through the crystal
material and back polarizer, reflects off the reflector, and back out
through the front of the display. Reflective operation thus uses external light energy and requires virtually no power.

In environments where only a small amount of light is available, a back lighted transmissive display is desirable. This display is assembled simply without the reflector material. Backlighting is generally done with a diffused incandescent or fluorescent light source.

Temperature range. The temperature range of currently available LCDs is limited to an operating range of about 0°C to +60°C. At severe low temperatures, LCDs usually require a heating system, possibly in conjunction with the back lighting.

AC drive signal. Liquid crystals require an ac drive signal, with no dc component. If dc voltages are applied, electrolysis and plating action can occur, decreasing the display life. However, preliminary experimentation that the Motorola passivated field-effect display can tolerate some dc bias on the ac excitation signal without seriously degrading the life of the device. For field-effect displays, this excitation signal may go as low as 2 V peak, typically at frequencies of 60 Hz to 10 kHz. The excitation signal for dynamic scattering displays is in the 7 to 30 V peak range, at 20 to 400 Hz.

8–7.3   Direct-drive LCD

Due to the relatively slow turn-on and turn-off times of LCDs, and the need for ac drive signals, simple multiplexing techniques have not been effective. It is usually simpler to direct drive LCDs, with each digit having its own counter, latch, decoder, and driver. The preferred method of generating the display ac drive signal is by means of an EXCLUSIVE OR gate where one input is the decoder segment signal, and the other input is the symmetrical square wave (with no dc component) excitation signal. Such an arrangement is shown in Figure 8-49a. Note that the segments are illuminated in accordance with the decimal number applied at the BCD input. For example, for a decimal 3, the BCD signal is 0011, and segments a, b, c, d, and g are illuminated. Segments e and f are not illuminated, and the display forms a numeral 3.

The excitation signal also feeds the LCD backplane. The output of the EXCLUSIVE OR is a square wave whose phase is related to the decoder output logic level. When the segment is to be de-energized, the segment drive signal is of the same phase and magnitude as the backplane signal; thus, there is no potential difference across the display. Conversely, when the segment is to be energized, the two plates will have two 180° out-of-phase signals across them, producing a square wave whose peak-to-peak voltage is twice the IC power supply value and whose average value is zero, as shown in Figure 8-49b.

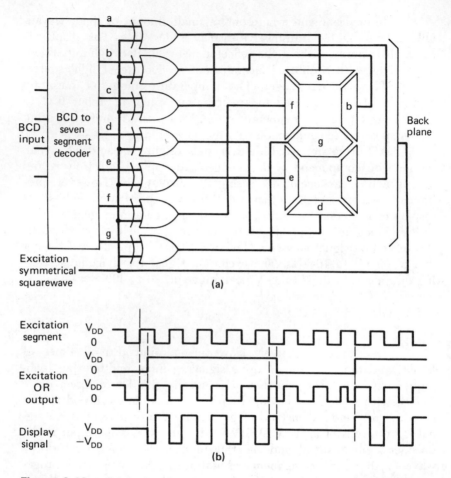

Figure 8-49.   Drive signal and waveforms for LCD using EXCLUSIVE OR gates. *(Courtesy of Motorola)*

CMOS, with its extremely low input power requirements and large power supply operating range, is ideally suited for driving LCDs, particularly for battery operated equipment. In the example of Figure 8-50, the BCD inputs are generated from the four cascaded up-counters (two MC14518 dual BCD up-counters) whose respective outputs feed the MC14543, a decoder designed specifically for driving LCDs. The decoder contains the required EXCLUSIVE OR gates, with the excitation signal applied to the Phase (Ph) input. When a 15 V power supply is used, a display excitation voltage of 30 V peak-to-peak is generated, a value compatible with both dynamic scattering and field-effect displays.

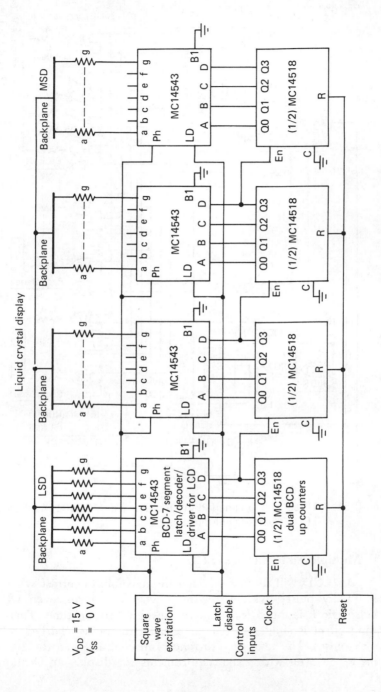

Figure 8-50. Four-digit direct drive LCD. (Courtesy of Motorola)

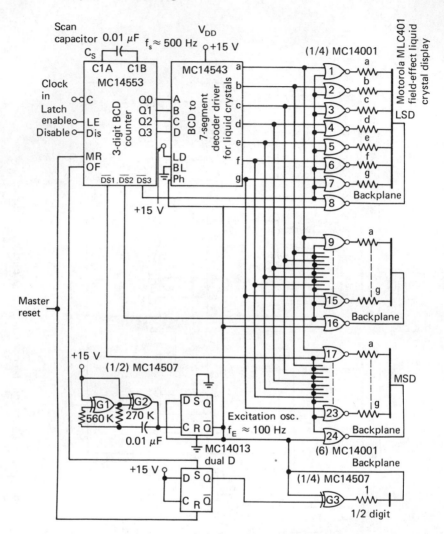

Figure 8-51. Three and one-half digit multiplexed MCC401 field-effect LCD. *(Courtesy of Motorola)*

8–7.4 Multiplexed Field-effect LCD

An LCD will ultimately reach its maximum contrast when it is continually pulsed with a low-duty cycle waveform, even when the pulse width is much less than the display turn-on time. This integrating effect is similar to that of a capacitor being charged by a series of pulses. The total time to reach full contrast is in the 100 to 400 ms range, well within the requirements for display applications.

Of the two basic types of LCDs, the dynamic scattering and the field effect, the excitation voltage is less for the field effect, with a corresponding lower threshold voltage (the voltage level at which the devices switches off to on). When the duty cycle of the excitation voltage is varied, the effective threshold varies inversely. Thus, as the number of multiplexed LCD digits increase (duty cycle decreases), the threshold, and therefore the required power supply, also must increase. For dynamic scattering liquid crystals, the required power supply is beyond the range of the CMOS ICs. Thus, the following example is limited to field-effect LCDs.

Figure 8-51 shows a 3½-digit LCD multiplexer which incorporates a Motorola MLC401 readout. The three-digit BCD counters (and respective latches), and the multiplexer and scan circuitry are all contained in one CMOS package, designated as the MC14553 three-digit BCD counter. The time division multiplexed BCD outputs are fed to the inputs of the MC14543 BCD-to-seven-segment latch/decoder/driver for liquid crystals. For other types of display (LED, incandescent, and so on), the digits can be strobed by simply scanning the digit drivers. Thus, all common digit segments can be tied to their respective decoder outputs. However, for LCDs, this approach cannot be used since the display will always see a signal across it, even when the backplane (digit select) is not strobed. The resultant segment signal is single-ended, and thus has an undesired dc component.

When gates are placed in series with the segment lines, the display can readily be strobed. For this example, MC14001 quad two-input NOR gates are used, with the control inputs being the negative-going digit select pulses from the MC14553 scan circuit. During the nonselect time when the scan output is high, the outputs of the NOR gates are low regardless of the state of the excitation signal. Zero voltage is then present across the MLC401 display, as shown in Figure 8-52. During the select low level, the gate outputs are controlled by the decoder output, and excitation signal. The selected digit then operates as a direct driven system.

The excitation frequency for the field-effect LCD can vary from 60 Hz to 10 kHz, with the 100 Hz used in this example being derived from the two CMOS gate ($G1$ and $G2$) FF oscillator. This toggle FF square wave circuit ensures that the excitation signal will have an exact 50 percent duty cycle with no dc component. The scan frequency, using capacitor C_S and the MC14553 internal oscillator, is set for approximately 500 Hz, resulting in a display refresh rate of approximately 170 Hz (500/3), a frequency well beyond the detectable flicker rate of 30 to 60 Hz. The effective threshold voltage

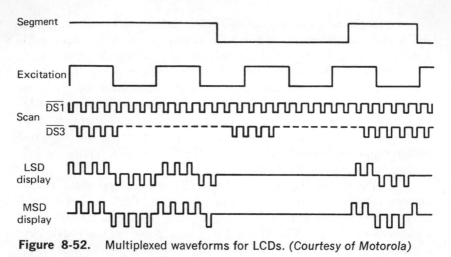

Figure 8-52.   Multiplexed waveforms for LCDs. *(Courtesy of Motorola)*

for a 3-digit scan (33 percent duty cycle) is approximately 6.5 V. The use of a 15 V power supply ensures that the device is driven well above threshold, with the resulting LCD voltage being approximately 30 V peak-to-peak.

8–7.5 Fluorescent Displays

Fluorescent displays, both the diode and triode types, are electrically similar to their vacuum-tube counterparts, the exception being that the anodes are coated with a phosphor. When a positive potential is placed across the anode and the directly heated cathode (or filament), the electrons hitting the anode will cause the phosphor to fluoresce and emit light. The light output peaks in the blue-green range which, when appropriately filtered, can display other colors.

Triode displays with their control grids are somewhat easier to multiplex since the strobing is performed at a lower power level. The diode fluorescent displays are multiplexed in the relatively high power filament circuit which can be powered by either ac or dc. The display digits can be packaged in individual tubes, or in a planar multiple-digit display contained in a single envelope with all like segments interconnected.

8–7.6 Six-digit Fluorescent Triode Display

Fluorescent displays can be energized by series or shunt switching a positive voltage to the anode (with respect to the cathode). *Series switching* is when the display anode is energized through a series semiconductor switch (bipolar or MOS). In *shunt switching*, the voltage is applied through a limiting resistor with the energized switch (across the display) clamping the display off.

The six-digit fluorescent triode display system of Figure 8-53 uses two sets of cascaded counters and decoders similar to that of Figure 8-51 to produce the six-digit count. Series switching with the MC14511 BCD to seven-segment latch/decoder/driver is used in Figure 8-53. The MC14511 units have output emitter followers capable of sourcing the required anode currents, and can thus drive the display directly. The cathodes of the display are set at approximately −18 V derived from the −28 V supply, and the 10 V series zener diode $D1$.

Digit scanning is done by turning on the grid control PNP transistors $(Q1–Q3)$ with the negative-going digit select outputs of one MC14533. This switches the grid from the OFF (−28 V) condition to the ON (approximately +15 V) value. The zener diode ensures that the grid-cathode is back-biased in the off mode, and satisfies the display's specification of −7 V minimum grid cutoff voltage. When energized, the display has approximately 32 V across it ($V_{oH} + V1 - V_z = 14 + 28 - 10 = 32$).

Limiting resistors $R1–R3$ are required when interfacing between a 15 V powered CMOS gate and bipolar transistor to prevent the current into the gate from exceeding 10 mA. For high temperature operation, emitter-base resistors are recommended for I_{CBO} considerations. In this example, filament power is supplied by a floating 1.5 V supply, referenced to the −18 V cathode, but it can also be powered by a 1.5 V transformer secondary.

Timing for the MC14553 counters is derived from one CMOS IC MC14572 with the Disable pulse obtained from the two-inverter, nonsymmetrical, astable multivibrator. Diode $D2$ allows the timing capacitor $C1$ to charge and discharge through two resistive paths; $R7$ for the long count duration time, and $R7$ in parallel with $R8$ for the short, positive-going, approximately 100 mS wide Disable (Dis) pulse. The Latch Enable (LE) and Master Reset (MR) pulses are obtained from the three, cascaded, half-monostable circuits.

8–7.7   Light-emitting Diodes (LED)

LED displays have taken over an appreciable part of the display business in the last several years, primarily due to their high reliability, long life, fast response time, operation from low voltage, dc, and ready adaptability to miniature displays. LEDs are

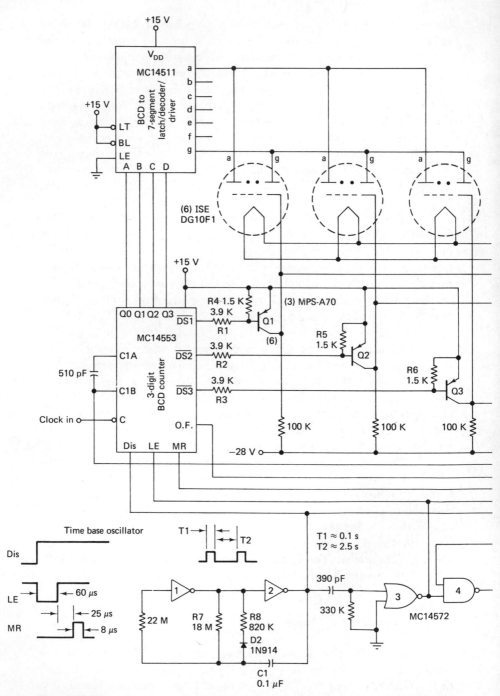

Figure 8-53. Six-digit fluorescent triode display. *(Courtesy of Motorola)*

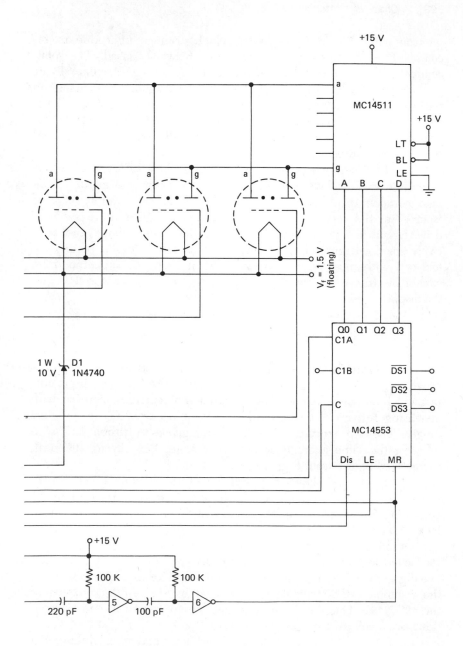

Figure 8-53. (Continued)

semiconductor PN junction diodes which produce light due to re-combination of holes and electrons when forward biased. The semi-conductor material is either gallium arsenide phosphide (GaAsP) or gallium phosphide (Gap), with the former being more prevalent in red display applications. Similar to any junction diode, the voltage drop across the forward-biased junction is relatively constant (ap-proximately 1.6 V for GaAsP) and is generally driven with a con-stant current source.

There are two possible connections when LEDs are used as seven-segment displays; common cathode and common anode. *Common cathode* displays require the drive circuit to supply current (source) to the segments. *Common anode* displays require segment drive circuit sink capability. Because of their low voltage and rela-tively small current requirements, LED displays can be readily in-terfaced with most families of ICs. When such ICs lack the drive capability, transistors can be easily interfaced between the ICs and the display.

8–7.8   Five-digit LED Display

Typical GaAsP currents are in the 5 to 30 mA per segment range, generally varying with the size of the display. The light out-put and luminous efficiency (light output/unit current) increase with increasing forward current until saturation occurs (100 to 150 mA range). These properties make it advantageous to strobe LEDs so that, for the same average current with lower duty cycle, the peak current and light output is greater.

Figure 8-54 shows a typical five-digit real-time LED display system. This system is designed for a peak forward current of about 40 mA (40/5 = 8 mA average). The MC75491 quad, and the MC75492 Hex MOS to LED drivers are ideally suited for this multiplexing application as only three interface ICs are required; two quads for the seven-segment-plus-decimal-point drivers, and one hex for the five-digit drivers and production circuit. All like anode segments of the common cathode displays are driven by the emitter outputs of the MC75491. This device must source the 40 mA LED current con-tinuously (each peak current sequentially).

The MC75492 digit drivers must be capable of sinking the total seven-segments-plus-decimal-point current of 320 mA (8 × 40 mA) for the 20 percent duty cycle. This resultant 64 mA average cur-rent is well within the 250 mA continuous maximum rating of the MC75492. Resistors $R1–R8$ in the collector circuits of the quad driv-ers set the LED current as follows:

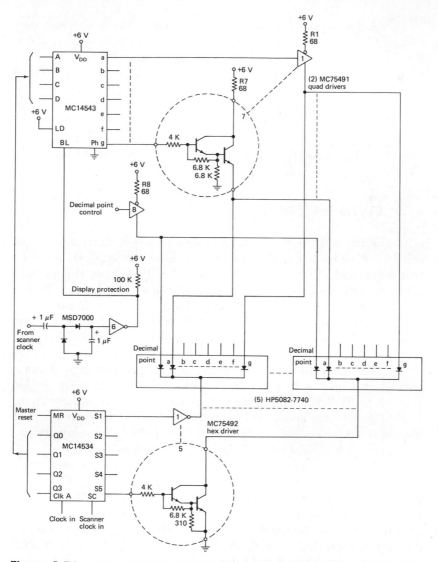

Figure 8-54. Five-digit LED display. *(Courtesy of Motorola)*

$$I_f = \frac{V_{DD} - V_{CE(\text{sat})}\,591 - V_D - V_{CE(\text{sat})}\,592}{R}$$

$$I_f \approx \frac{(6.0 - 0.8 - 1.7 - 0.9)\ \text{V}}{68\ \Omega}$$

$$I_f \approx \frac{(6.0\ \text{V} - 3.4\ \text{V}) \approx 39\ \text{mA}}{68\ \Omega}$$

A 6 V supply is chosen to allow a more nearly constant current LED drive, and to ensure adequate worst-case base current for the direct-coupled LED drivers. The approximately base currents for the quad and hex drivers are 170 and 320 μA, respectively. These values are compatible with the worst-case sourcing capability of the CMOS address outputs. The required active high outputs from the MC14543 are obtained by tying the Phase input to ground. The blanking input of the 14543 is used to blank the decoder output, and thus protect the LEDs and digit drivers if the scan circuit were to fail.

8–7.9 Gas Discharge Display

First of the gas discharge tubes to reach the market, and still in use with little noticeable change from the original introduction in the mid 1950s, was the Nixie ® tube. They have always been a highly readable device, with the drawback of not having the numerals in one plane. The ten preshaped cathodes (0 to 9) are physically stacked within the envelope, which causes the display to jump in and out as the digits are changed, and limits the useful angle of view.

The more recently introduced planar gas discharge displays have their numerals in one plane, facilitating wide viewing angles, from 1½ to 16 digits being contained in the one neon-filled envelope. Each digit has one anode and seven or more cathode segments. The like segments can be tied together as in most multidigit displays, or can be brought out individually (if the number of digits is small).

When a voltage greater than the ionization potential, generally 170 V, is applied between the selected anode and cathode, the gas will ionize and an orange glow will appear around the cathode. For multiplexed displays, a blanking period is required between the cathode select and the anode scan pulses. This ensures that the previous digit is completely deionized before the following digit is strobed and thus prevents erroneous readouts.

Since gas discharge displays required high ionization potentials, interfacing from low voltage logic ICs to this higher voltage requires level translation for shifting. This can be done by translating upward to the anode drivers with the cathode drivers referenced to the logic potential, or downward to the cathode drivers with the anode circuits at the logic potential. Level shifting can be done by directly-coupled high voltage transistors, or by capacitor coupling.

8–7.10   Twelve-digit TTL Multiplexed Gas Discharge Display

Figure 8-55 shows a twelve-digit TTL multiplexed planar gas discharge display system using capacitor coupling to the display anode drivers. The system uses the standard latch-decoder multiplexing technique, with the anode scan circuit derived from a counter-decoder. Twelve BCD words form the data inputs to the MC4035 quad latches. The similar outputs of the latches are connected in parallel to the four input lines of the time-shared MC7448 seven-segment decoder/driver.

Multiplexing occurs when the sixteen latches are sequentially enabled. The approximate 6 kHz scan frequency is generated by the two inverter (13, 14) astable multivibrator whose output clocks the MC7493 four-bit binary counter. The counter outputs provide the inputs to the MC8311 1-of-16 decoder. The thirteenth output pulse is fed back through an inverter to reset the counter. The resulting twelve sequential pulse outputs are used to both drive the PNP display anode transistors, and the twelve latch enable inputs through their respective inverters.

Display cathode blanking is obtained by differentiating the scan oscillator output through two cascaded, capacitor coupled inverters (15, 16). This signal is fed to the RBI output of the MC7448, resulting in a cathode drive signal that has an approximately 20-μS leading edge, and 50-μS trailing edge blanking interval, relative to the anode digit select pulse. The outputs of the MC7448 are resistive coupled to the NPN cathode segment drivers (Q1–Q7) to supply approximately 100 μA of base current. The 10-kΩ base-emitter resistors allow for high temperature operation. The collector resistors (R25–R31) limit the peak segment currents to approximately 0.8 mA.

To address a particular segment requires turning on the appropriate anode and cathode drive transistors. In the off condition, the collectors of these transistors are referenced to a voltage divider (R65–R67) through isolation resistors (R46–R57 and R58–R64) which allow using lower breakdown voltage devices. For the resistor values shown, the anode and cathodes are at approximately 115 V and 65 V, respectively, when not addressed, resulting in about 50 V across the display. This is well below the ionization potential. When addressed, the anode is switched to the supply voltage (less $V_{CE(sat)}$) and the segment limiting resistors to near ground, thus causing the display to ionize.

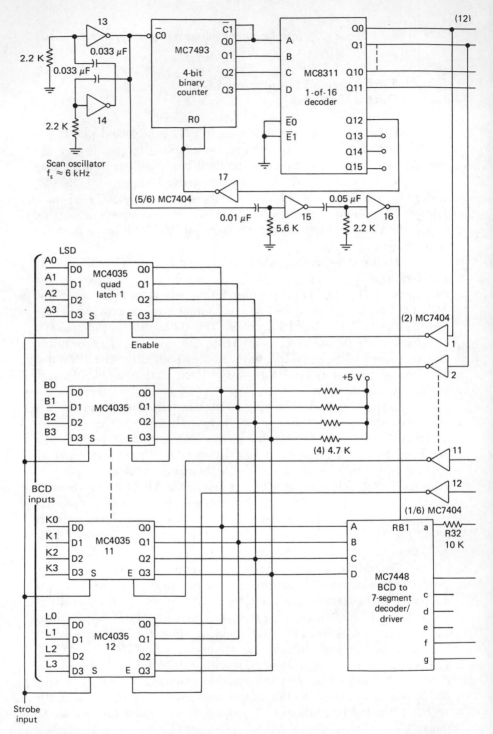

Figure 8-55. Twelve-digit TTL multiplexed display Panaplex II. *(Courtesy of Motorola)*

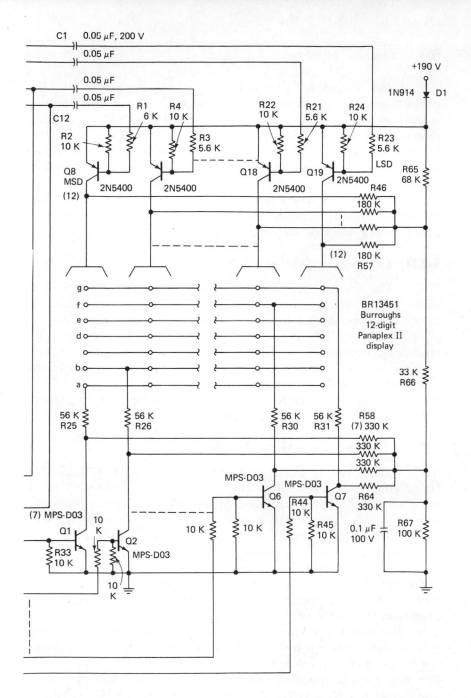

Figure 8-55. (Continued)

403

The anodes are scanned by the negative-going MC8311 pulses that are coupled through capacitors $C1$–$C12$ and their limiting resistors to turn on the PNP digit drive transistors ($Q8$–$Q19$). The RC time constant must be long enough to hold the transistors on for the anode select time. The base-emitter resistors are for ICBO considerations, and the time constant with that resistor added must be short enough to permit the coupling capacitors to recharge before the next scan cycle. Since the coupling capacitors are charged to nearly the full supply voltage, isolation diode $D1$ is used to protect the emitter base junctions when the supply is turned off.

8–7.11 Incandescent Display

The smaller direct-view incandescent displays have their seven helical coil segments fashioned from tungsten alloy. Their power requirements (1.5 to 5 V at 8 to 24 mA) when direct driven are comparable to LEDs (Sec. 8–7.7 and 8–7.8) which generally make incandescent compatible with LED drivers or decoder/drivers. When multiplexing to incandescent displays, maintaining an equivalent brightness to the direct drive requires greater filament peak power. Also, blocking diodes are required, one for each segment, to prevent erroneous display indications through sneak electrical paths.

8–7.12 Four-digit Incandescent Display

The circuit of Figure 8-56 shows interfacing of CMOS to a four-digit incandescent display. The system requires about 4.5 V at 24 mA (108 mW) per segment when direct driven. When multiplexing four digits, the instantaneous power has to be translated to approximately 9 V at 48 mA to maintain the same average power per segment, thus the need for interfacing devices.

The digit driver transistors ($Q1$–$Q4$) must be able to source about 350 mA ($7 \times 48\,\text{mA}$) and still have sufficient h_{FE} to interface with the CMOS decoder gates. The MPS–U95 small plastic PNP Darlington transistor easily meets this requirement. The MPS–U95 worst case, low temperature h_{FE} is approximately 10,000 at 350 mA, and the MPS–U95 can handle collector currents to about 2 A. Making the input current limiting resistors ($R1$–$R4$) equal to 18 kΩ results in a forced h_{FE} of approximately 800, and a MC14011 sink current of approximately 0.45 mA. The segment drive transistors ($Q5$–$Q11$) are inexpensive, plastic NPN transistors that readily interface with the positive-going decoder output pulses.

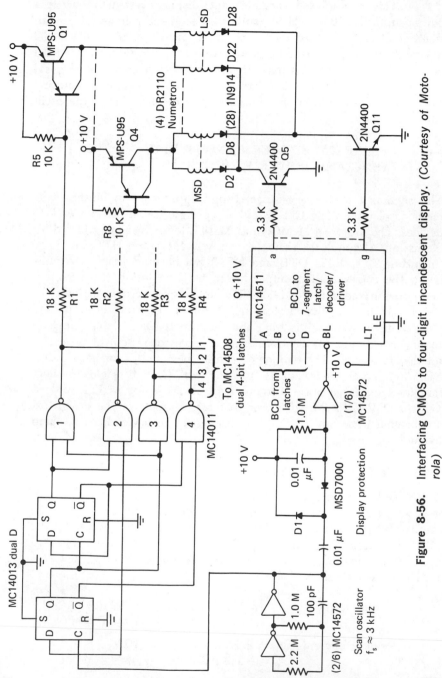

Figure 8-56. Interfacing CMOS to four-digit incandescent display. (*Courtesy of Motorola*)

405

This multiplexed incandescent display system requires a power supply of 10 V, which results in a segment voltage (after subtracting all of the semiconductor potential drops) of about 8 V at 40 mA peak. If the scan oscillator were to fail, this high current would be continuously applied to one digit, and soon degrade the display. The display protection circuit prevents this condition by monitoring the scan oscillator output and blanking the display should the oscillator fail.

8–8   INTERFACE BETWEEN INDUSTRIAL LOGIC AND POWER DEVICES

This section describes worst-case design approaches to illustrate the methods of interfacing CMOS and HTL logic to various power load devices. The Motorola McMOS and MHTL logic lines are selected for discussion. Although the examples apply to MOS and HTL, any logic (including TTL, RTL, DTL, and ECL) can be used simply by translating the corresponding output characteristics (using the examples here) to the specified condition. The technique here is to interface between logic and power devices by: (1) interpreting the output parameters of the logic devices from the datasheet, (2) defining the load requirements, and (3) characterizing the interface element necessary for the translation. The two logic families best suited for most industrial *high noise environments* are the MHTL and McMOS. Both families have large noise margins, typically 5 V for MHTL and 45 percent of V_{DD} for McMOS. For this reason, most of the design examples given here are for these families, but can be readily translated to any logic family.

8–8.1   MHTL Current Sink/Source Considerations

When interfacing between logic outputs and logic inputs, it is necessary to establish input and output loading factors (or the source and sink current capability of the device). For example, the MC670 (a triple three-input NAND HTL gate with passive pull-up) has an input loading factor of one, and an output loading factor of 10. The datasheets show:

output voltage $\quad V_{OL} = 1.5$ V max at $I_{OL} = 12$ mA

$V_{OH} = 12.5$ V min at $I_{OH} = 30$ μA

Reverse current $\quad I_R = 2$ μA max at $V_R = 16$ V

$I_F = -1.2$ mA max at $V_F = 1.5$ V

Thus, when driving ten loads, the device has a sink current capability of 10 I_F equal to 12 mA, with the low level output voltage V_{OL} guaranteed to be less than 1.5 V max (typically $V_{CE(sat)}$ about 0.2 V). Similarly, when sourcing ten loads, the specified minimum output voltage will exceed 12.5 V at a source current of 30 μA, a value somewhat higher than the ten reverse current loads (10 $\times$ 2 μA = 20 μA). This does not imply that the device can only source 30 μA, but states that the device is guaranteed to have a high level, minimum output voltage of 12.5 V at 30 μA. Normally, when driving other logic gates, the high level (one) driver output need only reverse bias the input diodes (or emitter-base junctions) and source the specified leakage current.

Passive pull-up.   When driving other than logic devices, such as discrete circuits, the following will suffice. Figure 8-57a shows the schematic of a passive pull-up gate. When R_{L1} is connected as illustrated, the sink current is dictated by the equation shown, and the maximum specified sink current of 12 mA which, in turn, is limited to minimum h_{FE} of the output transistor Q_2.

With the output at the logic one level (Q_2 not conducting), the source current I_{source} is expressed by the equation shown. The limiting value of source current would be when R_{L2} is zero (the output shorted), and would be 15 V/15 kΩ, or 1 mA. The specified short-circuit current I_{SC} is 1.5 mA, and is due to the -33 percent resistor tolerance of the 15 kΩ Q_2 collector resistor (67% $\times$ 15 kΩ = 10 kΩ).

Thus, when directly interfacing with the base-emitter load of an NPN transistor, the nominal source current (or base current for that transistor) would be:

$$\frac{15 \text{ V} - 0.7 \text{ V}}{15 \text{ k}\Omega} \approx 1 \text{ mA for the passive pull-up gate.}$$

Active pull-up.   The sink current limitations of the active pull-up gate shown in Figure 8-57b are the same as for the passive pull-up, as shown by the equations. (Note that V_{OL} in the passive pull-up is replaced by $V_{D3} - V_{CE}$ in the active pull-up, and is approximately 1.5 V max.)

The sourcing capability of the active pull-up gate is quite different from that of the passive pull-up, as shown in Figure 8-57c. When the gate output is high (logic 1), Q_2 is cut off and the circuit operates as an emitter-follower. For worst-case design, the value of h_{FE}(min) in the equations of Figure 8-57c can be considered as ten.

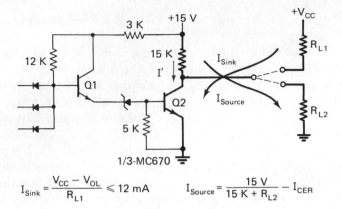

$$I_{Sink} = \frac{V_{CC} - V_{OL}}{R_{L1}} \leqslant 12 \text{ mA} \qquad I_{Source} = \frac{15 \text{ V}}{15 \text{ K} + R_{L2}} - I_{CER}$$

Passive pull-up HTL gate

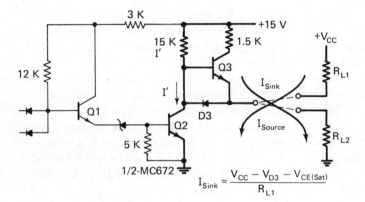

$$I_{Sink} = \frac{V_{CC} - V_{D3} - V_{CE(Sat)}}{R_{L1}}$$

Active pull-up HTL gate

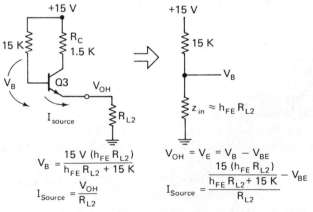

$$V_B = \frac{15 \text{ V} (h_{FE} R_{L2})}{h_{FE} R_{L2} + 15 \text{ K}}$$

$$I_{Source} = \frac{V_{OH}}{R_{L2}}$$

$$V_{OH} = V_E = V_B - V_{BE}$$

$$I_{Source} = \frac{\frac{15 (h_{FE} R_{L2})}{h_{FE} R_{L2} + 15 \text{ K}} - V_{BE}}{R_{L2}}$$

Active pull-up source current derivation

Figure 8-57. Sinking and sourcing MHTL gates. *(Courtesy of Motorola)*

For a particular sourced load, the high level output voltage can be readily calculated. The limiting source current occurs when the load is zero (output short circuited) or simply, the emitter current resulting from the common-emitter configuration:

$$I_{\text{source(max)}} = I_E \approx \frac{V_{CC} - V_{CE(\text{sat})Q3}}{R_C} = \frac{15 - 0.2}{1.5 \text{ k}\Omega} \approx 10 \text{ mA}$$

The specified short-circuit current is 15 mA maximum to take into account the tolerance variation of the collector limiting resistor R_C.

Direct coupling the active pull-up. When the output of the active pull-up is directly coupled to the base-emitter junction of an NPN common-emitter transistor, the source current will be nearly that of the short circuit current, and will be:

$$I_{\text{source}} \approx \frac{15 - 0.2 - 0.7}{15 \text{ k}\Omega} \approx 9.4 \text{ mA}$$

However, the high level output will be clamped to the base-emitter voltage drop of the load transistor. By resistive coupling to the load, the high level output voltage V_{OH} can be maintained at a higher voltage level, if desired, dictated by the size of the coupling resistor and the base current requirements of the load transistor.

Other loads. By extension, any type of load can be interfaced using the data of Figure 8-57. Of course, the sink or source current requirements of the load must be within the capability of the logic device. However, to interface with larger loads, *power simplifications* must be used as described in the following paragraphs of this section.

8–8.2 McMOS Current Sink/Source Considerations

Figure 8-58 shows the source/sink capabilities of the CMOS logic. Note that the CMOS inverter is as discussed in Sec. 2–1.7. Due to the extremely high input impedance of CMOS, the basic CMOS inverter has the capability of interfacing with many CMOS gates (a typical fan-out is 50 or greater). However, when interfacing with other loads, the CMOS gate is current limited mainly by the channel resistance and, to a lesser extent, by the forward transadmittance. The dc resistance between drain and source when the device is turned on is generally labeled "ON resistance, R_{ON}" or $r_{DS(on)}$.

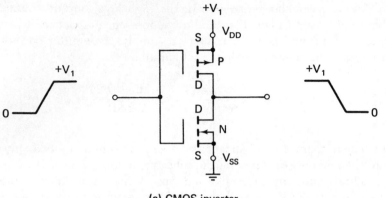

(a) CMOS inverter

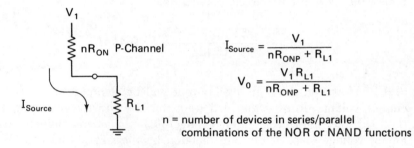

$$I_{Source} = \frac{V_1}{nR_{ONP} + R_{L1}}$$

$$V_0 = \frac{V_1 R_{L1}}{nR_{ONP} + R_{L1}}$$

n = number of devices in series/parallel
combinations of the NOR or NAND functions

(b) Sourcing current

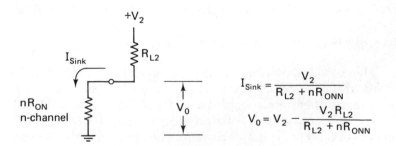

$$I_{Sink} = \frac{V_2}{R_{L2} + nR_{ONN}}$$

$$V_0 = V_2 - \frac{V_2 R_{L2}}{R_{L2} + nR_{ONN}}$$

(c) Sinking current

Figure 8-58.   CMOS source-sink capabilities.

Equivalent source and sinking circuits. The equivalent circuits for sinking and sourcing current into an external load is shown in Figures 8-58b and 8-58c. Normally, when interfacing CMOS to CMOS, because of the extremely small load currents, the logic outputs will be very near their absolute maximum states (V_1 or 0 V). With other types of loads, the current and the resulting output voltage are dictated by the simple voltage divider of R_{ON} and the load resistor R_L where R_{ON} is the total series and/or parallel resistance of the devices comprising the NOR or NAND function.

An illustration of the sink/source currents at various supply voltages with the specified logic level outputs is shown in Table I of Figure 8-59. From this information, the ON resistance can be extrapolated for worst-case design.

As an example, the CL/CP series with guarantees of $V_{DD} = 10$ V, and $V_{OH} = 9.5$ V, and the source current $I_{OH(min)} = -0.2$ mA, the maximum R_{ON} would be:

$$R_{ON(max)} = \frac{V_{DD} - V_{OH}}{I_{OH(min)}} = \frac{10 - 9.5}{0.2} = 2.5 \text{ k}\Omega$$

Table 1. Source/sink characteristics

Characteristic	Symbol	V_{DD}	AL series			CL/CP series			Unit
		V_{dc}	Min	Typ	Max	Min	Typ	Max	
Output drive current P-channel (I_{Source})	I_{OH}	5.0	−0.5	−1.7	—	−0.2	−1.7	—	mA dc
		10	−0.5	−0.9	—	−0.2	−0.9	—	
		15	—	−3.5	—	—	−3.5	—	
N-channel (I_{Sink})	I_{OL}	5.0	0.4	0.78	—	0.2	0.78	—	
		10	0.9	2.0	—	0.5	2.0	—	
		15	—	7.8	—	—	7.8	—	

Table 2. Approximate value of ON resistance

V_{DD} (Volts)	V_{OH} (Volts)	Source $R_{ON(Typ)}$	Source $R_{ON(Max)}$	V_{OL} (volts)	Sink $R_{ON(Typ)}$	Sink $R_{ON(Max)}$
5.0	2.5	1.7 K	12.5 K	0.4	500	2 K
10	9.5	500	2.5 K	0.5	420	1 K
15	13.5	430	—	1.5	190	—

Figure 8-59.   Source-sink characteristics and ON resistance values. *(Courtesy of Motorola)*

Similarly, for the sink current condition, also at 10 V, with $I_{OL(\text{min})} = 0.5$ mA:

$$R_{ON(\text{max})} = \frac{V_{OL} - V_{SS}}{I_{OL(\text{min})}} = \frac{0.5 - 0}{0.5} = 1 \text{ k}\Omega$$

Continuing these calculations for the other conditions, the maximum ON resistances can be approximated as shown in Table 2 of Figure 8-59. As shown, the ON resistance decreases with increasing supply voltage.

Although the minimum currents are not shown on the datasheet for the 15 V case, the maximum ON resistance can be no greater than for the 10 V example, and can be assumed for worst-case approximation to be 1 kΩ and 2.5 kΩ for sink and source current cases, respectively.

Generally, for most CMOS devices, the maximum current is limited to 10 mA. With a 15-V supply, the device is capable of supplying greater than 10 mA, but should be limited to that current by means of a resistor in series with the load. With 5 V and 10 V supplies, the CMOS outputs can be short-circuited, and can thus directly drive transistor or diode junctions.

8–8.3 Load Characteristics

Loads can be resistive, reactive, and linear or nonlinear. With linear resistive loads (those loads that do not change resistance with power or time), the determination of the required power gain is simple. With nonlinear loads, the transient conditions must be taken into account to ensure that the specifications of the interface devices are not exceeded. Typical nonlinear loads are lamps, whose cold filament surge currents are many times greater (5 to 15) than the hot filament, and dc motors whose *start* and *stall* currents are often several times greater than *run* currents.

Transient response of loads. As shown in Figure 8-60, the transient response associated with nonlinear loads must be considered to adequately define the interface device, and to ensure that the ratings are not exceeded.

Non-h_{FE}-limited response. Figure 8-60.1 shows the typical transient current voltage and device dissipation of a lamp load being driven by an interface device that is not h_{FE} limited. (That is, the interface can supply all the current that the load demands.) Under these conditions, the output voltage will quickly swing from the OFF state V_1 to ON state $V_{CE(\text{sat})}$ in a time dictated by the transient response of the device.

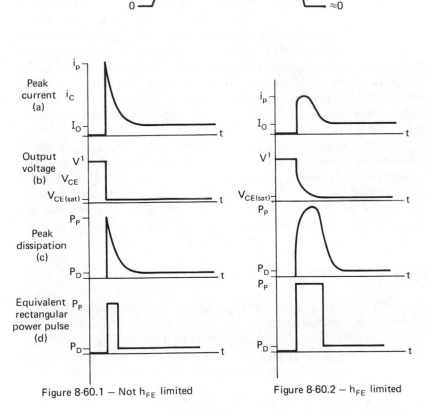

Figure 8-60.1 — Not h_{FE} limited   Figure 8-60.2 — h_{FE} limited

Figure 8-60. Transient response of nonlinear load. *(Courtesy of Motorola)*

The peak current is a function of the lamp's cold, low resistance filament, and will decay exponentially as the lamp filament heats and resistance increases. This thermal response can last for tens of milliseconds, depending on the lamp rating and filament size. The peak device dissipation, the product of voltage and current, is similar to that power waveform of Figure 8-60.1c. This dissipation can be equated to the rectangular power pulse of Figure 8-60.1d to simplify average power calculations. The peak dissipation resulting from these transient conditions must then be related to the interface device transient thermal response, and Safe Operating Area (SOA) to ensure that average dissipation does not exceed the maximum ratings.

h_{FE} **limited response.** Figure 8-60.2 shows the response where the interface can not supply all the required load current. This condition results in even greater device dissipation. For example, as shown by the transient waveforms of Figure 8-60.2, the peak current is less than in the not-limited example, but the device leaves the saturation state during the switching transitions, due to insufficient input drive current. The resulting power dissipation pulse can be excessive and destructive. For this reason, it is always good practice to design the interface device (discrete or IC) with adequate current gain under transient conditions. Obviously, the device must be able to sustain these peak currents. In cases where the designer has no control of the current gain and forced h_{FE} of the device, as in some IC high current logic-devices, it is important to ensure that the device rating are not exceeded.

The surge currents are not limited only to lamp loads. Surge currents are inherent in start-up and stall currents of dc motors. Also, as with any inductive load (motors or relays), it is mandatory that the reactive load line be within the device's safe operating area, and that the device be protected with inductive voltage suppressors (such as clamp diodes, zeners, resistor-capacitor networks, and so on) so as not to exceed the breakdown voltage of the device.

8–8.4 Interface Device Characterization

To ensure that the interface device has adequate current gain under worst-case design often requires extrapolation of the somewhat fragmentary published information on the datasheet. As an example, a transistor might have $h_{FE(min)}$ and $h_{FE(typ)}$ specified at a current other than the current in question, with no typical h_{FE} versus current and temperature curves published. There also might be another $h_{FE(typ)}$ specified at a different current. By extrapolating the two typical h_{FE} points, the $h_{FE(typ)}/h_{FE(min)}$ ratio, the desired $h_{FE(min)}$ can then be roughly determined.

If the circuit is required to operate at some much lower temperature than the 25°C at which the device was characterized, then the above calculated $h_{FE(min)}$ must also be modified for consideration of low temperature operation.

Generally, h_{FE} falls off by an approximate 50 to 67 percent factor at −55°C. Once the minimum h_{FE} is determined, the designer must decide which forced h_{FE} and/or base overdrive factor to use to ensure that the transistor is saturated under worst-case conditions (including worst-case conditions of the driver logic device). The probability of all blocks in the circuit being worst-cased simultaneously must also be considered, particularly when this imposes undue current requirements on the logic driver. In that case, it might be necessary to make tradeoffs on the worst-case design with the current capability of the driver.

The example cited is for a bipolar transistor but the reasoning applies to any interface device, be it a thyristor with its maximum gate current requirements, or an IC high-current logic device with its limited drive current. With these design guidelines as a basis, it becomes a relatively simple task to interface the industrial logic with the power devices. The following circuits illustrate these techniques where examples of various loads and power levels, both ac and dc, using bipolar transistors and thyristors as power devices, are shown.

8–8.5 MHTL Interface Circuits

Low current passive pull-up circuit. One of the simplest circuits for interfacing between HTL logic and a nonlinear load in the 1- to 5-W range is shown in Figure 8-61. This circuit illustrates the passive pull-up MC670 gate turning on a lamp driver transistor (activating the lamp) when the logic output goes low (active state low).

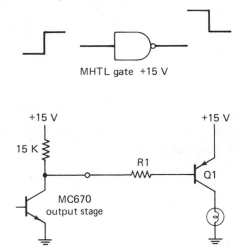

Condition	Load	Q1	R1
(a)	Lamp CM330 14 V @ 80 mA	2N4403	1.2 K
(b)	Lamp 1816 13 V @ 350 mA	2N6296 Darlington	2.7 K

Figure 8-61. MHTL logic passive pull-up with low active. (*Courtesy of Motorola*)

The emitter of the PNP driver transistor Q_1 is tied to the same +15 V supply as the gate. When the logic output is high, Q_1 is biased OFF, and the lamp is OFF. When the logic switches low, the logic output must be able to sink the base current of Q_1, and still be within the logic maximum specifications (12 mA). Thus, the limiting resistor R_1 is selected to furnish adequate base current to Q_1 under worst-case conditions: surge currents, minimum h_{FE}, and minimum operating temperature.

For the first example, Figure 8-61 condition (a), the 80 mA lamp load with its high surge current and a forced gain of about 7 was selected, resulting in a base current of approximately 11.5 mA, which is less than $I_{sink(max)}$ of 12 mA. This results in a base current limiting resistor equal to:

$$R_1 = \frac{V_{CC} - V_{CE(sat)} - V_{BE1}}{I_B} = \frac{15 - 0.3 - 0.7}{11.5} \approx 1.2 \text{ k}\Omega$$

An inexpensive, small-signal, plastic PNP transistor (2N4403) was selected as the interface driver Q_1. The pertinent specifications are: $IC_{max} = 600$ mA continuous (well within the quiescent lamp current of 80 mA), and $h_{FE(min)} = 100$ at I_C of both 50 mA and 150 mA. The h_{FE} versus I_C curve can be presumed flat between the two specified points, resulting in a minimum h_{FE} of 100 to 80 mA.

Low temperature operation ($-30°$C, the low limit of MHTL) may produce a lower of h_{FE} to approximately 70 percent of the 25°C figure, resulting in a minimum h_{FE} of about 70. This worst case h_{FE} results in a base overdrive factor of:

$$\frac{h_{FE(min)}}{h_{FE(forced)}} = \frac{70}{7} = 10,$$

ensuring that Q_1 is saturated for steady-state conditions.

However, the cold filament surge current may be greater than ten times steady state (approximately 800 mA) and lasting for tens of milliseconds. Although this peak current exceeds the maximum continuous current of 600 mA, it is within the published *SOA* pulse curve of 1 second at 1 A. This 1 S pulse is predicted on the lamp's thermal response being less than that period.

With a base current of 11.5 mA and a minimum h_{FE} of 70, a collector current of 800 mA (11.5 × 70) can result before the transistor comes out of saturation. Thus, under worst-case conditions, where the lamp surge current is 10 times steady state, Q_1 is saturated and dissipation is minimal. For this particular circuit, a surge current of 570 mA, lasting for 15 ms, was measured with the collector voltage driven sharply into saturation (Figure 8-61, conditions a and b).

For high temperature operation, particularly when the circuit is deactivated (active logic high, Q_1 off), a resistive shunt path for I_{CBO} leakage current is provided through R_1 and the 15 kΩ pull-up resistor.

High current passive pull-up circuit. For larger lamp currents to still be within the logic driver's 12 mA sink capability, the 350 mA lamp of Figure 8-61 condition (b) requires a higher h_{FE} of approximately 70. The 2N6296, a PNP metal-packaged TO-66 Darlington with built-in base-emitter resistors and maximum continuous collector current of 4 A, was selected. After extrapolating the published $h_{FE(typ)}$ versus collector current and temperature curves, and the minimum specified h_{FE} at 2 A, an $h_{FE(min)}$ at 350 mA and $-30°C$ was approximated to be 250.

If the base current is approximately 5 mA, a forced h_{FE} for steady-state conditions should be 70 (350/5), with a resulting base overdrive factor of about 3.5 (250/70). The limiting resistor R_1 is then:

$$R_1 = \frac{V_{CC} - V_{OL} - V_{BE \text{ Darlington}}}{I_B} = \frac{15 - 0.2 - 1.3}{5 \text{ mA}} \approx 2.7 \text{ k}\Omega$$

This results in a saturated circuit under steady-state conditions. However, with minimum h_{FE}, low temperature, surge current operation, the transistor could pull out of saturation with an associated increase in dissipation. Even under these worst-case conditions, the transistor is still operating within its *SOA*.

Passive pull-up, logic output high. Figure 8-62 shows the circuit where the passive pull-up gate activates the load when the logic output goes high. The load can be lamps or a small dc motor.

The circuit is deactivated when the logic output is low (saturated), clamping the interface transistor Q_1 base-emitter off. Although the published $V_{OL(max)}$ $V_{CE(sat)}$ for the MC670 is 1.5, the actual saturation voltage is about 0.4 V, with typical values being in the 0.1 to 0.2 V range at 5 to 10 mA sink currents. This low value of V_{OL} ensures proper clamping (Q_1 turning off).

When the logic output goes high (that gate output transistor cut off), the internal 15 kΩ collector resistor then supplies the base current, and is about 1 mA (15 V/15 kΩ). For larger base current requirements, the pull-up resistor R_1 in parallel with the 15 kΩ collector

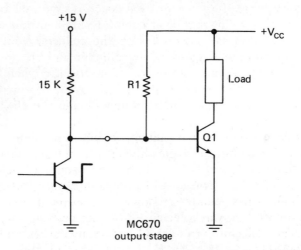

Condition	Load	V_{CC}	Q1	R1
(a)	Lamp 14 V @ 80 mA	15 V	2N4401	8.4 K
			MPSA13 Darlington	∞
(b)	Lamp 13 V @ 350 mA	15 V	2N6294 Darlington	2.7 K
(c)	D.C. motor 24 V @ 65 mA	24 V	2N4401	8.2 K

Figure 8-62. MHTL logic passive pull-up with high active. *(Courtesy of Motorola)*

resistor furnishes the current. The parallel resistor combination must be within the sink current capability of the gate.

The table of Figure 8-62 describes the transistor types and pull-up resistors for the various loads. As in the previous examples, the transistors and resistors are chosen to satisfy the *surge current* requirements of the load.

For Figure 8-62 condition (a), using the MPS-A13 Darlington transistor, no pull-up resistor is required since this device has adequate h_{FE} to be saturated by the sourcing of the MC670. Also, for the Figure 8-62 circuits, the supply voltage need not be the +15 V, but can be of any value required by the load.

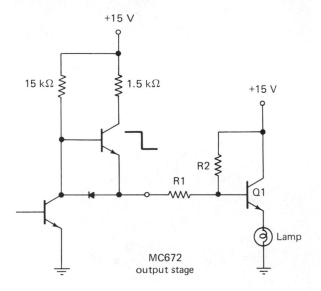

Figure 8-63.   MHTL logic, active pull-up, activation low. (*Courtesy of Motorola*)

Condition	Load	Q1	R1	R2
(a)	Lamp 14 V @ 80 mA	2N4403	2.2 kΩ	10 kΩ
(b)	Lamp 13 V @ 350 mA	2N6296	3.3 kΩ	INF

When using active pull-up gates, the sinking current criteria of the gate is almost identical to the passive pull-up. The basic difference between the two is that $V_{OL(max)}$ for the active pull-up gate (MC672) is 1.5, whereas the MC670 (passive) is about 0.2 V.

Active pull-up gate, logic output low.   Figure 8-63 shows an active pull-up gate that activates the load when the logic goes low. The base current limiting resistor R_1 is chosen to be within sink capability of the logic device, and still provide adequate drive to the interface PNP transistor under worst-case conditions. When the logic goes high, Q_1 is at cut off and the load is de-energized. Resistor R_2 is required for leakage current bypass considerations. When using the

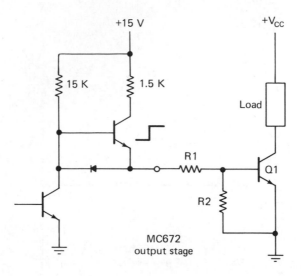

Figure 8-64.   MHTL logic, active pull-up, activation high. *(Courtesy of Motorola)*

Condition	Load	V_{CC}	Q1	R1	R2
(a)	Lamp	+15 V	2N4401	2.2 K	4.7 K
	14 V @ 80 mA			0	4.7 K
(b)	D.C. motor 24 V @ 65 mA	+24 V	2N4401	2.2 K	4.7 K

2N6296 Darlington transistor as the interface, condition (b) of Figure 8-63, R_2 is not required as the Darlington has built-in base-emitter shunt resistors for I_{CBO} effects.

Active pull-up gate, logic output high.   An example of how the load is activated when the MC672 logic goes high is shown in Figure 8-64. Again, the base limiting resistor R_1 is designed to furnish adequate drive to the NPN transistor. The limiting value of R_1 is zero ohms, condition (a) of Figure 8-64, where the base current (source current of the MC672) is approximately the short-circuit current of the logic device. Under this condition, the high logic output is clamped to the base-emitter voltage of Q_1 (about 0.7 V), and cannot be used to drive some other circuit. By making R_1 finite, the high logic level is dictated by the equivalent emitter-follower driver and its load ($R_1 +$ Q_1), and may be predictably high enough to drive other logic circuits.

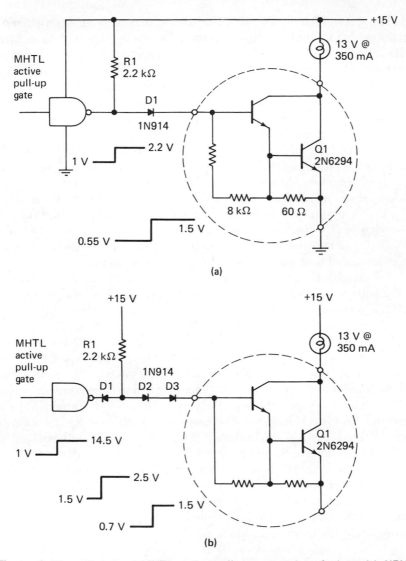

Figure 8-65. High level MHTL active pull-up output interfacing with NPN Darlington. *(Courtesy of Motorola)*

The load (and Q_1) is de-energized when the logic output goes low. Under these conditions, the active pull-up transistor is at cutoff, the logic driver is saturated and base drive to Q_1 is removed, cutting Q_1 off. Resistor R_2 is necessary to reduce leakage current of Q_1, and should be designed for maximum operating temperature I_{CBO} considerations.

Active pull-up, high gain. Figure 8-65a shows an active

pull-up device driving an NPN Darlington transistor for high current gain. In this example, a base current of 6 mA is required to ensure adequate drive. The MHTL gate cannot completely source this current, and still ensure a V_{OH} compatible with the Darlington. However, by adding the pull-up resistor R_1 to source the major part of the current, the circuit is made reliable. Diode D_1 is added to ensure that the maximum low level logic output, $V_{OL(max)}$, is 1.5 V, and will not turn on the two cascaded emitter-base junctions of the Darlington.

The logic high level output is clamped to about 2.2 V, which is the two emitter-base voltage drops, plus the diode drop. This makes the level incompatible for driving other logic circuits. A technique for overcoming this condition is shown in Figure 8-65b. Diode D_1 isolates the high logic output from the interface circuit, allowing the output to swing to V_{CC} (less one emitter-base drop, or $15 - 0.5$ V $= 14.5$ V). Diode D_3 is required to compensate for diode D_1 drop (when the logic goes low), and thus ensure that the Darlington is cut off.

High voltage supply operation. When the load is powered by a voltage greater than the MHTL supply of 15 V, and when the MHTL low level output is required to activate this load, then the simple circuits of Figures 8-61 and 8-63 will not suffice. This is due to the logic V_{OH} being less than the base voltage of the PNP transistor, and thus possibly turning the transistor on when it should be off.

This problem can be overcome by the circuit of Figure 8-66a. Here, transistor Q_1 acts as a *level translator* that is referenced to the MHTL supply. When the logic level goes low, Q_1 is turned on, Q_2 and the load are energized. The base-emitter resistor R_2 is required for leakage current considerations of Q_1 when driven with an active pull-up logic device. R_2 is not needed for passive pull-up devices.

In some applications, where other loads are connected to the MHTL output, pulling it down (as shown in Figure 8-66b), a coupling zener diode D_1 is required to prevent the loaded V_{OH} from turning Q_1 on. With the component values shown, the logic output can drop approximately 7 V before the load is erroneously energized.

8–8.6 McMOS Interface Circuits

McMOS (CMOS) devices can sink and source currents from fractions of 1 mA to about 10 mA, depending on how high (or low) an output is acceptable. In many cases, where the load power requirements are relatively low, the CMOS gate can be directly interfaced to the load driver, assuming that ample power gain is available.

Figures 8-67a and 8-67b show a typical CMOS gate directly interfacing with small-signal plastic Darlington transistors driving an 80-mA lamp load. With the component values shown, the respective sink and source currents are approximately 350 μA, resulting in V_{OL}

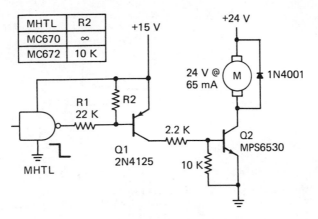

MHTL activation: low

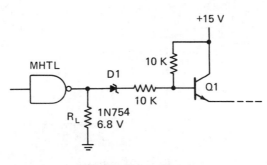

Zener diode coupling

Figure 8-66.   MHTL interfacing with load powered by greater than +15 V supply. *(Courtesy of Motorola)*

and V_{OH} (due to R_{ON}) of 0.04 and 14.8 V, respectively.

In addition to driving bipolar transistors, the CMOS gates can directly drive sensitive gate SCRs (Figure 8-67c). This circuit shows a small signal, plastic TO-92 case 2N5060, (with a maximum gate trigger circuit I_{GT} at low temperatures, $-65°C$, of 350 μA) being triggered directly by the CMOS gate high-level output. The 39 kΩ current limiting resistor supplies the approximate 350 μA (which is more than adequate for $-55°C$ McMOS operation), and still maintains V_{OH} (approximately 14.8 V) at acceptable levels for driving other CMOS logic.

8-8.7 Thyristor-Controlled AC Loads

Figures 8-68 through 8-71 show how CMOS and HTL can be interfaced to higher-power ac loads using thyristors as the ac control element.

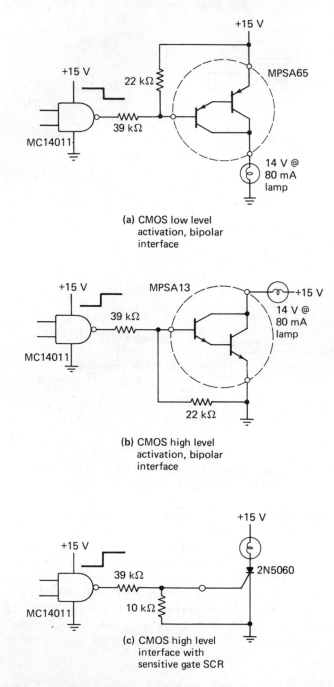

(a) CMOS low level
activation, bipolar
interface

(b) CMOS high level
activation, bipolar
interface

(c) CMOS high level
interface with
sensitive gate SCR

Figure 8-67. Interface techniques between CMOS and low power loads.
(Courtesy of Motorola)

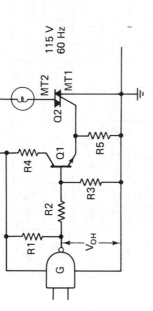

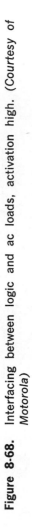

Condition	Logic device G	Load	Q1	Q2	R1	R2	R3	R4	R5	V_{OHtyp}
(a)	MHTL MC672	300 W	2N4401	MAC10-4	∞	2.2 K	10 K	150 2 W	∞	9 V
(b)	MHTL MC670	300 W	2N4401	MAC10-4	5.6 K	0	∞	150 2 W	∞	1.7 V
(c)	McMOS MC14011	25 W	2N4401	MAC92A-4	∞	12 K	10 K	560 1 W	1 K	14.6 V
(d)	McMOS MC14011	300 W	2N4401	2N6346	∞	2.2 K	10 K	110 4 W	∞	12.8 V

Figure 8-68. Interfacing between logic and ac loads, activation high. (Courtesy of Motorola)

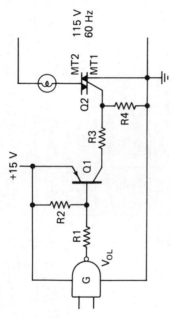

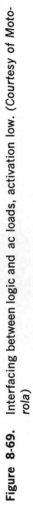

Condition	Logic device G	Load	Q1	Q2	R1	R2	R3	R4	$V_{OL,typ}$
(a)	MHTL MC672	300 W	MPS3638A	MAC10-4	2.2 K	10 K	150 2 W	∞	1.1 V
(b)	MHTL MC670	300 W	MPS3638A	MAC10-4	2.2 K	∞	150 2 W	∞	0.25 V
(c)	McMOS MC14011	25 W	MPS3638A	MAC92A-4	12 K	10 K	560 1 W	1 K	0.15 V
(d)	McMOS MC14011	300 W	2N4403	2N6346	3.9 K	10 K	110 Ω 4 W	∞	0.50 V

Figure 8-69. Interfacing between logic and ac loads, activation low. (Courtesy of Motorola)

426

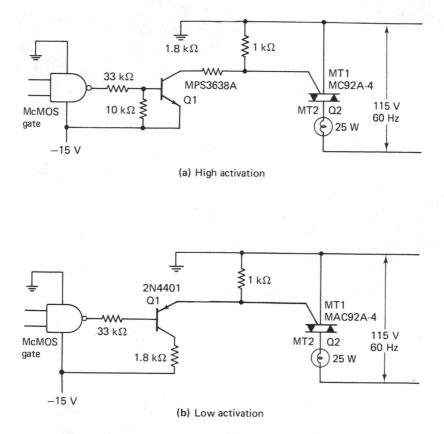

(a) High activation

(b) Low activation

Figure 8-70.   McMOS interfacing with sensitive gate triac using a negative supply. *(Courtesy of Motorola)*

Triac control, logic output high.   Figure 8-68 shows the interface configurations for high level activation of both MHTL and McMOS gates with triac-controlled ac loads. The active logic 1-level (high level) output turns on the effective common-emitter transistor Q_1, the emitter current of which (as determined by collector resistor R_4) triggers the triac. This current is designed for maximum I_{GT} at $-40°C$ (gate sensitivity falls off with decreasing temperature). The I_{GT} for the three triacs used, MAC10-4, MAC92A-4, and 2N6345, are approximately 90, 25, and 125 mA, respectively.

The MAC92A-4 is a sensitive gate triac, and typically requires a 1 kΩ resistor between the gate and Main Terminal 1 to desensitize the triac to noise transients. The other two triacs have internal resistors (built-in), and require no desensitizing.

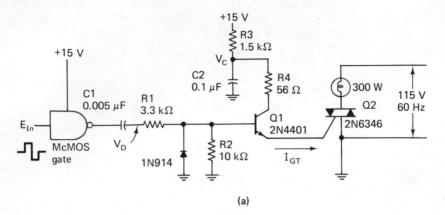

(a)

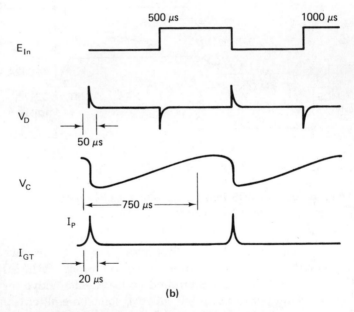

(b)

Figure 8-71. McMOS driving triac using capacitor storage techniques. *(Courtesy of Motorola)*

The NPN transistor Q_1 must have adequate current gain for worst-case conditions to interface between the logic sourcing capability, and the triac triggering requirement. The input network of Q_1 is dictated by which logic device is used. The MC670 requires a pull-up resistor R_1 for increased current sourcing, and is directly coupled to the Q_1 base. The MC672 and MC14011 can source adequate current through a limiting resistor R_2. The sourcing current in the 300 W circuits is approximately 3 mA. For the 25 W circuit, condition (c) of Figure 8-68, the sourcing current is about 1 mA. For the CMOS gates, the effect of R_{ON} is apparent in the value of V_{OH} for the two different sourcing current cases (1 and 3 mA) being 14.6 and 12.8 V.

Triac control, logic output low.   Figure 8-69 shows the interface circuits when an active low-level logic is required to activate the load. These circuits are essentially the complement of the Figure 8-68 circuits, in that the logic zero sinks the base current of the PNP transistor Q_1 and drives it into saturation. The collector current of Q_1, determined by R_3, supplies the positive gate current to the triac Q_2, turning Q_2 and the load on.

The gate sink currents are approximately the same as those in Figure 8-68 (1 and 3 mA). V_{OL} for the CMOS gates is approximately 0.15 and 0.5 V, for sink currents of 1 and 3 mA respectively. R_{ON} (sink) is less than R_{ON} (source).

Triac control with negative supply.   Triacs can be triggered with both positive and negative gate currents, with four combinations of gate current and Main Terminal 2 ($MT2$) voltages possible. The most sensitive mode is when the gate and $MT2$ are positive with respect to $MT1$ (this is known as quadrant 1). The least sensitive condition is when the gate is positive and $MT2$ is negative (known as quadrant 4).

When gate sensitivity for logic current drive capability is a criteria for switching an ac load, it is recommended to use negative gate trigger current with its associated nominal sensitivity (known as quadrants 2 and 3). The circuits of Figure 8-70a and 8-70b show both high and low CMOS logic activating a 25-W load. To achieve the negative gate current (current flowing out of the gate) requires that the logic and interface transistor high supply line be grounded, and the low line attached to a negative supply. The maximum negative gate current at $-40°C$ required to fire the triac is specified as 8 mA, whereas the quadrant 4 current is 25 mA.

Triac control using capacitor storage.   The relatively large power supply drain of the Figure 8-70 circuit can be excessive in some applications. The circuit of Figure 8-71 using capacitor storage techniques overcomes this problem. Storage capacitor C_2 completely charges to V_{CC} (+15V) in five R_3C_2 time constants. Shortly after C_2 is charged, Q_1 is fired by the positive-going, differentiated pulse derived from the input square wave. Capacitor C_2 is discharged through R_4 and Q_1, causing triac Q_2 to fire and energize the ac load.

To get maximum power to the load, the triac should be fired early in its conduction angle. As an example, if the triac is fired at a conduction angle of 18°, greater than 99 percent of the power is delivered; at 30°, approximately 97 percent. When the input square wave trigger is 1 kHz, the greatest change in conduction angle on a cycle-to-cycle basis for a 60 Hz load is approximately:

$$\text{conduction angle} = \frac{360°}{(1 \text{ kHz}/60 \text{ Hz})} \text{ or } 21.6°$$

and the minimum output power is approximately 98 percent of the maximum available. The period of the trigger $(1/1 \text{ kHz} = 1 \text{ ms})$ is greater than the five time constants of the storage network, $5 \times (1.5 \text{ k}\Omega \times 0.1 \text{ }\mu\text{F}) = 750 \text{ }\mu\text{S}$. This ensures that the capacitor is completely charged.

To ensure rapid triac turn-on, it is recommended that the triac be fired with a fast rise time high-current pulse (within the triac ratings). The minimum specified gate pulse width for the 2N6346 triac is 2 μS. Thus, Q_1 should be on for greater than this minimum time, as determined by the differentiating time constant, $(R_1 + R_{ON}$ CMOS$) \times C_1$, to allow the storage capacitor to discharge through R_4, Q_1, and the input of Q_2. Current limiting resistor R_4 sets the peak current and, with C_2, determines the pulse width. The circuit waveforms of Figure 8-71b show the gate trigger current waveform to be approximately 200 mA peak, and 20 μS wide.

The average $+15$ V power supply drain is about 4 mA relative to the 125 mA drain of the dc coupled example of Figure 8-68, condition (d).

8–8.8   High-current Logic Drivers

There are several high-current industrial logic IC drivers that readily interface wtih compatible loads. In the MHTL family, the MC679 dual lamp driver can sink 150 mA, as shown in Figure 8-72. The McMOS family is represented by MC14009, MC14049, MC14010, and MC14050 which are hex inverter/buffers and noninverting hex/ buffers. The MC14009/10 and MC14049/50 can typically sink 35 and 40 mA, respectively. Also shown in the table of Figure 8-72 are other families of high-current logic drivers, MTTL and Linear, capable of interfacing directly with higher power loads.

Figure 8-73 shows examples of low-power industrial logic interfacing with these higher-power devices.

Product family	Device number	Description	V_{CC} or V_{DD}	$V_{I(max)}$	I_O	Remarks
MHTL	MC679	Dual lamp drivers	15 V	30 V	150 mA max	Open collector
	MC699	Dual power AND gate	5 to 20 V	30 V	500 mA	Sink or source I_O
McMOS	MC14009	Hex inverter/buffer	V_{DD} = 3 to 18 V	$V_{CC} \leqslant V_{DD}$	$I_{OL(typ)}$ = 35 mA @ V_{CC} = 15 V V_{OL} = 1.5 V	Requires two supplies $V_{CC} \leqslant V_{DD}$
	MC14010	Non-inverting hex buffer				
	MC14049	Hex inverter/buffer	3 to 18 V		$I_{OL(typ)}$ = 40 mA@ V_{CC} = 15 V V_{OL} = 1.5 V	One supply $I_{O(max)}$ = 45 mA
	MC14050	Non-inverting hex buffer				
MTTL	MC7406	Hex inverter buffer/driver	5 V	30 V	$I_{Sink(max)}$ = 40 mA	Open collector
Linear	MC55325	Dual memory driver	5 V	24 V	I_O = 600 mA @ V1 = 15 V	Contains 2 sink switch pairs and 2 source pairs
	MC75450	Dual peripheral	5 V	30 V	300 mA	TO-116 14 pin case
	MC75451	Positive "AND" driver				8 pin plastic case
	MC75452	"NAND" driver				Open collector
	MC75453	"OR" driver				
	MC75454	"NOR" driver				
	MC75491	Multiple LED driver Quad driver	10 V max	—	I_{Source} or I_{Sink} = 50 mA	MOS to LED Darlington drivers
	MC75492	Hex driver			I_{Sink} = 250 mA	

Figure 8-72. Motorola high-current logic drivers. (Courtesy of Motorola)

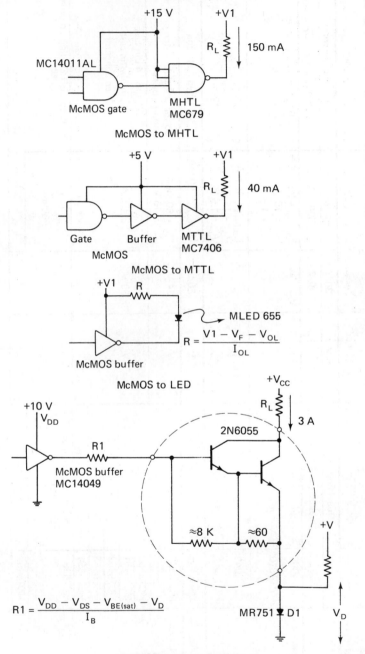

Figure 8-73. Typical high-current logic driver circuits. *(Courtesy of Motorola)*

McMOS to MHTL.   Figure 8-73a shows a typical McMOS gate driving the MC670 lamp driver. To energize the load, the McMOS output goes high, driving the MHTL output low. To turn the load off, the gate must be able to sink the MC679 forward current I_F of 1.2 mA maximum, and still have $V_{OL(max)}$ less than the MHTL threshold of 6.5 V minimum.

With a +15 V supply, the MC14011AL has a specified $V_{OL(typ)}$ of approximately 0.4 V at an I_{OL} of 1.2 mA, and a temperature of 85°C (at a constant I_{OL}, V_{OL} increases with increasing temperature). Although worst-case I_{OL} is not specified, from the previous calculations of $R_{ON(max)}$ in table 2 of Figure 8-59, R_{ON} for the sink condition can be assumed to be no greater than 1 kΩ. The simple voltage divider calculations of R_{ON} with the MHTL input pull-up resistor of 15 kΩ results in a $V_{OL(max)}$ of about 1 V, and well within the 6.5-V threshold.

McMOS to MTTL.   Figure 8-73b shows the interfacing between McMOS and MTTL MC7406 30-V, 40-mA, hex inverter buffer/driver. Since the MTTL is a 5 V device, the McMOS also operates from this supply. The forward current I_F of the MC7406 is 1.6 mA, and is well within the worst-case of the McMOS buffer sink capability of 2.1 mA at a supply of 5 V, V_{OL} of 0.4 V (which is the max TTL V_{OL}) and a temperature of 1.25°C.

McMOS to LED.   The high-current logic drivers are well suited for driving light emitting diodes (LED). Figure 8-73c shows a visible red LED being driven by a McMOS buffer. The sinking current of the gate is within the typical LED current of 10 mA. As shown by the equation, the value of limiting resistor R is dictated by the sinking current I_{OL}, the supply V_1, the LED forward voltage drop V_F of 1.6 V (typical), and the low-level output voltage V_{OL} of the driver.

McMOS to power Darlington.   Higher-current loads can be directly interfaced with McMOS buffers when there is adequate power gain in the interface device. Figure 8-73d shows such a circuit driving a 3-A load, using the 2N6055 8-A Darlington transistor. Worst-case h_{FE} at −55°C and collector current of 3 A is approximately 500. Using a base overdrive factor of 3-to-1 results in a base current of approximately 18 mA, a current well within the sourcing capability of the McMOS buffer MC14049. For high temperature operation, it is recommended that diode D_1 be placed in the emitter of the Darlington. This ensures reverse biasing of the Darlington in the off state, and thus prevents thermal runaway.

8–8.9   Optoelectronic Couplers

An optoelectronic coupler is formed when an LED is packaged with a photodetector. Typically, the LED is of the gallium arsenide, infrared light emitted type, whereas the detector can be a single photo transistor or Darlington. Optoelectronic couplers are designed for applications requiring electrical isolation, and have input/output isolation voltages as great as 2500 V, with isolation resistances being typically in the 10^{11} ohm range. Due to the high isolation properties of these couplers, and a typical band width capability from dc to about 300 kHz (in some cases), optoelectronic couplers can be used where it is required to isolate higher voltage and power loads from the low level logic.

The coupling from logic to load is simplified as the optoelectronic coupler can replace a dc coupled system (with its associated general requirement for level translation), and a pulse transformer or coupling transformer (with their limited bandwidths). Isolation is complete between the logic and the load in that the power supplies and their circuit grounds can be completely independent of each other.

Motorola manufactures a complete series of optoelectronic couplers and some of these are listed with their characteristics in Figure 8-74. Examples of optoelectronic control circuits interfacing between CMOS logic and loads (both ac and dc) are shown in Figures 8-75 through 8-78.

Optoelectronic coupler	Photo detector type	Isolation voltage (min) (V)	DC current transfer ratio % (min)	Detector working voltage (min)	Bandwidth (typ) (kHz)
4N25	Photo-transistor	2500	20	$BV_{CEO} =$ 30 V	300
4N26		1500	20		
4N27		1500	10		
4N28		500	10		
4N29	Photo-Darlington transistor	2500	100	$BV_{CEO} =$ 30 V	30
4N30		1500	100		
4N31		1500	50		
4N32		2500	500		
4N33		1500	500		

Figure 8-74. Motorola optoelectronic couplers. *(Courtesy of Motorola)*

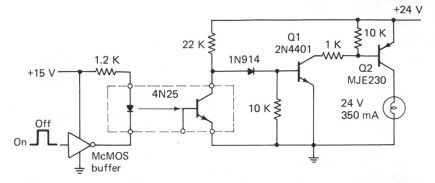

Figure 8-75.   Dc control using the optoelectronic coupler 4N25. *(Courtesy of Motorola)*

Dc control.   Figure 8-75 shows a 4N25 coupler interfacing between a McMOS buffer and a 350 mA dc powered lamp load. The lamp is normally de-energized when the 4N25 is energized. A high level on the input of the McMOS inverter energizes the 4N25, clamping the base-emitter junction of Q_1 off. Transistor Q_1 removes drive to Q_2, thus de-energizing the load. When a logic 0 is applied to the 4N25, the clamp to Q_1 is removed, and Q_2 is driven into conduction through the load, thus energizing the load. With the values shown, about 10 mA of input coupler current can control a completely isolated 350 mA load.

When greater sensitivity and current gain are required, the 4N29-33 series of optoelectronic couplers can be used. These couplers use photo-Darlington transistors as the detector, and have minimum dc transfer ratios I_C/I_F (the Darlington collector current to LED forward current) of as great as 500 percent.

Half-wave ac control.   Figure 8-76 shows a circuit for isolated, half-wave ac control, using the 4N26 coupler. The detector is used as a static series switch in the gate circuit of the 2N5064 SCR. When the logic input goes high, the coupler is energized, allowing the 2N5064 SCR principle current to flow. This current turns on the power SCR 2N6402, and the load for that positive half-cycle of the ac line voltage. Conversely, when the logic goes low, the coupler is de-energized, removing the 2N5064 gate current. The load current ceases at the next zero excursion of the ac source. Thus, with only about 5 mA of isolated dc control current, a half-wave load of up to 16 A can be controlled using the 2N6402.

Full-wave, normally-off control.   Figure 8-77 shows a circuit for isolated full-wave control of an ac load using the 4N26 and the

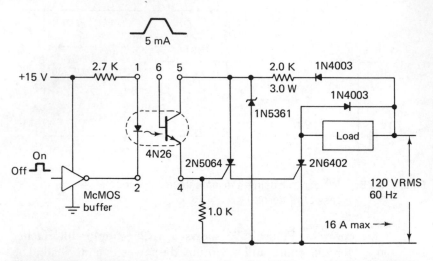

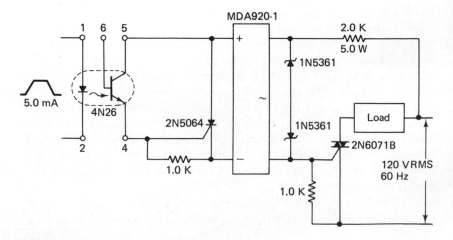

Figure 8-76. Half-wave ac control using the 4N26 optoelectronic coupler. *(Courtesy of Motorola)*

2N6071B sensitive gate triac. For full-wave operation of the 2N5064, a bridge rectifier is required to convert the bidirectional ac line voltage into a unidirectional (full-wave rectified) voltage. The SCR is placed across the bridge dc output. When the SCR conducts, a path for the triac gate trigger current is completed through the bridge and the SCR. Thus, a high-level logic input energizes the optoelectronic coupler, which triggers the SCR, and provides gate current for the triac.

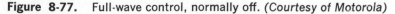

Figure 8-77. Full-wave control, normally off. *(Courtesy of Motorola)*

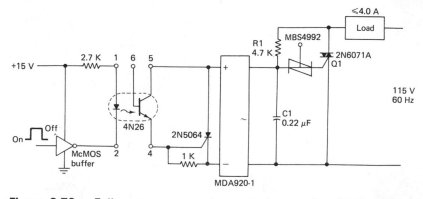

Figure 8-78.   Full-wave, zero crossing control, normally off. *(Courtesy of Motorola)*

The load turns on immediately after application of the logic signal (less propagation delays). When turn-on occurs other than at zero crossings of the ac line, it is possible for electrical noise to be induced in the circuit.

Full-wave, zero-crossing control.   Figure 8-78 shows a circuit for overcoming the induced electrical noise problem. The circuit consists of triac Q_1, with the trigger circuit R_1, C_1, S_1 and a clamp circuit consisting of the 4N26, 2N5064 SCR and bridge rectifier. To energize the load, the logic input goes low, the coupler is de-energized, and the clamp is removed from across the trigger capacitor C_1, allowing C_1 to charge through timing resistor R_1. When the voltage across C_1 reaches the triggering voltage of S_1 (approximately 8 V), S_1 fires, allowing C_1 to dump the charge into the gate of the triac. This turns both the triac and load on. Capacitor C_1 is chosen small enough to fire the triac early in the conduction cycle (near zero crossing), thus minimizing electrical noise, and maximizing power delivered to the load.

To de-energize the load, the logic high input energizes the coupler which, with the SCR and bridge rectifier, clamps the trigger capacitor, thus inhibiting the triac from firing.

8–8.10   Power Transistor Controlled DC Loads

Industrial logic devices can interface with any power load level, assuming that the circuit has the required number of stable power gain stages. Previous examples described low dc power circuits, and thyristor-controlled ac circuits. By cascading additional power gain stages to the low-level examples, higher-power loads can be readily driven. The following circuits of Figures 8-79 through 8-80 describe the interfacing between industrial logic devices, using both low- and high-level activation, and a 24-V, 18-A, dc motor.

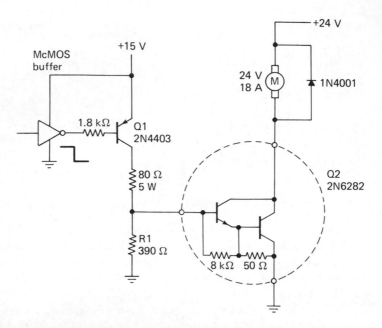

Figure 8-79.   McMOS low-level activation of motor load. (*Courtesy of Motorola*)

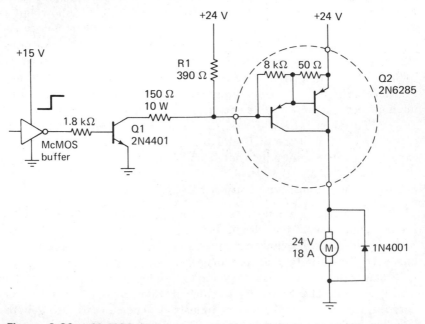

Figure 8-80.   McMOS high-level activation of motor load. (*Courtesy of Motorola*)

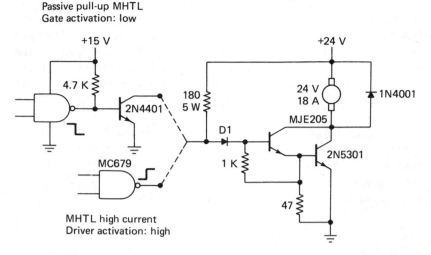

Passive pull-up MHTL
Gate activation: low

Figure 8-81. Clamped Darlington motor control. *(Courtesy of Motorola)*

Low-level activation. Figure 8-79 shows how low-level logic output activation of the motor circuit can be accomplished with McMOS hex/buffers. When the output is low, the buffer sinks the approximate 8 mA base current for Q_1. Transistor Q_1 then furnishes base current for the Darlington Q_2. Worst-case h_{FE} is approximated to be 210. By using a base current of 150 mA, a forced h_{FE} of 110 results, which ensures a 2:1 base overdrive factor. This current is derived from the +15-V logic power supply which, in some applications, might be excessive.

High-level activation. Figure 8-80 shows high-level activation of the motor circuit. The circuit of Figure 8-80 minimizes current drain, and is essentially a complement of the Figure 8-79 circuit. In the circuit of Figure 8-80, the buffer *sources* the 8 mA base current to Q_1 which, in turn, sinks the 150 mA base current to Q_2. This current is furnished by the +24 V motor supply, thus reducing the drain on the logic supply by about 150 mA. For high temperature operation (where junction temperatures exceed 100°C), the addition of R_1 across the Darlington input is recommended. This additional leakage current shunt ensures that Q_2 will not go into a thermal runaway condition.

Clamped Darlington motor control. Figure 8-81 shows another motor control circuit. The approach here is to supply drive current through a pull-up resistor, with motor control obtained by means of a transistor clamp across the input of the Darlington. (The previous circuits supplied Darlington base current through a switched transistor.) The clamp driver, Figure 8-81a, is a passive pull-up gate. The clamp, Figure 8-81b, is a high current driver.

The circuit of Figure 8-81 uses two discrete transistors for the Darlington. The MJE205 (a 5-A plastic case transistor) drives the 2N5301 (a 30 A, TO-3 case transistor). The worst-case h_{FE}s of the two transistors are 19 and 9, at collector currents of 2 and 18 A, for the MJE205 and 2N5301, respectively. An input current of about 100 mA is required by the Darlington. When the shunt clamp is de-energized, input current flows, the Darlington saturates, and the motor is powered.

The clamp, when energized (motor de-energized), must be able to sink the input current and maintain a saturation voltage less than the on voltage of the Darlington. Both circuits, Figure 8-81a and 8-81b, meet these requirements. Diode D_1 is used for high temperature operation to further ensure proper clamping.

9

Noise Considerations
in Logic Circuits

This chapter compares the noise immunities of the five major logic families used today in industrial logic systems designs: CMOS, TTL, ECL, HTL, and DTL. In effect, the chapter answers the question "What is the noise immunity of each family, and how does it compare with that of other logic families?" Also included in the chapter are general discussions of common noise sources, precautions against noise, noise specifications, and standard noise tests. Sections 9–1 through 9–5 apply particularly to CMOS, TTL, HTL, and DTL families. Section 9–6 through 9–9 describe the special noise problems of the ECL family.

9–1 NOISE SOURCES

Sources of electrical noise may be classified as either external or internal to a digital logic system. *External noise* may be generated by electric motors, arcing relay contacts, circuit breakers, and so on. This noise is usually in the form of randomly-generated spikes of electromagnetic interference (EMI) that couples inductively to the power, ground, and signal interconnections within the logic system. *Internal noise* may be generated on the signal lines by *crosstalk* (capacitative coupling between adjacent signal lines), on transmission lines from *reflections* (due to impedance mismatch), and on the power and ground lines by *current surges* that occur during switching.

External noise, although it may be thousands of volts, has a minimum effect on logic circuits since the external noise source is not connected directly to the logic. External noise is induced in the logic wiring, which has entirely different impedances than the noise generators. Thus, fortunately, external noise is attenuated long before it reaches the logic currents.

The effect of internal noise on logic circuit operation depends on the *noise margin, speed,* and *impedance* of the circuit. Unless long lines are being driven (Chapter 8), the effects of crosstalk and transmission-line reflections are minimal in all logic families, except superfast ECL. (For that reason, separate sections of this chapter describe ECL noise problems.) Power and ground noise is due primarily to poor power and ground returns in the system. (Internally generated noise on the power and ground lines seems to be a significant effect only in logic with very fast rise and fall times).

9–2   NOISE PRECAUTIONS

There are two approaches for dealing with electrical noise in logic modules in a noisy industrial environment: (1) keeping the noise out of the system, and (2) minimizing the influence of noise that does get into the system. Chapter 2 describes some practical suggestions to minimize the electrical noise problem (wiring and construction tips, physical layout of logic ICs, and so on). Here are some additional system design and construction practices commonly used to keep noise out:

1. Segregate logic wiring from field wiring. (The term "field" is used here to designate high-power interface circuitry (either semiconductor or electromechanical) which is external to (and possibly controlled by) the logic module. Do not design input converters and output drivers so that field wiring uses the same connectors that carry logic signals. Arrange to use *opposite ends of printed boards* for logic and field-wiring connections, and never allow both types of wiring to be adjacent or to be bundled together.

2. Do not mix logic grounds with field grounds. This does not mean that logic grounds should float. However, heavy currents should not pass through the logic system ground on the return path to the power supply. A good scheme is to switch the ac line with an

optically-isolated triac. Dc solenoid drivers might seem difficult to isolate, but proper use of ground-isolating resistors and auxiliary chassis tie points can force most of the load current outside the logic system ground.

3. Use high-density packaging. Computer-type modular construction minimizes lead lengths in the logic, reducing the coupling between logic wiring and nearby field wiring. Dense packing also cuts resistance and inductance in the logic ground system, minimizing interference from residual noise currents that may be present. Select a logic family with a strong complement of MSI functions (or possibly even LSI, if that is practical). This will improve packing density.

4. Where logic and power circuits must be adjacent, use shielding. For example, a group of printed circuit boards carrying field circuits can be shielded from general-purpose logic modules by inserting unetched copper-clad boards in the sockets that separate the two groups.

5. Filter the line voltage where it enters the logic power supply, and at the supply output terminals. Bypass capacitors, typically 100 pF silver mica to remove high-frequency transients, and 0.1 μF ceramic to remove lower frequency transients, should be used wherever power supply lines enter the circuit board.

9–2.1 Selecting a Logic Family for Noise Considerations

One of the best ways to minimize the effects of noise that may enter a logic system, in addition to the previously mentioned precautions, is to design with a family that has good noise-rejection characteristics. Thus, the *inherent noise immunity* of the logic family should be considered, in addition to the usual considerations such as cost, speed, availability, and compatibility. Here are some considerations that should be taken into account when selecting a logic family to operate in a hostile industrial noise environment:

1. Slower speed or longer propagation delay logic families are usually less susceptible to noise, since noise is generally more intense at higher frequency (MHz range). Metal-to-metal contacts are nearly ideal step generators, and wiring resonances often generate high frequency noise peaks.

2. Logics requiring fast rise and fall time inputs are more inherently sensitive to high-frequency noise than slower rise time logic families that switch by level sensing rather than edge sensing.

3. Using complex MSI and LSI functions in a logic system can reduce the number of circuit components and interconnection lines into which noise can be coupled. Also, complex functions usually reduce the total system cost, since fewer individual components must be assembled.

4. Inherently fast rise and fall time logic generates large current spikes internally on system power and ground lines. Slower logic greatly reduces this problem.

9–3   NOISE SPECIFICATIONS

Before comparing the noise immunity characteristics of the different logic families, a discussion of the different types of noise immunity specifications is in order. There are three basic ways to specify noise immunity: (1) dc noise margin, (2) ac noise immunity, and (3) noise-energy immunity.

9–3.1   Dc Noise Margin

The worst-case signal line dc noise margin (most commonly found on datasheets) of a logic family is shown in Figure 9-1, and is defined as follows:

$$V_{NSL} \text{(min)} = V_{IL} \text{(}max\text{)} - V_{OL} \text{(max)} \quad \text{(called low-level signal line noise margin)}$$

$$V_{NSH} \text{(min)} = V_{OH} \text{(min)} - V_{IH} \text{(min)} \quad \text{(called high-level signal line noise margin)}$$

where:

V_{OH} (min) = minimum high-out voltage for worst-case source current load (I_{OH})

V_{IH} (min) = minimum high-input voltage to guarantee the appropriate output logic level $(V_{OH}$ or $V_{OL})$

V_{IL} (max) = maximum low-input voltage to guarantee the appropriate output logic level $(V_{OH}$ or $V_{OL})$

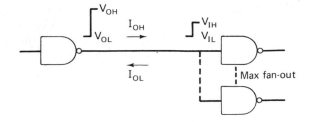

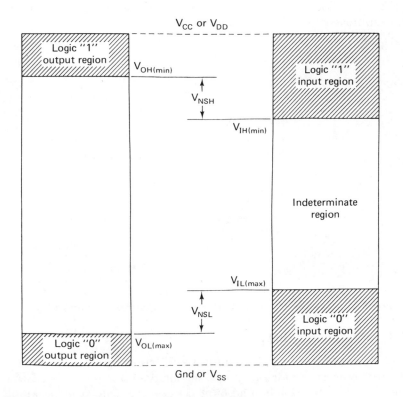

Figure 9-1.   Dc noise margin. *(Courtesy of Motorola)*

$$V_{OL}\,(\text{max}) = \text{maximum low-output voltage for worst-case sink current load } (I_{OL})$$

A dc noise margin specification is a good rule-of-thumb, but it does not give the designer all the necessary information to completely characterize the inherent noise immunity of a logic family. The dc noise margin predicts only the effects of a steady-state variation in signal line voltage levels.

9–3.2   Ac Noise Immunity

Ac noise immunity adds another dimension to the immunity specification by demonstrating the relationship of both the amplitude and the pulse width of the noise affecting circuit operation. A typical ac noise immunity plot, such as in Figure 9-2, shows the effects that propagation delay has on noise immunity. Note that as the noise pulse width approaches the propagation delay of the circuit, the voltage amplitude required to affect the circuit becomes quite high. Also note that as pulse width increases, the noise amplitude approaches the dc noise margin.

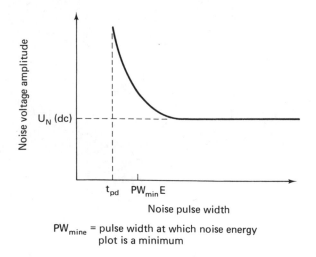

PW$_{mine}$ = pulse width at which noise energy
plot is a minimum

Figure 9-2.   Typical ac noise immunity plot. *(Courtesy of Motorola)*

9–3.3   Noise-energy Immunity

A noise-energy immunity specification gives the designer the best criteria to judge the noise immunity of a logic family. Noise-energy immunity takes into account not only the noise voltage amplitude and pulse width, but also the impedance that coupled noise "sees" on a line. A typical noise-energy immunity plot is shown in Figure 9-3, and follows the equation:

$$E_N = \frac{V_N^2}{R_0} \, (PW)$$

where:

V_N = noise voltage amplitude required to cause a circuit malfunction

R_0 = line impedance (the parallel combination of the circuit output and input impedances of the signal line, or the return impedance on the power and ground lines)

PW = noise pulse width

The plot of Figure 9-3 shows the noise energy required to affect a circuit reaches minimum value at the point $(PW_{min}E)$ where the ac noise immunity begins bending upward. This minimum value is the point at which noise-energy margin can be most meaningfully specified. The noise-energy curve also could be generated from the ac noise immunity curve by knowing the characteristic impedance of the line from which the curve was generated.

9–4   NOISE IMMUNITY TEST

Since noise immunity must be approached from a systems standpoint, standard testing for noise immunity must be made under normal logic system operation conditions. Testing of a single gate without considering typical input and output loading factors would not give a true noise figure of merit. Also, testing a large logic system could be impractical.

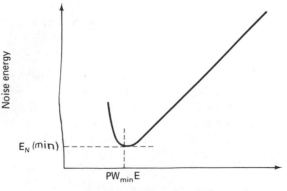

Figure 9-3.   Typical noise energy plot. *(Courtesy of Motorola)*

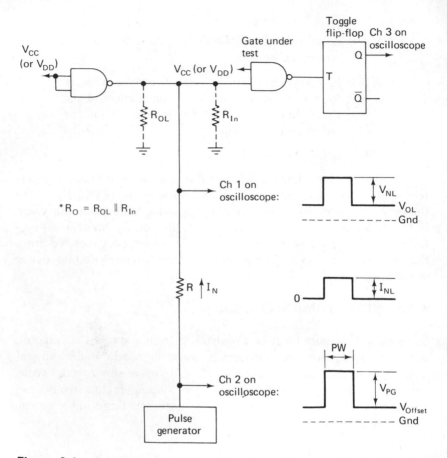

Figure 9-4.   Low-level signal line noise immunity test circuit. *(Courtesy of Motorola)*

Test circuits that give the best tradeoffs between the two extremes are shown in Figures 9-4 through 9-7. The test circuits include two gates and an FF from the same logic family. Such tests give the logic designer a good idea how well a logic gate will reject noise under the condition of a typical family input and output loading, propagation delays, and voltage thresholds.

9–4.1   Signal Line Tests

The signal line low-level and high-level noise immunity tests are shown in Figures 9-4 and 9-5, respectively. Since the noise margin, line impedance, and propagation delay characteristics of a gate vary between the high and low-level logic states, both states must be

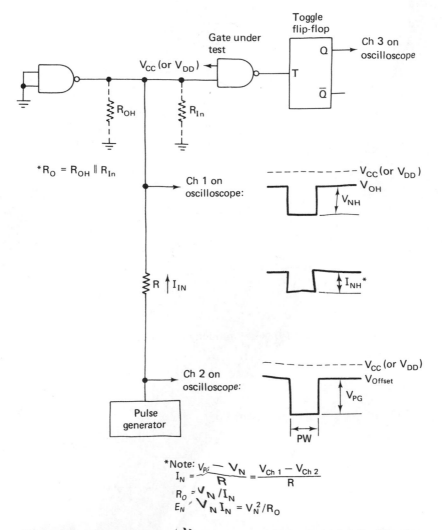

Figure 9-5.   High-level signal line noise immunity test circuit. (*Courtesy of Motorola*)

tested. Both positive and negative-going spikes of noise are generated at the interconnection between the gates using a pulse generator. The pulse generator dc offset is used to maintain quiescent dc voltage levels at the interconnection. An oscilloscope is used to measure the pulse voltage amplitude, V_N, and current amplitude, I necessary to toggle the FF with various noise pulse-widths. Fr these measurements, the line-impedance and noise-energy marg a typical gate-to-gate signal-line interconnection can be calc

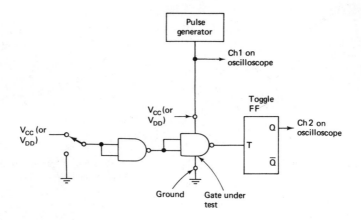

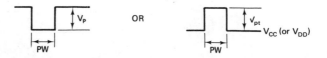

Figure 9-6. Power supply noise immunity test circuit. (Courtesy of Motorola)

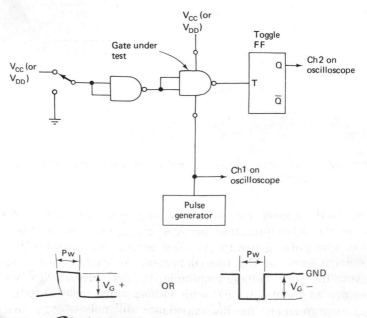

9-7.   Ground line noise immunity test circuit. (Courtesy of Motorola)

9–4.2   Power Supply and Ground Line Tests

The power supply and ground line noise immunity tests are shown in Figures 9-6 and 9-7, respectively. Logic family noise-energy margin concepts that apply to the signal line tests are not directly applicable to the testing of the power and ground lines. This is particularly true in the case of high-impedance MOS logic circuits. Since the impedance, looking into the V_{DD} and V_{SS} terminals, of a CMOS gate is typically $10^9\Omega$, the energy required to change the output voltage sufficiently to cause a malfunction is negligible. However, the energy required to raise the voltage level of the low impedance power and ground busses in a practical system can be very large.

It is more meaningful to test the *noise voltage level* on the gate power or ground lines that will cause a malfunction. The energy required to produce this voltage is dependent on the power supply output impedance, bypassing precautions taken, and wiring layout and ground return rules used. The system designer must determine the adequacy of the procedures used. As with the signal line immunity, the supply and ground line immunity is a function of noise pulse width. Depending on the logic state and family type, the logic gate may be sensitive to both positive and negative-going pulses on the power and ground lines.

9–5   TEST RESULTS FOR CMOS, TTL, HTL, DTL

The following paragraphs describe test results for both signal-line noise immunity, and power / ground line noise immunity.

9–5.1   Signal-line Noise Immunity

Figures 9-8 and 9-9 show low-level and high-level signal line *ac noise* immunity, respectively. Figures 9-10 and 9-11 show signal line *noise energy* immunity for the same logic families. As discussed in Sec. 9-3.3, the most comprehensive comparisons are taken from the plots of relative noise energy immunity (Figures 9-10 and 9-11). In these illustrations, the combined effects of voltage threshold, line impedance, and propagation delay on device noise immunity become quite clear.

CMOS vs HTL.   Due to the greater energy scale required, the 15-V noise energy characteristics of the CMOS and HTL families are plotted alone in Figure 9-10. The advantage of low line-impedanc becomes clear in this plot. Note that the HTL, with its low im dance (about 140Ω), low-level state, shows a minimum noise-ene

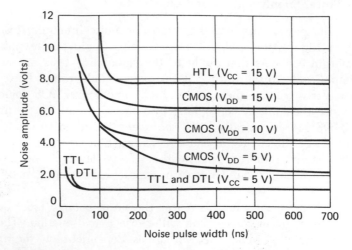

Figure 9-8.   Low-level signal line ac noise immunity. *(Courtesy of Motorola)*

immunity of 60 nanojoules (nJ). This is nearly double that of the best minimum for CMOS. However, the relatively high-impedance (about 1.6 kΩ) of HTL in the high-level state results in a minimum noise immunity less than that of the 15-V CMOS. (CMOS has a slightly lower impedance, about 1 kΩ, and comparable noise-amplitude margins in the high-level state.) The HTL does not have the additional advan-

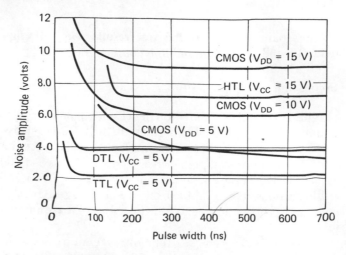

Figure 9-9.   High-level signal line ac noise immunity. *(Courtesy of Motorola)*

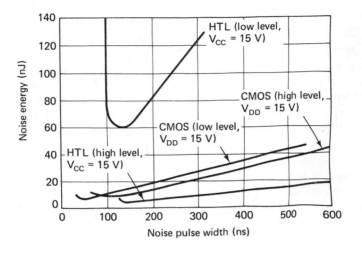

Figure 9-10.   Single line noise energy immunity (CMOS, HTL at 15 V). *(Courtesy of Motorola)*

tage of slower response time, causing the HTL to reach its minimum energy at a wider noise pulse width, and thus be less susceptible to higher frequency noise than CMOS.

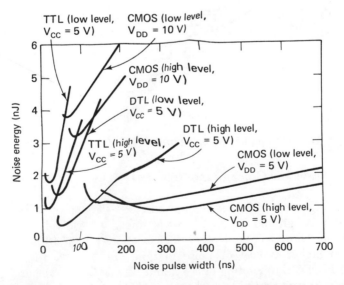

Figure 9-11.   Signal line noise energy immunity (CMOS at 5 V and 1⁰ TTL, DTL at 5 V). *(Courtesy of Motorola)*

CMOS vs TTL/DTL. Since the pulse width value indicates the noise frequency to which the device is most sensitive, it is clear that not only should the minimum noise energy value be considered, but also the pulse width at which the minimum energy immunity occurs. Keep this in mind when comparing the noise energy immunity plots of Figure 9-11. Low line impedances of TTL devices offer a slight advantage in minimum noise-energy immunity compared to the other devices at 5 V. However, TTL high-speed capability does make it much more susceptible to higher noise frequencies than CMOS, which reaches minimum energy immunity value at a much wider pulse width.

At the higher 10-V power supply level, CMOS has slower response time than TTL and DTL. Also, at 10 V, CMOS has higher minimum noise energy immunity values. Note that the high-level passive load (about 1.8 kΩ) of the DTL family give it the poorest value of noise energy immunity.

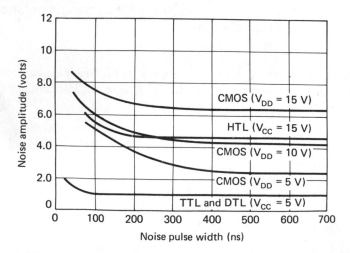

Figure 9-12. Ground line ac noise immunity. *(Courtesy of Motorola)*

9–5.2 Power and Ground Line Noise Immunity

Figures 9-12 and 9-13 show the results of the test for ac noise immunity on ground lines and power lines, respectively. In general, the graphs show: (1) the CMOS devices operating at 10 or 15 V have much higher typical dc noise margin, and slower response time than TTL/DTL (operating at 5 V); and (2) CMOS power and ground noise margin at 15 V is higher than HTL (also at 15 V).

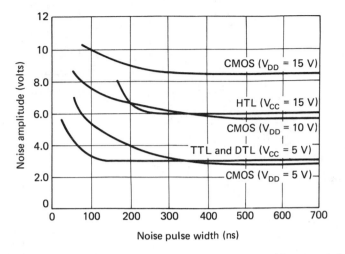

Figure 9-13.   Power supply line ac noise immunity. *(Courtesy of Motorola)*

Even when operating at 10 V, CMOS noise margin is comparable to that of the 15-V HTL devices, although the HTL devices have a noticeable slower response time to noise pulses on the power supply line. CMOS dc noise margin at 5 V, compared to TTL/DTL at 5 V, is about the same on the power supply line, and about twice as great on the ground line. However, due to the slower response at 5 V, CMOS shows much higher noise immunity than TTL/DTL as noise pulse width decreases.

Effect of noise spikes.   As discussed in Sec. 9–4, a device may be sensitive to both positive and negative noise spikes on the power and ground lines. The CMOS devices show a sensitivity to only negative-going noise spikes on the power line, and only to positive-going spikes on the ground line. The other logic families show various degrees of sensitivity to both positive and negative spikes on the power and ground lines, depending on the output logic state of the device under test. Only the worst-case conditions are plotted in the graphs of Figures 9-12 and 9-13.

Line impedance versus noise immunity.   Thus far, only device power and ground line noise immunity have been discussed. The designer alone has complete control over the *impedances of the power and ground lines of the system,* and these line impedances determine the energy. Significant results of noise immunity comparisons are tabulated in Figure 9-14.

Logic family	Power supply (volts)	Typical quiescent power dissipation (mW)	Typical propagation delay (ns)		Signal line DC noise margin				Typical* power supply line AC noise margin (volts)	Typical* ground line AC noise margin (volts)	Typical* signal line impedance (ohms)		Typical* noise energy minimum Logic state			
					V_{NL} (volts)		V_{NH} (volts)						Low		High	
			t_{PLH}	t_{PHL}	Min	Typ*	Min	Typ*			Low	High	E_{NL} nJ	@ PW ns	E_{NH} nJ	@ PW ns
DTL (Gate: MC849)	5.0	8.0	20	50	0.7	1.2	0.7	3.8	3.0	1.0	49	1.8 K	1.4	45	0.4	40
TTL (Gate: MC7400)	5.0	10	8.0	12	0.4	1.2	0.4	2.2	3.0	1.0	30	140	1.7	20	1.0	25
HTL (Gate: MC672)	15	25	85	130	5.0	7.5	4.0	7.0	6.0	4.5	140	1.6 K	60	125	5.0	145
CMOS (Gate: MC14011)	5.0	$25 \cdot 10^{-6}$	35	100	1.5	2.2	1.5	3.4	2.8	1.0	1.7 K	4.8 K	1.0	155	0.9	280
	10	$50 \cdot 10^{-6}$	20	35	3.0	4.2	3.0	6.0	5.7	4.3	670	1.5 K	3.7	70	3.1	90
	15	$150 \cdot 10^{-6}$	8.0	15	4.5	6.3	4.5	9.0	8.5	6.4	460	1 K	7.2	50	8.5	75

*Typical values are from experimental results of testing a small sample quantity of parts and may not reflect the manufacturer's specifications.

Figure 9-14. Tabulated noise immunity results. *(Courtesy of Motorola)*

9–6   TEST CIRCUITS FOR ECL

The following discussion applies specifically to Motorola MECL. However, the same problems apply to all ECL units. Transient noise can enter a MECL circuit at one or more of four different points: at the input to a circuit; on V_{CC}, which is normally at ground; on V_{EE}, which is normally −5.2 V; and on the termination voltage V_{TT}, which is normally −2 V. The response of the MECL circuit to noise depends on where the noise enters. For this reason separate tests are used to measure the response of a circuit to induced noise.

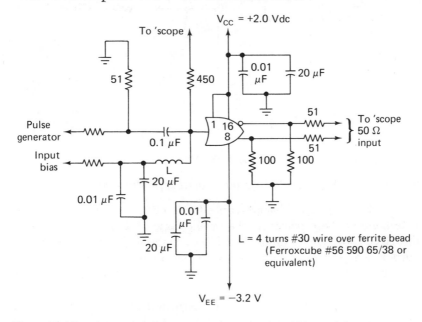

Figure 9-15.   Input noise test circuit. *(Courtesy of Motorola)*

9–6.1   Input Noise Test

Figure 9-15 shows a test circuit for measuring noise immunity to a transient signal on an ECL input line. As is characteristic of MECL testing, the test fixture puts +2 V on V_{CC} and −3.2 V on V_{EE}, to permit terminating the MECL outputs to ground through a 50-Ω oscilloscope probe. This has proven to be the most accurate method for testing MECL 10,000 and MECL III because possible problems with high impedance scope probes are eliminated.

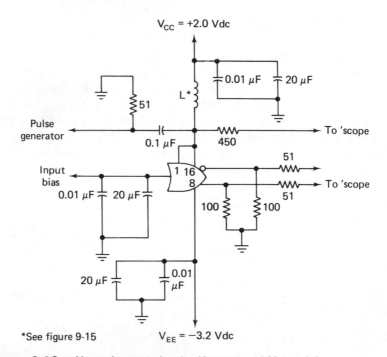

$V_{CC} = +2.0$ Vdc

0.01 μF   20 μF

L*

51

Pulse generator

0.1 μF   450   To 'scope

51

Input bias

0.01 μF   20 μF

1 16
8

51   To 'scope

100   100

20 μF   0.01 μF

*See figure 9-15 $V_{EE} = -3.2$ Vdc

Figure 9-16. V_{CC} noise test circuit. *(Courtesy of Motorola)*

Both V_{CC} and V_{EE} are capacitor bypassed to ground to eliminate noise at these points. If the oscilloscope 50-Ω input will handle the HIGH logic level (about +1.1 V) accurately, the MECL circuit outputs may be connected directly to the oscilloscope inputs. If not, 2:1 attenuation resistors and 100-Ω pull-down resistors may be used as shown in Figure 9-15.

The input of the test circuit is arranged to use an accurate bias, provided by a power supply, to set the input logic levels. The pulse generator is terminated by a 51-Ω resistor, and is capacitor-coupled to the input pin, to induce a noise pulse. An inductor, made of four turns of wire wound around a ferrit bead, isolates the noise from the input bias source.

A 450-Ω resistor is used in series with the 50-Ω oscilloscope input to isolate the input from the scope. This results in a 10:1 amplitude attenuation, but still gives an accurate picture of the input noise.

9–6.2 V_{CC} Noise Test

Figure 9-16 shows a test circuit for measuring noise immunity on the V_{CC} line. The test is similar to that for input noise. A power supply is connected at the circuit input to set the HIGH and LOW

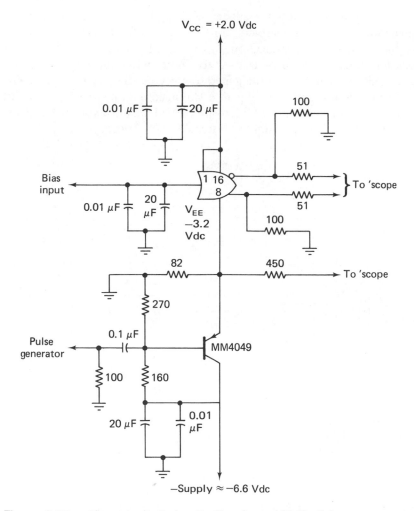

Figure 9-17. V_{EE} noise test circuit. *(Courtesy of Motorola)*

logic levels. The pulse insertion network is connected to V_{CC} pins 1 and 16. Since good MECL design practice is to tie both V_{CC} pins to a common point, tests are not conducted for signals inserted on only one V_{CC} pin.

9–6.3 V_{EE} Noise Test

Figure 9-17 shows a test circuit for measuring noise immunity on the V_{EE} line. This test is somewhat more complex than the tests for input and V_{CC}. A PNP transistor is connected as an emitter-follower to provide a low-impedance V_{EE} path. The high-speed pulse generator is capacitor-coupled into the base of the transistor. Because of the large signals required on V_{EE} to affect the circuit outputs, the emitter-follower transistor is used to allow better termination of the pulse generator than with the generator driving the V_{EE} pin. V_{EE} noise amplitude is approximately equal to the maximum pulse output of the pulse generator used. A very high-speed MM4049 transistor or equivalent should be used to retain the fast pulse edge speed.

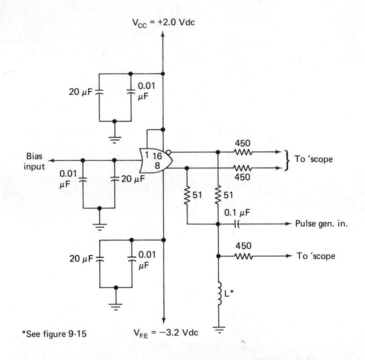

Figure 9-18. V_{TT} noise test circuit. *(Courtesy of Motorola)*

9–6.4 V_{TT} Noise Test

Figure 9-18 shows a test circuit for measuring noise immunity on the V_{TT} line. Power supplies are connected to V_{CC}, V_{EE}, and the circuit input. Bypass capacitors are used to keep noise from these points. The output of the test circuit is arranged to insert noise on V_{TT}. For this test, the 51-Ω termination resistors are connected to ground through an inductor. The induced pulse is capacitor-coupled to the V_{TT} point.

Circuit outputs are isolated from the 50-Ω oscilloscope inputs by 450-Ω attenuation resistors. These resistors prevent the MECL circuit from driving both the 51-Ω resistors to V_{TT}, and the 50-Ω oscilloscope input.

9–7   TEST CONDITIONS FOR ECL

The test limits chosen to determine noise immunity of typical MECL 10,000 parts are based on the specified voltages shown on datasheets. The circuit output is allowed to move to the input threshold points before the noise amplitude is considered a maximum. The HIGH logic level is allowed to drop to V_{IHA}, 0.895 V, and the LOW level to rise to V_{ILA}, 0.525 V (the voltage given for V_{CC} at +2 V).

These two points are chosen because the outputs of a circuit with these input levels are defined by the *threshold voltages* on the datasheets. Output test limits should be determined when evaluating parts for specific test information.

The input bias points are set at normal test voltages. These are +1.11 V for a HIGH logic level, and +0.31 V for a LOW logic level. With these inputs, the outputs of the circuits under test averages the following levels:

$$V_{OH} \text{ "OR"} = 1.1 \text{ V};$$

$$V_{OH} \text{ "NOR"} = 1.099 \text{ V};$$

$$V_{OL} \text{ "OR"} = 0.29 \text{ V, and}$$

$$V_{OL} \text{ "NOR"} = 0.265 \text{ V}.$$

Using the limits described, the following *transitions are allowed on the outputs* before maximum noise is reached:

$$\Delta V_{OH} \text{ "OR"} = 0.205 \text{ V};$$

$$\Delta V_{OH} \text{ "NOR"} = 0.204 \text{ V};$$

$$\Delta V_{OL} \text{ "OR"} = 0.235 \text{ V}; \text{ and}$$

$$\Delta V_{OL} \text{ "NOR"} = 0.26 \text{ V}.$$

This technique for reading *output transition amplitudes* is used because it is more convenient than reading dc values on an oscilloscope.

The tests are conducted by setting the power supplies and circuit inputs into the quiescent conditions. A pulse width is selected. Then the amplitude of the pulse generator is increased until the output reaches the predetermined limit. The pulse width is then changed and the test repeated.

A very high speed pulse generator is required to generate the narrow pulses used in the tests. The generator must be capable of 500 MHz operation and has pulse edges of about 0.5 ns rise time. This presents worst-case noise conditions, because such an edge speed is much faster than would be expected in a MECL system.

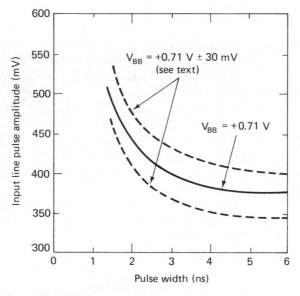

Figure 9-19.   Input signal ac noise immunity. *(Courtesy of Motorola)*

9–8   TEST RESULTS FOR ECL

Figure 9-19 shows typical noise immunity for a MECL 10,000 gate. A pulse is applied to the circuit input and the pulse amplitude and width are measured to determine the tolerance limits previously defined. The average is shown by the solid line which levels at about 370 mV for a 6-ns pulse width. The curve is relatively flat to about 3 ns, then rises sharply as the circuit becomes limited by the bandwidth.

The test results show that there is some deviation from the single line, due to variations in V_{BB} from the exact center voltage of 0.71 V. When this happens there will be two curves as shown by the dotted lines in Figure 9-19. V_{OH} "OR" and V_{OL} "NOR" follow one curve, and V_{OL} "OR" and V_{OH} "NOR" follow the other curve. For example, if V_{BB} is 30 mV low, V_{OH} "OR" and V_{OL} "NOR" will follow the upper dotted curve.

This behavior can be explained by the MECL transfer curves shown in Figure 9-20. The heavy vertical lines show the nominal input test levels. The outputs must move to the heavy horizontal lines. Under conditions greater than about 6-ns pulse width, the circuit follows the transfer curve. A low level input has to move from +0.31 V to about

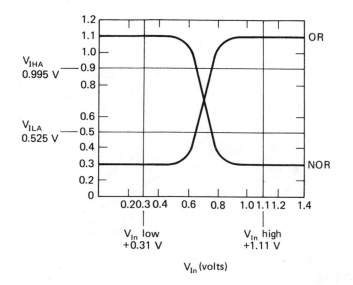

Figure 9-20. MECL 10,000 transfer characteristics. *(Courtesy of Motorola)*

+0.67 V before the output noise limit is reached. Similarly, a high input must move from 1.11 V to about +0.74 V before the maximum output change is reached. This gives about 370 mV noise margin, as shown in Figure 9-19. If V_{BB} is slightly to one side of the 0.71 V nominal value, the transfer curves shift slightly to the right or left. This causes the dotted curves of Figure 9-19.

9–8.1   Noise Immunity to Transients on the V_{CC} Line

Figure 9-21 shows noise immunity to transients on the V_{CC} supply line. The tests are made by applying a negative pulse to V_{CC} when the tested outputs are at a HIGH logic level, and a positive pulse when at a LOW level output. The circuits are most susceptible to negative noise spikes on V_{CC} affecting the HIGH logic levels.

For wide pulses there is a one-to-one loss of noise margin with pulse amplitude. At 6 ns, there is about 215 mV of noise immunity. This compares with the 204 mV maximum output transient allowed by the previously defined test limits. For narrower pulses the response time is shown in Figure 9-21.

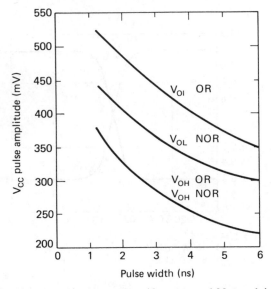

Figure 9-21. V_{CC} ac noise immunity. *(Courtesy of Motorola)*

9–8.2   Noise Immunity to Transients on the V_{EE} Line

Figure 9-22 shows noise immunity to transients on the V_{EE} supply line. The tests are performed by applying a pulse on the V_{EE} line, using the circuit shown in Figure 9-17. The pulse polarity is chosen so that there is a positive pulse on V_{EE} when the output is LOW, and a negative pulse with a HIGH output. The amount of noise

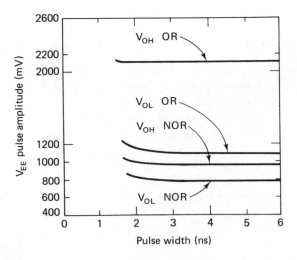

Figure 9-22. V_{EE} ac noise immunity. *(Courtesy of Motorola)*

immunity in the MECL circuit is dependent on whether the OR or NOR output is tested, and the logic state of the output.

From Figure 9-22 it may be seen that the OR outputs have more noise immunity than the NOR outputs (as would be expected from the V_{BB} coupling). HIGH logic levels are more immune than LOW levels due to the voltage tracking ratio of the output logic levels with V_{EE}.

The LOW level NOR output is most affected by V_{EE} noise. This is due to a combination of V_{OL} tracking and V_{BB} coupling. However, the noise immunity to this worst case condition remains greater than 700 mV. The greatest immunity occurs on the HIGH level OR output and is greater than 2 V. In general, the MECL circuits are more immune to noise on V_{EE} than on any other point in the system.

9–8.3 V_{TT} Noise Immunity

The noise immunity of MECL circuits to noise on the V_{TT} termination voltage is dependent on the output impedance of the MECL gate and on the response time of the MECL circuit. The test circuit of Figure 9-18 shows that the pulse generator signal is connected to the outputs through 50-Ω termination resistors. Since the resistors have a better bandwidth than the pulse signal, there is no noise immunity change as a function of pulse width. Full amplitude output transitions are present down to the measuring limit of about 1.5 ns.

Figure 9-23 shows waveforms associated with V_{TT} noise immunity measurements. The upper trace shows the noise pulse coupling through the circuit output, and the MECL gate recovering from

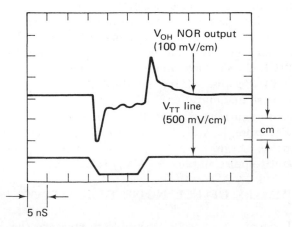

Figure 9-23. V_{TT} ac noise immunity with 0.5 ns noise edge speed. *(Courtesy of Motorola)*

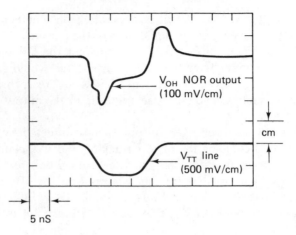

Figure 9-24. V_{TT} ac noise immunity with 0.2 ns noise edge speed. *(Courtesy of Motorola)*

the pulse, then restoring the logic level. The response time of the MECL 10,000 output circuit is about 2 ns.

The lower oscilloscope trace shows the pulse generator noise on the V_{TT} line. About 400 mV is required to cause a 204 mV transient on the output. The ratio of noise amplitude on the V_{TT} line to the noise on the output is dependent on the output impedance of the MECL circuit. As a result, the noise immunity to fast pulse edges is independent of the OR or NOR output tested and of the polarity of the noise pulse. For all outputs, about 400 mV of V_{TT} noise with 0.5 ns rise and fall times is required before the test limits are reached.

Since the V_{TT} noise immunity depends on the MECL circuit response time, the 0.5 ns pulse rise times are a worst-case test condition. By increasing the pulse rise and fall times for the noise on V_{TT}, the MECL circuits become more immune to V_{TT} noise. With the outputs at a HIGH logic level and a negative pulse with 2 ns edges on V_{TT}, the MECL circuits have about 800 mV noise immunity, as shown in Figure 9-24. Because of the emitter-follower outputs of MECL circuits, the noise immunity for a LOW level output is limited by the output transistors turning off, and the outputs following the V_{TT} voltage level. For example, with V_{TT} at ground in the test circuit, and a low level noise margin output test limit of 0.525 V, the maximum tolerable positive pulse amplitude on V_{TT} is 525 mV.

9–9   SUMMARY OF ECL NOISE IMMUNITY

The MECL circuits are most susceptible to noise on the V_{CC} supply line. For this reason, MECL design rules commonly suggest operating MECL circuits with a positive ground on V_{CC} and −5.2 V on V_{EE}. By having a low impedance system ground, noise on the V_{CC} line between two circuits can be controlled and minimized.

In some cases, it is desirable to operate MECL circuits with ground and +5 V supplies. MECL circuits operate very well in this mode when care has been taken to keep noise on the +5 V supply line to a minimum. The MECL circuits are most immune to noise on the V_{EE} supply line. With standard capacitor bypassing techniques (Chapter 2), noise on the V_{EE} line is controlled to safe system levels.

The amount of noise present in a digital system depends on many factors, such as power supplies and system environment. A primary source of noise is crosstalk, which is proportional to signal speed and signal amplitude. The 800 mV logic swing and relatively slow rise and fall times of MECL 10,000, along with its ability to operate in a transmission line environment, serve to reduce crosstalk in a MECL system.

One factor contributing to power supply noise is the amount of current "spiking" inherent in a logic family. MECL circuits have emitter-follower outputs and differential-amplifier switches. Consequently, MECL generates very little noise. The high ratio of noise immunity to internally-generated noise in MECL is a feature leading to reliable system operation.

Miscellaneous
Logic Circuits

This chapter is devoted to special logic circuits not covered in previous chapters. The information in this chapter shows the logic designer how to implement circuits often needed in system design from basic gates and logic elements. Such circuits include oscillators, multivibrators, multipliers, detectors, delays, clocks, and latches.

10–1 MOS DEVICES USED AS ASTABLE
AND MONOSTABLE OSCILLATORS

It is possible to connect complementary MOS devices in simple circuits to form astable and monostable oscillators (or multivibrators). The following paragraphs describe a few examples. All of the circuits described involve the use of MOS NAND and NOR gates, or MOS inverters. (Inverters are formed when all inputs of a NAND or NOR gate are tied together.)

10–1.1 Astable Oscillators

Figure 10-1 shows an astable multivibrator circuit that uses two COS/MOS inverters. This simple circuit requires only one resistor and one capacitor. Operation of the circuit is as follows:

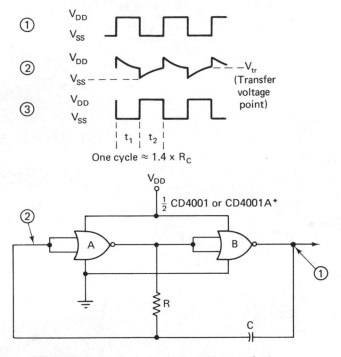

One cycle ≈ 1.4 x R_C

* This circuit can also be implemented by use of other COS/MOS devices such as NAND gates or inverters in place of NOR gates.

Figure 10-1. Astable multivibrator implemented with two NOR gates connected as inverters. *(Courtesy of RCA)*

When waveform 1 at the output of inverter B is high (logic 1), the input to inverter A is also high. Under these conditions, the output of inverter A is low, and capacitor C is charged. Resistor R is returned to the output of inverter A to provide a path to ground for discharge of capacitor C.

As long as the output of A is low, the output of inverter B is high. As capacitor C discharges through R, the voltage shown as waveform 2 approaches and passes through the transfer voltage point of inverter A. At the instant of crossover, the output of A becomes high, the output of B becomes low (logic 0), and C charges in the opposite direction.

The crossover process is repeated at a rate determined by the time constant of R and C. Because of the input-diode protection circuits included in the RCA COS/MOS, the generated drive waveform is clamped between V_{DD} and V_{SS}. Consequently, the time to complete one cycle is *approximately* 1.4 times the RC time constant, because *one time constant* is used to control the switching of *both states* of the multivibrator circuit.

$$T = - RC \ln \left[\frac{V_{tr}}{(V_{DD} + V_{tr})} + \ln \frac{(V_{DD} - V_{tr})}{2 V_{DD} - V_{tr}} \right]$$

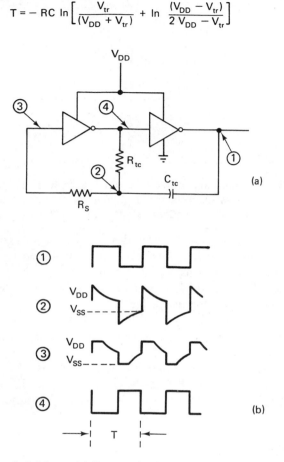

Figure 10-2.   Astable multivibrator (independent of supply voltage) implemented with two COS/MOS inverters. *(Courtesy of RCA)*

The time constant of the Figure 10-1 circuit is *voltage dependent,* making the $(1.4 \times RC)$ figure an approximation. However, assuming that the transfer voltage point V_{tr} varies as much as 33 to 67 percent of V_{DD}, the time constant multiplication factor will vary from about 1.4 to 1.5 times RC. Thus, the maximum variation in the time period is about 9 percent with a ± 33 percent variation in V_{tr} (from unit to unit).

Nonvoltage-dependent oscillator.   The oscillator can be made independent of supply voltage variations by use of a resistor R_S in series with the input lead to inverter A, as shown in Figure 10-2. Resistor R_S should be at least twice the value of R_{tc}. This will allow the voltage waveform generated at the junction of R_S, R_{tc}, and C_{tc} to rise to $V_{DD} + V_{tr}$. The waveform is still clamped at the input between V_{DD} and V_{SS}, as shown by the waveforms of Figure 10-2b.

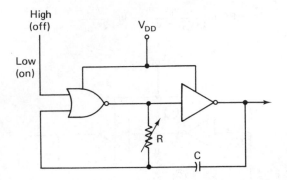

Figure 10-3.   Astable multivibrator in which a NOR or NAND COS/MOS gate is used as the first inverter to permit gating of the multivibrator. *(Courtesy of RCA)*

The use of resistor R_s provides several advantages in the circuit. First, because the RC time constant controls the frequency, the overall maximum variations in time period are reduced to less than 5 percent with variations in V_{tr}. Resistor R_s also makes the frequency independent of supply voltage variations.

The time period T for one cycle can be computed using the equations shown in Figure 10-2. The frequency of the oscillator is the reciprocal of the time period, or $1/T$.

For example, assume that V_{DD} is 10 V, V_{tr} is 5 V, R is 0.4 MΩ, C is 0.001 μF, and R_s is 0.8 MΩ (twice the value of R). $RC = (0.4 \times 10^6)(0.001 \times 10^{-6}) = 0.0004$ second, or 0.4 ms. The period T is found by:

$$T = 0.4 \text{ ms} \times \left[1. \frac{5}{10 + 5} + 1. \frac{(10 - 5)}{(20 - 5)} \right]$$

$$= 0.4 \text{ ms} \times 2.66 = 1.064 \text{ ms}$$

The frequency is approximately 1 kHz.

The astable multivibrator of Figure 10-1 or 10-2 can be gated ON and OFF by means of a NOR or NAND gate as the first inverter. Such an arrangement is shown in Figure 10-3.

10–1.2   Variable Duty Cycle Astable Oscillators

A true square-wave pulse is obtained only when V_{tr} occurs at the 50 percent point. The duty cycle can be controlled if part of the resistance in the RC time constant is shunted out with a diode, as shown in Figure 10-4.

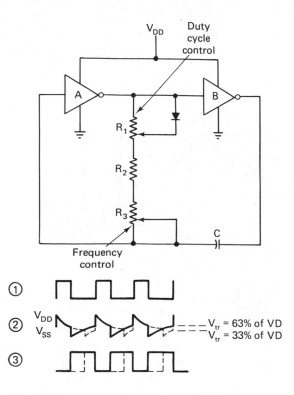

Figure 10-4.   Variable duty-cycle astable multivibrator using COS/MOS inverters. *(Courtesy of RCA)*

Because adjustment of this diode shunt to obtain a specific pulse duty factor causes the frequency of the circuit to vary, a frequency control R_3 is added for compensation. It may be necessary to reverse the diode to obtain the desired duty factor. The frequency of any of the circuits (Figures 10-1 through 10-3) can be made variable by use of a potentiometer for resistor R.

10–1.3   Monostable Circuits

Figure 10-5 shows a compensated monostable multivibrator type of circuit that can be triggered with a negative-going pulse (V_{DD} to ground). In the quiescent state, the input to inverter A is high, and the output is low. Thus, the output of inverter B is high. When a negative-going pulse or spike is introduced into the circuit as shown in the waveforms, capacitor C_1 becomes negatively charged to ground,

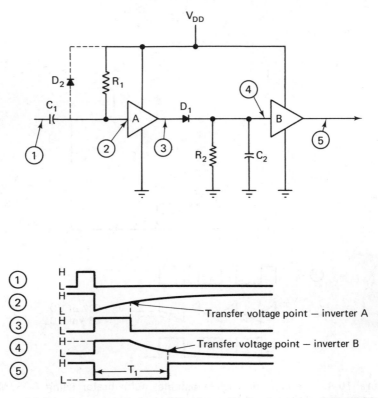

Figure 10-5. Compensated monostable multivibrator using COS/MOS inverters. *(Courtesy of RCA)*

and the output of inverter A goes high. Capacitor C_2 then charges to V_{DD} through the diode D_1 and inverter A, and the output of inverter B becomes low.

As capacitor C_1 discharges negatively, it charges to the opposite polarity through R_1 to V_{DD} (waveform 2). The output of inverter A remains high until the voltage waveform generated by the charge of C_1 passes through the transfer voltage point of inverter A. At that instant, the output of inverter A goes low.

Diode D_1 temporarily prevents the discharge of capacitor C_2, which was charged when inverter A was high (waveform 3). Capacitor C_2 then commences to discharge to ground through R_2 (waveform 4). The output of inverter B remains low until the waveform is generated by the discharge of C_2 passing through the transfer voltage point of inverter B. At that point, the output returns to its high state (waveform 5).

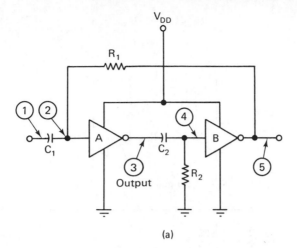

(a)

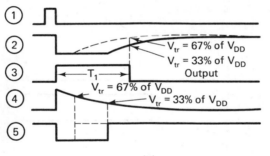

(b)

Figure 10-6.   Monostable multivibrator triggered by a negative-going input pulse. *(Courtesy of RCA)*

When two inverters are fabricated on the same chip their transfer voltage point will be similar. This is an advantage in that any variations in V_{tr} are cancelled out.

When $R_1 = R_2$ and $C_1 = C_2$, the period T is *approximately* equal to the RC time constant. However, from a practical standpoint, the period will usually be somewhat greater than the time constant. For example, if $R_1 = R_2 = 1M\Omega$, and $C_1 = C_2 = 0.001\ \mu F$, the period will be about 1 to 1.1 ms.

Alternate monostable circuits. Figures 10-6 and 10-7 show variations of the monostable circuit, together with the associated waveforms. The circuit of Figure 10-6 triggers on the negative-going swing of the input pulse, in the same manner as the circuit of Figure 10-5.

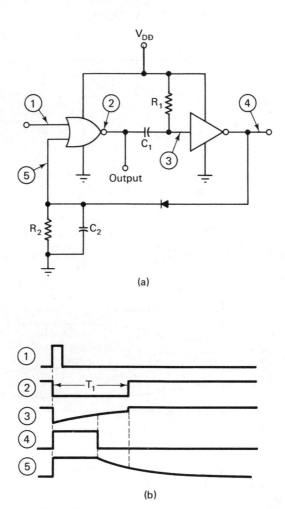

Figure 10-7. Monostable multivibrator triggered by a positive-going input pulse. *(Courtesy of RCA)*

The output pulse is positive-going and is taken from the first inverter (waveform 3). No external diode is needed.

The circuit of Figure 10-7 triggers on the positive-going swing of the input pulse, and then locks back on itself until the RC time constants complete their discharge.

Note that the circuits of Figures 10-6 and 10-7 *cannot be re-triggered until* they return to their quiescent states.

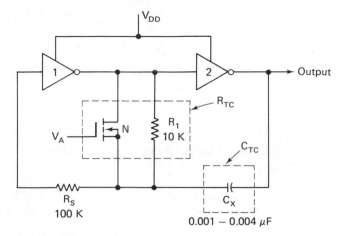

$0 \leqslant V_A \leqslant V_{DD}$

Inverters and N-channel device are available in a
single COS/MOS package (CD4007 or CD4007A).

	Period (μS)		
V_A	V_{DD} = 5 V	V_{DD} = 10 V	V_{DD} = 15 V
0	120	54	48
5	115	45	41
10	– –	32	30
15	– –	111	24

Figure 10-8.   Voltage-controlled oscillator using COS/MOS inverters.
(Courtesy of RCA)

10–1.4   Voltage-controlled Oscillator

Figure 10-8 shows a circuit similar to that of Figure 10-2.
However, in the Figure 10-8 circuit, C is made variable by C_x, and R
is made variable by adjustment of V_A (which is applied to the gate of
an N-channel MOS device). The value of R varies from approximately
1 to 10 kΩ. These limits are determined by the parallel combination of
R_1 and the N-channel MOS device resistance (which varies from about
1 kΩ when fully ON, to about $10^9 \Omega$ when fully OFF).

When $V_A = V_{ss}$, the N-channel device is OFF, and $R =$ the
parallel combination of R_{OFF} and R_1, or about 10 kΩ, since R_{OFF} is
much greater than R_1.

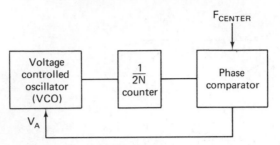

Figure 10-9.   Voltage-controlled oscillator used as a phase-locked loop. *(Courtesy of RCA)*

When $V_A = V_{DD}$, the N-channel device is ON, and $R =$ the parallel combination of R_{ON} and R_1, or about 1 kΩ, since R_{ON} is much smaller than R_1.

The oscillator center frequency is varied by adjustment of C_X. The table in Figure 10-8 shows a comparison of the output waveform period as a function of V_{DD} and V_A.

10–1.5   Phase-locked Voltage Controlled Oscillator

The voltage-controlled oscillator of Figure 10-8 can be operated as a phase-locked oscillator by the application of a frequency-controlled voltage to the gate of the N-channel device. Figure 10-9 shows the block diagram of an FM discriminator using the phase-locked voltage-controlled oscillator (VCO).

The VCO block is the same circuit as Figure 10-8. The output of a standard phase comparator is fed into the gate of the N-channel device (V_A). If the two inputs to the phase comparator are different, the change of V_A causes the output frequency of the VCO to change. This change is divided by 2^N (through a counter), and fed back to the phase comparator.

10–1.6   Voltage-controlled Pulse-width Circuit

Figure 10-10 shows a further modification of the Figure 10-2 circuit to provide a modulated pulse width. That is, the pulse width can be controlled by a voltage at V_A. The frequency will not be affected by the V_A voltage, provided the value of R_X is high. The values shown provide pulse periods and pulse widths, as indicated by the table.

Note that the period is determined by component values and V_{DD}, whereas pulse width is primarily dependent on V_A. For example, with the resistance values shown, C at 1500 pF, V_A at 5 V, and V_{DD} at 10 V, the period is 35 μS, and the pulse width is 17.7 μS. If V_A is changed to 10 V, then pulse width changes to 16.2 μS, but the period remains at 35 μS.

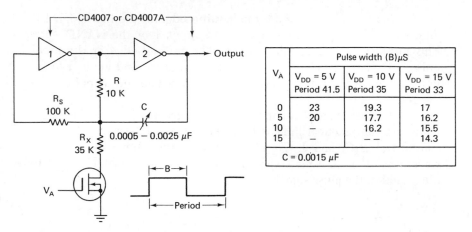

	Pulse width (B)μS		
V_A	V_{DD} = 5 V Period 41.5	V_{DD} = 10 V Period 35	V_{DD} = 15 V Period 33
0	23	19.3	17
5	20	17.7	16.2
10	–	16.2	15.5
15	–	– –	14.3
C = 0.0015 μF			

Figure 10-10. Voltage-controlled pulse-width circuit using COS/MOS inverters. *(Courtesy of RCA)*

10–1.7 Frequency Multiplier

Figure 10-11 shows a frequency multiplier (doubler) using MOS IC devices. A 2^N multiplier can be realized by cascading the basic circuit with N–1 and other identical circuits. That is, a 2^3 multiplier is formed when three basic circuits are cascaded.

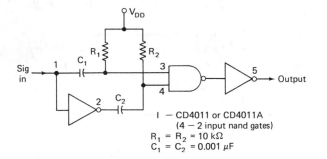

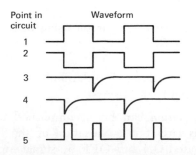

Figure 10-11. Frequency doubler circuit using COS/MOS devices. *(Courtesy of RCA)*

With the basic circuit, the leading edge of the input signal is differentiated by R_1 and C_1, applied to input 1 of the NAND gate. This produces a pulse at the output. The trailing edge of the input pulse, after having been inverted, is differentiated, and applied to input 2 of the NAND gate. This produces a second output pulse from the NAND gate.

In theory, any number of basic circuits can be cascaded. However, the practical limit is between 5 and 10. Also, the RC time constant of R_1C_1 and R_2C_2 sets a limit on the frequency. For best waveforms, the RC time constant should be approximately equal to the period of the pulse signals.

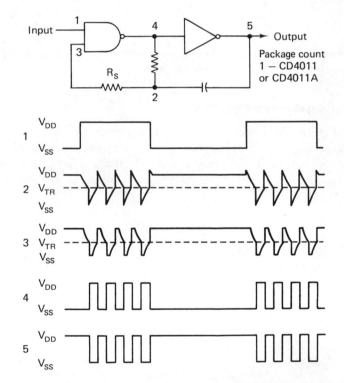

Figure 10-12.   Pulse modulator circuit using COS/MOS devices. *(Courtesy of RCA)*

10–1.8   Envelope Detection (Modulation/Demodulation)

Pulse modulation can be accomplished using the circuit of Figure 10-12. This circuit is a variation of the Figure 10-2 circuit. The oscillator is gated ON and OFF by signal input 1 at the NAND gate. The number of pulses at the output (waveform 5) during the gated period (waveform 1) is dependent on the RC time constants.

Demodulation or envelope detection of pulse modulated waves is performed by the circuit shown in Figure 10-13. The carrier burst is inverted by inverter A. The first negative transition at point 2, turns on diode D to provide a charging path for C through the N-channel resistance to ground. On the positive transition of the signal at point 2, diode D is cut off, and C discharges through R. The discharge RC time constant is *much greater* than the time of burst duration. Thus, point 3 never reaches the switch point of inverter B until the burst has ended.

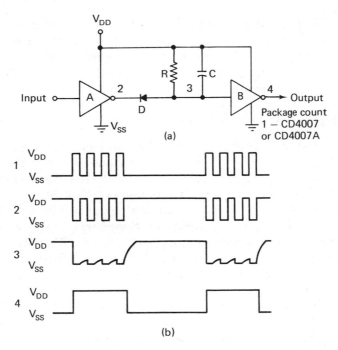

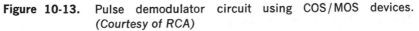

(b)

Figure 10-13. Pulse demodulator circuit using COS/MOS devices. *(Courtesy of RCA)*

10–2 CLOCK WAVEFORM CIRCUITS

One of the more important areas in designing with logic circuits (particularly ICs) is the generation and distribution of waveforms suitable for triggering FFs (clock pulses). This is particularly true when using JK FFs, since these devices often require a clock pulse fall time between 10 and 100 ns.

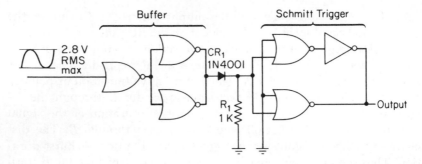

Figure 10-14.   Clock waveform generator. *(Courtesy of Motorola)*

In general, it is best to drive FFs with a buffer or gate element in IC form, rather than directly from circuitry composed of discrete components. In cases where this is not feasible, the driving circuit should be designed to have output characteristics similar to the type of IC being driven.

10–2.1   Clock Waveform Generator

Figure 10-14 shows a basic clock waveform generator suitable for most FFs. The circuit uses NOR gates and an inverter connected to form a Schmitt trigger. The input is an ac sinewave, whereas the output is a rectangular pulse. If it is assumed that the input is 2.8 V (RMS) with the value of R_1 at 1 kΩ, the output will be a 1.5-V pulse. When the 1N4001 diode is used, the circuit will trigger at approximately 1.5 V on the positive-going slope, and at 1.1 V on the negative-going slope of the input waveform. This 0.4-V differential insures a stable output since the uncertainty region ($V_{OFF} - V_{ON}$) is less than 0.4 V in all cases.

The actual output level of the circuit is set by the characteristics of diode CR_1. The primary purpose of R_1 is to provide an input load for the Schmitt trigger portion of the circuit.

10–2.2   Clock Waveform Generator With FF Input

Figure 10-15 shows another circuit for producing clock waveforms. Again, a Schmitt trigger is formed with gates and an inverter. However, the trigger is driven by two D-type FFs, which accept input pulse rise and fall times up to 100 μS. This method is particularly useful in ripple counter applications, but can also be used with shift registers. Note that since two FFs are used, there is one output pulse for each four input pulses.

As a general rule, the clock input to a ripple counter or similar circuit should have a rise and fall time of less than 100 μS, while a clocked counter or shift register requires fall times of approximately

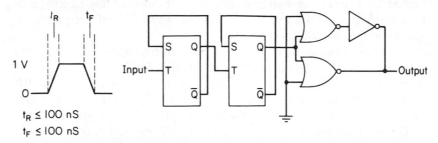

Figure 10-15. Clock waveform with FF input circuit. *(Courtesy of Motorola)*

100 ns effectively to combat "race" problems (where clock pulses overlap input data pulses, causing the circuit to latch, or move rapidly back and forth between states).

10–3 CRYSTAL OSCILLATOR FOR LOGIC CIRCUITS

In designing logic circuits, it is often convenient to have a pulse source, locked in frequency by a crystal. A typical application for such a pulse source is that of clock pulses for a counter.

Figure 10-16 shows a crystal-controlled pulse oscillator implemented with logic gates. In this circuit, two NOR gates are con-

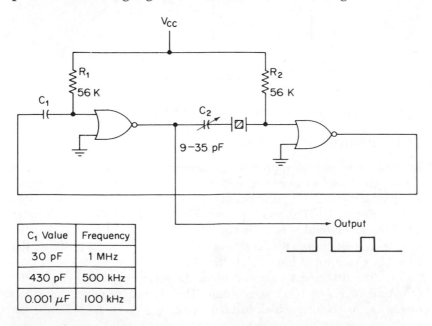

C_1 Value	Frequency
30 pF	1 MHz
430 pF	500 kHz
0.001 μF	100 kHz

Figure 10-16. Crystal oscillator for logic circuits. *(Courtesy of Motorola)*

nected in a cross-coupled configuration. In effect, the circuit is a free-running multivibrator whose square wave output frequency is locked by the crystal.

The resistors R_1 and R_2 serve as biasing elements, in addition to being a part of the circuit time constants. With the crystal placed in the circuit as shown, R_1 and C_1 determine the period (and thus the frequency) of the output waveform. Since R_1 also establishes the bias of the gate input, and must be fixed for a given V_{CC}, frequency is controlled by selection of C_1 (and the crystal). Note that three values of C_1 are given in Figure 10-16, one for each of three frequencies. These values are based on the assumption that R_1 is 56 kΩ. Trimmer capacitor C_2 permits exact adjustment of the frequency.

Any type of NOR gate can be used. However, the beta of the transistors used in the amplifier-inverter portion of the gate should be sufficiently high to sustain oscillation. A beta value of 50 to 100 (which is typical for a NOR gate in IC form) should provide sufficient feedback to sustain oscillation.

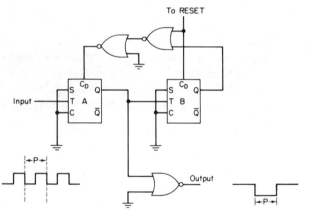

Figure 10-17.   Period selector. *(Courtesy of Motorola)*

10–4   PERIOD SELECTOR

In some logic circuit applications, particularly period counters, it is convenient to select one and only one period of an incoming signal, and produce an output pulse based on the period.

Figure 10-17 shows a period selector implemented with gates and FFs. The circuit is initially reset by application of a fixed dc voltage, or a reset pulse, to the direct reset (or preclear) inputs of the FFs. The Q output of both FFs is then low.

The first negative transition of the incoming signal causes the Q output of FF A (Q_A) to go high. The second negative transition causes Q_A to go low, which in turn causes Q_B to go high. Thus, the high condition of Q_A exists only during one complete period of the input to the circuit. This high state is inverted by the NOR gate, and becomes the output pulse.

At first glance it appears that the $\overline{Q}$ output of FF A $(\overline{Q_A})$ could be used directly and eliminate the need for inversion. However, most JK FFs behave in the following manner: As the toggle input is clocked, the negative transitions actually cause the $\overline{Q}$ output (when high) to attempt to go low. Since the direct clear is held high, $\overline{Q}$ immediately returns to the high state. Consequently, the $\overline{Q}$ output produces a negative spike at each negative transition of the toggle input. On the other hand, a high on the direct clear insures that the $\overline{Q}$ output remains low during the toggle input transitions. Thus, it is desirable to use the Q output and invert, rather than cope with the negative spikes on the $\overline{Q}$ output.

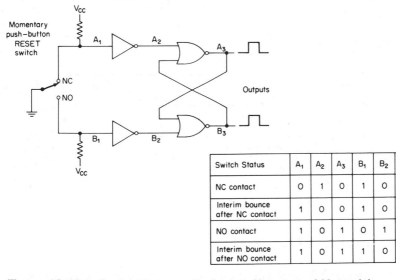

Switch Status	A_1	A_2	A_3	B_1	B_2	B_3
NC contact	0	1	0	1	0	1
Interim bounce after NC contact	1	0	0	1	0	1
NO contact	1	0	1	0	1	0
Interim bounce after NO contact	1	0	1	1	0	0

Figure 10-18. Contact bounce eliminator. *(Courtesy of Motorola)*

10–5   CONTACT BOUNCE ELIMINATOR

When the contacts of mechanical switches close, they may "bounce" several times. This bouncing can result in a series of pulses being generated on the line. This can cause malfunctions in high-speed logic circuits.

The circuit of Figure 10-18 eliminates any contact bounce, and is completely independent of the duration of contact bounce. The circuit of Figure 10-18 requires a double-pole switch, such as a momentary push-button-type switch. In effect, the circuit is a bistable

MV (or an *RS* FF) that is triggered by the momentary switch. As shown in the table of Figure 10-18, once the switch arm makes contact with either the normally closed (NC) contact, or the normally open (NO) contact, no amount of bounce can change the state of the output.

One restriction in the circuit is that the switch can not rebound completely between the NC and NO contact. (Switches of this type are generally considered "choppers" or "vibrators.") The circuit output is, in effect, a momentary ON or OFF condition (similar to true switch action, without bounce).

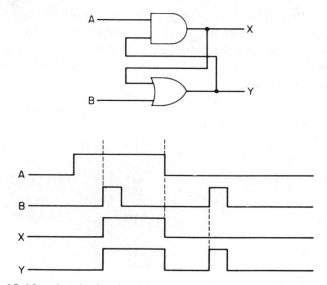

Figure 10-19.   Latch circuit with noninverting gates. *(Courtesy of Motorola)*

10–6   LATCH CIRCUITS

While "latching" in most counters and shift registers is undesirable, it is often necessary to implement a circuit that will latch upon a given command. Figures 10-19 through 10-21 show three latch circuits that can be used with almost any logic configuration.

The circuit of Figure 10-19 uses an AND gate cross-coupled with an OR gate. This circuit is a temporary "memory" as shown by the timing diagram. However, since complementary outputs are not available, the circuit is somewhat limited in use.

The circuit of Figure 10-20 is a capacitively coupled, edge-triggered latch, while the circuit of Figure 10-21 is a direct-coupled latch. The capacitively coupled version has the inherent disadvantage

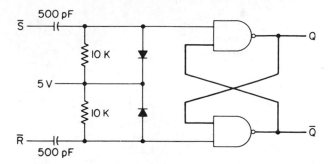

Figure 10-20. Latch circuit with capacitive coupling. *(Courtesy of Motorola)*

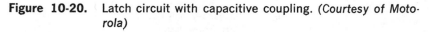

Synchronous
Truth Table

t_n				t_n+1
S1	S2	C1	C2	Q
0	0	0	0	U
1	X	1	X	Q_n
X	1	X	1	Q_n
0	1	1	0	Q_n
0	0	X	1	1
1	X	0	0	0
X	1	0	0	0

Single Trigger
Truth Table
(S2 and C1 tied
together)

t_n		t_n+1
S1	S2	Q
0	0	U
1	0	0
0	1	1
1	1	Q_n

0 = Dynamic transition from Hi to Low state, when
    referred to S2 or C1
0 = Low state, when referred to S1, C2 or Q
1 = High state   X = Don't care
Q_n = No change   U = Intermediate state

Figure 10-21. Latch circuit with direct coupling. *(Courtesy of Motorola)*

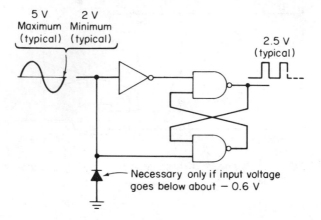

Figure 10-22.   Pulse-forming multivibrator. *(Courtesy of Motorola)*

of low-speed operation, relatively low noise margin, and presenting a heavy capacitive load at the inputs, but is more economical than the direct-coupled type.

10–7   PULSE-FORMING MULTIVIBRATOR

In many logic applications, it is often desired to shape an analog signal so that a pulse waveform is obtained. Figure 10-22 shows two NAND gates and an inverter connected as an MV to serve this purpose. The circuit is similar in characteristics and applications to a Schmitt trigger, but differs in that switching occurs at some nominal input voltage level (determined by the threshold of the gates) rather than at the zero voltage level.

10–8   ASTABLE MULTIVIBRATOR

The circuit of Figure 10-23 is essentially a ring counter made up of inverters (or gates connected as inverters). The frequency of operation is determined by the propagation delay of the gates. Note that the equation shown in Figure 10-23 provides the approximate frequency, and should be used only as a starting point for design. The external capacitor C_x modifies the delay associated with two of the gates, and thus changes the frequency. The output is essentially a square wave, even though only one capacitor is used.

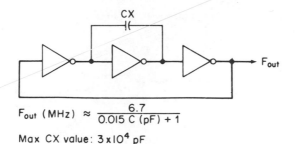

$$F_{out} \text{ (MHz)} \approx \frac{6.7}{0.015 \, C \text{ (pF)} + 1}$$

Max CX value: 3×10^4 pF

Figure 10-23.   Astable multivibrator using inverters (or gates connected as inverters). *(Courtesy of Motorola)*

10–9   MONOSTABLE MV CIRCUITS

Monostable MVs may be grouped into two general classifications. One group consists of circuits that stretch an input pulse (output pulse width is some constant value, plus the input pulse width). The second group has an output pulse width that is independent of the input pulse width (pulse-shaping monostable MV).

Aside from this basic difference, it should be noted that pulse-shaping monostable MVs have finite allowable duty cycle (defined as the ratio of pulse width to period), whereas the pulse-stretching monostable MV normally will operate at any duty cycle up to 100 percent. In addition, a second input pulse (arriving during the timed interval) will completely recycle the pulse-stretching monostable MV. The effect of the second pulse on the pulse-shaping monostable MV varies from negligible to a significant increase of output pulse width according to circuit design. Complete recycling of the pulse-shaping MV normally does not occur, however.

10–9.1   Pulse-stretching Monostable MV Circuits

A simple pulse-stretching circuit is shown in Figure 10-24. The circuit values shown are typical for power gates and standard gates, but must be used only as starting points for design. As capacitor size increases, the minimum input pulse width must increase in order to insure that the capacitor is fully discharged. If an even greater output pulse width is required, the circuit of Figure 10-24 can be operated in series with another identical circuit.

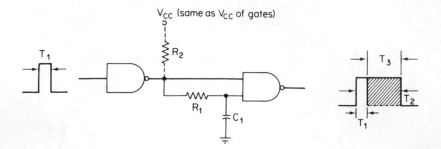

$T_3 \text{(nS)} \approx 0.55\, C\text{(pF)} + 60$

	Typical power gate	Typical standard gate	
R_1	21 Ω	68 Ω	
R_2	6 Ω	∞	
Minimum input pulse width	0.2 C	0.5 C	$\left\{ \begin{array}{l} \text{Time in NS} \\ C \text{ in pF} \end{array} \right\}$
Maximum* capacitance value	50 μF	15 pF	

* May be increased to infinity by tripling value of R_1 and minimum input pulse width

Figure 10-24. Pulse-stretching monostable multivibrator. *(Courtesy of Motorola)*

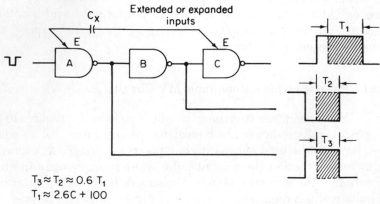

$T_3 \approx T_2 \approx 0.6\, T_1$
$T_1 \approx 2.6C + 100$
Time in NS, C in pF

Figure 10-25. Pulse-stretching monostable multivibrator with multiple outputs. *(Courtesy of Motorola)*

The pulse-stretching circuit of Figure 10-25 represents a variation of the ring oscillator described in Sec. 10-8. The circuit of Figure 10-25 may be found useful in applications that require two timed intervals. Note that this circuit requires gates with an "extended" or "expanded" input, in addition to the normal input (Sec. 1–17.1). Generally, the need for such extended inputs limits the use of this circuit to applications where the output loading of gates B and/or C require the use of buffer or power gate elements. (Most low-power gates do not have the extended input feature, nor do all power gates.)

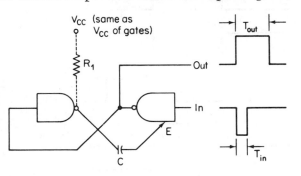

$T_{out} \approx 2.75\,C$

Time in NS, C in pF

R_1 (typical)	Allowable duty cycle
∞	10%
620	40%

Figure 10-26.   Pulse-shaping   monostable   multivibrator.   *(Courtesy   of Motorola)*

10–9.2   Pulse-shaping Monostable MV Circuit

The circuit of Figure 10-26 may be used as a pulse-shaping monostable MV, provided that the input signal is always shorter than the output. The allowable duty cycle is governed by component values as shown. Again, note that these values are "typical."

10–10   ECL GATE MONOSTABLE MV CIRCUITS

High-speed monostable MV circuits can be formed by using ECL gates. Operating frequencies of 70 MHz and pulse widths of 4 ns are possible with typical IC ECL gates. Two ECL MV circuits are shown

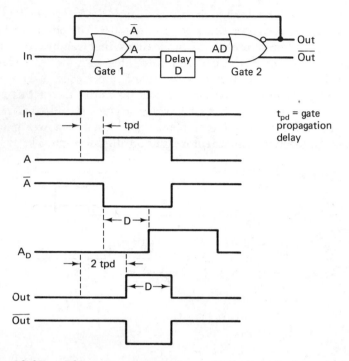

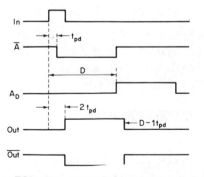

Figure 10-27.   ECL gate monostable multivibrator (independent pulse width). *(Courtesy of Motorola)*

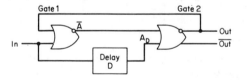

Figure 10-28.   ECL gate monostable multivibrator (fast recovery). *(Courtesy of Motorola)*

in Figures 10-27 and 10-28. The two circuits are identical with the exception of the delay element connections. In Figure 10-27, the delay element is connected so that the output pulse is independent of propagation delays, but in so doing, the duty cycle is slightly limited. In Figure 10-28, a higher-duty cycle is possible, but the output pulse width is somewhat dependent on propagation delay (designated as t_{pd}).

With either circuit, both the OR and NOR outputs must be available from the gates. It is assumed that the basic gate propagation delay is about 2 ns (typical for high-speed ECL gates in IC form). It is also assumed that the input to the FF is a typical -1.6 V for a logic 0, and -0.75 V for a logic 1. If the input is not compatible with these levels, it can be converted by either of the two methods shown in Figure 10-29.

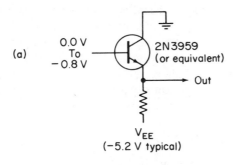

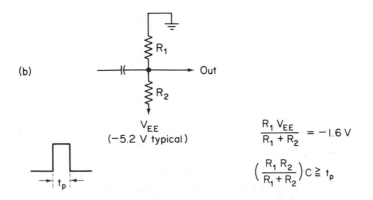

Figure 10-29. Circuits for converting logic voltage levels to values compatible with typical ECL gates. *(Courtesy of Motorola)*

10–10.1   Converting Logic Voltage Levels

The circuit of Figure 10-29a is simply an emitter follower, and the output follows the input, reduced by the voltage drop across the transistor diode. The circuit of Figure 10-29a is good for inputs in the range between about 0 V and −0.8 V.

The circuit of Figure 10-29b is a capacitively coupled voltage divider. The values of R_1 and R_2 should be selected to give a logic 0 (−1.6 V) at the output, if the equation shown is used. The value of C can be determined by the fact that the RC time constant $(R_1R_2)/(R_1 + R_2) \times C$ should be several times greater than the duration of the input pulse.

10–10.2   Input Pulse Duration

In the circuits in both Figure 10-27 and Figure 10-28 it is necessary for the input pulse to be of sufficient duration to allow the output to latch in the 1-state. By inspection of either diagram, it can be seen that two propagation delays are required for the output to assume a 1-state. The output feeds back to the input to hold the 1-state for the required time. Thus, the monostable input pulse must be at least twice as long as two propagation delays of the gate (typically 4 ns).

10–10.3   Inverted Output

In both circuits, the inverted output is available (either or both OR and NOR outputs can be used simultaneously). Using the inverted output makes it possible to reshape high duty cycle inputs by using a short delay. This is shown in Figure 10-30. The only disadvantage in using the inverted output in this manner is that the output is effectively delayed by D ns, as shown.

10–10.4   Reshaping Waveforms

Figure 10-31 shows several examples of reshaping waveforms by means of monostable MVs. Because of the intended high operating frequencies of these circuits, performance is affected by the physical layout. Thus, good high-frequency design procedures should be followed, all interconnections being kept as short as possible. (Such procedures are described in the author's *Handbook of Simplified Solid-State Circuit Design,* published by Prentice-Hall, Inc., Englewood Cliffs, New Jersey.) If the circuits are implemented with IC gates, it is standard practice to connect all unused input leads to the V_{EE} voltage source. This prevents undesired inputs from entering the gate.

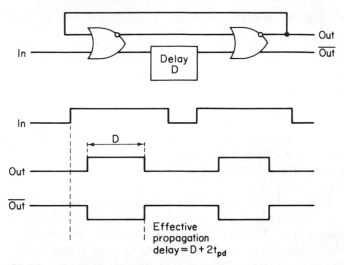

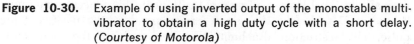

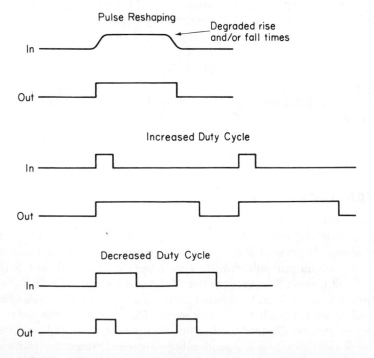

Figure 10-30. Example of using inverted output of the monostable multivibrator to obtain a high duty cycle with a short delay. *(Courtesy of Motorola)*

Figure 10-31. Examples of waveshaping with monostable multivibrator. *(Courtesy of Motorola)*

10–10.5   Independent Pulse Width

The circuit of Figure 10-27 has the advantage that the output pulse width is determined solely by the delay element D. Circuit operation can best be understood by reference to the timing diagram of Figure 10-27.

Two propagation delays after the input pulse starts, the NOR output of gate 2 goes high because of the coincidence of 0 inputs (a NOR gate produces a one only when both inputs are at 0). This condition exists until D ns later, when a one signal from the delay element reaches the input of gate 2, switching the NOR output back to a zero.

It may also be verified from the timing diagram that the output pulse is precisely equal to D, and is not affected by the propagation delays. An additional time $D + 1t_{pd}$ past the output pulse is required for the delay element to return to zero. If the input is to be normally in the 1-state, and momentarily at 0, this places a minimum 0 time of D ns before the circuit can respond to the positive-going pulse. This restriction also limits the duty cycle of the monostable MV to less than 50 percent without using the inverting output, unless the inverting output is substituted as the MV output. The maximum duty cycle of the Figure 10-27 circuit is equal to:

$$\frac{D}{2D + 3t_{pd}}$$

The maximum operating frequency of the Figure 10-27 is equal to:

$$\frac{1}{2D + 3t_{pd}}$$

10–10.6   Fast Recovery

The circuit of Figure 10-28 differs from the Figure 10-27 circuit in that the delay element requires only a short recovery time. The timing diagram of Figure 10-28 illustrates operation of the circuit.

The output pulse width is no longer equal to D, but is $(D - 1t_{pd})$ in duration. Two propagation delays are required to switch the output to the high state, whereas only one propagation delay is required to switch back to the low state. This is due to the delay line being connected directly to the input of gate 1. The pulse traveling through the delay line is considerably shortened, since the pulse is of the same duration as the input pulse. The maximum duty cycle of the Figure 10-28 circuit is equal to:

$$\frac{D - 1t_{pd}}{D + 2t_{pd}}$$

It is possible to obtain the maximum duty cycle only when the input pulse is at a duration of $2t_{pd}$ (or less). If the input pulse is longer than this, the duty cycle is reduced to:

$$\frac{D - 1t_{pd}}{D + t_{pd}}$$

The maximum operating frequency of the Figure 10-28 circuit is equal to:

$$\frac{1}{D + 2t_{pd}}$$

10–10.7 Delay Elements for Monostable MVs

The circuits of Figures 10-27 and 10-28 require fixed delay elements D. There are three types of delays to be considered.

Commercial delay lines are available from numerous suppliers in practically any delay desired, as well as variable delays. These delay lines have temperature characteristics and impedances to match almost any gate. Thus, logic designers will do well to consider commercial delay lines for all delay applications.

Coaxial cable can also be used to provide delay. Most coaxial cable has a propagation delay of about 1.5 ns/ft. This makes cable useful for short pulse widths, but somewhat impractical for pulses longer than several ns. For example, if a 10-ns delay is required, the coaxial cable is about seven feet long. The coaxial method of delay does have the advantage of infinite resolution, since the cable may be cut to any desired length.

In order to minimize line ringing (undesired pulse changes caused by mismatch) with either coaxial cable or commercial delay lines, it becomes necessary to terminate the element into its characteristic impedance Z_0. In the case of delay elements connected to a gate, it is often desirable to use a resistive divider network that will match the impedance, as well as provide a fixed voltage at the gate input. Such a circuit is shown in Figure 10-32. The value of the fixed voltage

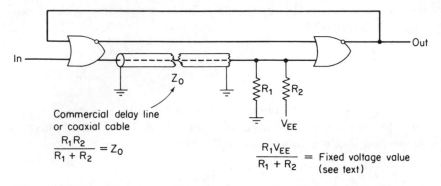

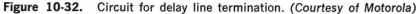

Commercial delay line
or coaxial cable

$$\frac{R_1 R_2}{R_1 + R_2} = Z_0$$

$$\frac{R_1 V_{EE}}{R_1 + R_2} = \text{Fixed voltage value (see text)}$$

Figure 10-32. Circuit for delay line termination. *(Courtesy of Motorola)*

should be just below the threshold level of the gate. For example, if the gate input is -1.6 V for a 0 input (with positive logic), a typical fixed voltage value is -2 V. The equations for finding the resistance values of the divider network are shown in Figure 10-32.

Gates can also be used as delay elements. Since each gate has a propagation delay associated with it, the delay (and consequently the output pulse width) of the monostable MV is equal to the propagation delay of one gate, multiplied by the number of gates used. Figure 10-33 shows how four NOR gates (with individual propagation delays of 3.5 ns) can be connected to provide a delay of 14 ns. In the case where only NOR outputs are available, there must be an even number of gates to obtain the correct number of inversions. If both OR and NOR outputs are available (as is the case with most ECL gates), both OR and NOR outputs can be mixed to make an even number of inversions.

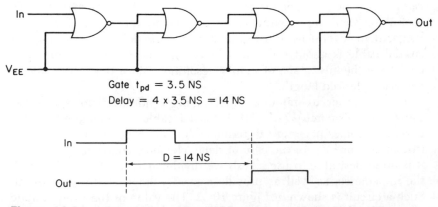

Gate $t_{pd} = 3.5$ NS
Delay $= 4 \times 3.5$ NS $= 14$ NS

Figure 10-33. Circuit for using NOR gates as fixed delay elements. *(Courtesy of Motorola)*

If gates are used as delay elements with the independent pulse width monostable MV (Figure 10-27), the requirement for an even number of inversions can be avoided by connecting the input of the delay element to the NOR output of the first gate, instead of the OR output. This can not be done with the fast recovery circuit (Figure 10-28), since the delay element input is connected directly to the input pulse of the monostable MV.

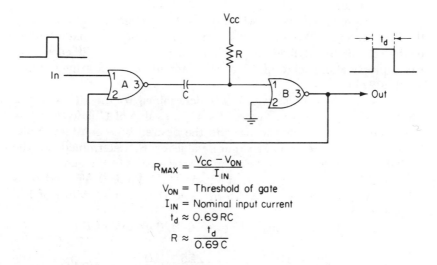

$$R_{MAX} = \frac{V_{CC} - V_{ON}}{I_{IN}}$$

V_{ON} = Threshold of gate

I_{IN} = Nominal input current

$t_d \approx 0.69\,RC$

$$R \approx \frac{t_d}{0.69\,C}$$

Figure 10-34.   One-shot multivibrator controlled by RC time constant. *(Courtesy of Motorola)*

10–11   ONE-SHOT MV CONTROLLED BY RC TIME CONSTANT

The circuit of Figure 10-34 is a one-shot MV composed of two NOR gates. The output pulse width is controlled by the RC time constant of the external resistor and capacitor. Operation of the circuit is as follows:

In the steady state, prior to an input pulse, a steady current flows through R, applying a voltage level equal to a logic 1 to $B1$. This results in a logic 0 at $B3$, which is fed back to input $A2$. Since both A inputs are at 0, $A3$ is at 1. There is little charge stored in C, since both plates are at about the same potential.

If a positive-going pulse (0 to 1) is applied to $A1$, $A3$ goes low and C begins to charge. With $B1$ at 0, and a permanent 0 at $B2$, $B3$ switches to 1. This 1 is fed back to $A2$, and maintains $A3$ at a low level until C charges to the point where $B1$ reaches the 1 threshold level. Then output $B3$ is switched to 0, completing the generation of the one-shot pulse. The 0 at B3 is fed back to A2, and the one-shot has returned to its original steady state.

The presence of this feedback loop makes the duration of the one-shot output *relatively* independent from the duration of the trigger input. The feedback insures that the output of gate A remains at 0 after the trigger input has reverted to 0. Thus, the duration of the 1 output from the one-shot is determined by the values of R and C, not the time duration of the trigger.

The minimum pulse width of the trigger input is governed by the propagation delays of the gates used, and is approximately equal to the sum of the individual propagation delays. Thus, if each gate has a propagation delay of 12 ns, the minimum input trigger pulse width is 24 ns.

The output pulse width t_d is determined by the RC constant. The equation is shown in Figure 10-34. The value of C is fixed, while the value of R is selected to provide the desired time constant. Note, however, that there is a maximum value of R, determined by the available voltage (V_{CC}) and the input characteristics of the gate.

For example, assume that V_{CC} is 3.6 V, V_{on} is 0.865, and I_{in} is 0.5 mA. Also assume that C is 75 μF, and that a pulse width of 225 ms is desired.

Using the equations of Figure 10-34, the value of R is:

$$R \approx \frac{225 \times 10^{-3}}{0.69 \times (75 \times 10^{-6})} \approx \frac{225 \times 10^{-3}}{51.75 \times 10^{-6}} \approx 4.3 \text{ k}\Omega$$

$$\text{The value of } R_{max} = \frac{(3.6) - (0.865)}{0.5 \times 10^{-3}} = 5.4 \text{ k}\Omega$$

Since the value of 4.3 kΩ is less than the maximum, the 4.3 kΩ resistor can be used for R.

Index